Mechanics of
Composite Materials
and
Structures

Mechanics of Composite Materials and Structures

Madhujit Mukhopadhyay

Universities Press

Universities Press (India) Private Limited

Registered Office
3-5-819 Hyderguda, Hyderabad 500 029 (A.P.), India
e-mail: hyd2_upilco@sancharnet.in

Distributed by
Orient Longman Private Limited

Registered Office
3-6-752 Himayatnagar, Hyderabad 500 029 (A.P.), India

Other Offices
Bangalore / Bhopal / Bhubaneshwar / Chennai
Ernakulam / Guwahati / Hyderabad / Jaipur / Kolkata
Lucknow / Mumbai / New Delhi / Patna

© Universities Press (India) Private Limited 2004

ISBN 81 7371 477 0

Printed in India at
Orion Printers Praivat Limited
Hyderabad 500 004

Published by
Universities Press (India) Private Limited
3-5-819, Himayatnagar, Hyderabad 500 029

Dedicated

to

Ratna, Kuttus, Chuchu, Shaon and Arko

Contents

Preface   xv

Chapter One   Introduction to Composite Materials   1

1.1   Definitions   1
1.2   History of Fibre Reinforced Composites   2
1.3   Constituent Materials   3
    1.3.1   Fibres   3
        1.3.1.1   Glass fibres   4
        1.3.1.2   Carbon fibres   5
        1.3.1.3   Aramid fibres   5
        1.3.1.4   Boron fibres   6
        1.3.1.5   Ceramic fibres   6
    1.3.2   Polymeric Matrix   6
        1.3.2.1   Polyester resins   7
        1.3.2.2   Epoxy resins   7
        1.3.2.3   Vinyl ester resins   8
        1.3.2.4   Phenolic resins   8
        1.3.2.5   High performance resins   8
    1.3.3   Prepegs   8
    1.3.4   Fillers and Other Additives   8
1.4   Lamina and Laminate   9
1.5   General Characteristics of FRPs   9
1.6   Why FRPs?   10
1.7   Micromechanics and Macromechanics   11
1.8   Properties of Typical Composite Materials   11
1.9   Application of Composites   12
    1.9.1   Marine Field   12
    1.9.2   Aircraft and Space   14
    1.9.3   Automotive Field   15
    1.9.4   Sporting Goods   15
References and Suggested Readings   15

Chapter Two   Processing of FRP Composites   17

2.1   Contact Moulding   17
    2.1.1   Mould Preparation   17
    2.1.2   Spray-up   17
    2.1.3   Hand Lay-up   17

viii *Mechanics of Composite Materials and Structures*

2.2	Compression Moulding Methods	18
	2.2.1 Matched Die Moulding	18
	2.2.2 Forming Methods Employing Gas Pressure	18
	2.2.3 Low Pressure, Closed Mould System	19
	2.2.4 Pultrusion	22
2.3	Filament Winding	23

Chapter Three Micromechanical Analysis of Composite Strength and Stiffness 25

3.1	Introduction	25
3.2	Volume and Weight Fractions	26
3.3	Assumptions and Limitations	27
3.4	Longitudinal Strength and Stiffness	28
3.5	Transverse Modulus	30
3.6	Inplane Shear Modulus	33
3.7	Poisson's ratio	35
	3.7.1 Poisson's Mismatch Effect	36
3.8	Problems on Micromechanical Analysis	36
	References and Suggested Readings	44
	Exercise 3	44

Chapter Four Elastic Properties of the Unidirectional Lamina 46

4.1	Introduction	46
4.2	Stress–Strain Relationship	46
	4.2.1 Stress–Strain Relation of a Monoclinic Material	48
	4.2.2 Stress–Strain Relation of a Specially Orthotropic Material	48
	4.2.3 Stress–Strain Relation for a Transversely Isotropic Material	49
	4.2.4 Stress–Strain Relation of an Isotropic Material	49
4.3	Engineering Constants	50
4.4	Stress–Strain Relations of a Thin Lamina	53
4.5	Examples: Series A	56
4.6	Transformation of Stress and Strain	58
	4.6.1 Transformation of Stress	58
	4.6.2 Transformation of Strain	59
4.7	Transformation of Elastic Constants	61
4.8	Transformed Reduced Compliance	62
4.9	Transformation Relations of Engineering Constants	64
4.10	Examples: Series B	65
4.11	Restrictions on Elastic Constants	75
4.11	Typical Elastic Properties of a Unidirectional Lamina	77
	References and Suggested Readings	77
	Exercise 4	78

Chapter Five Analysis of Laminated Composites — 81

- 5.1 Laminates — 81
- 5.2 Basic Assumptions — 81
- 5.3 Strain–Displacement Relationship — 82
- 5.4 Stress–Strain Relations — 83
- 5.5 Equilibrium Equations — 84
- 5.6 Laminate Stiffness — 86
- 5.7 Determination of Lamina Stresses and Strains — 88
- 5.8 Coupling Effects — 90
- 5.9 Types of Laminate Configuration — 92
  - 5.9.1 Symmetric Laminates — 93
    - 5.9.1.1 Symmetric laminates with isotropic plies — 94
    - 5.9.1.2 Symmetric laminates with specially orthotropic plies (symmetric cross-ply laminates) — 95
    - 5.9.1.3 Symmetric laminates with generally orthotropic layers (symmetric angle-ply laminate) — 95
    - 5.9.1.4 Symmetric laminates with anistropic layers — 95
  - 5.9.2 Balanced Laminates — 96
  - 5.9.3 Anti-symmetric Laminates — 96
    - 5.9.3.1 Anti-symmetric cross-ply laminates — 97
    - 5.9.3.2 Anti-symmetric angle-ply laminate — 98
  - 5.9.4 Quasi-Isotropic Laminate — 99
  - 5.9.5 Unsymmetric Laminates with Isotropic Layers — 99
- 5.10 Examples: Series A — 100
- 5.11 Laminate Engineering Constants — 116
- 5.12 Examples: Series B — 118
- *References and Suggested Readings* — 122
- Exercise 5 — 122

Chapter Six Analytical Methods of Laminated Plate — 124

- 6.1 Introduction — 124
- 6.2 Classical Laminate Plate Theory (CLPT) — 124
  - 6.2.1 Basic Assumptions — 124
  - 6.2.2 Equilibrium Equations of Laminated Plates — 125
  - 6.2.3 Bending of Composite Plates — 128
    - 6.2.3.1 A rectangular plate with all edges simply supported: Navier's solution — 128
  - 6.2.4 Bending of Symmetric Angle-Ply Laminates — 130
  - 6.2.5 Bending of Anti–Symmetric Cross-Ply Laminates — 132
  - 6.2.6 Bending of Anti–Symmetric Angle-Ply Laminates — 133

x   Mechanics of Composite Materials and Structures

6.3	Bending of Rectangular Plates with Two Simply Supported Edges	135
6.4	Shear Deformation in Laminated Plates	138
6.5	Higher Order Shear Deformation Theory	142
	6.5.1 Strain–Displacement Relations	142
	6.5.2 Generalised Stress–Strain Relations	144
6.6	Determination of the K_s-Parameter	146
6.7	Free Vibration of Laminated Plates (Based on CLPT)	148
	6.7.1 Free Vibration of Unsymmetrical Angle-Ply Laminated Plate (Based on CLPT)	151
6.8	Stability Analysis of a Rectangular Angle-Ply Plate under Uniform Biaxial Compression (Based on CLPT)	152
6.9	Layerwise Theory	155
	6.9.1 Displacements and Strains	155
	References and Suggested Readings	158
	Exercise 6	159

Chapter Seven   Analysis of Composite Beams   160

7.1	Definitions	160
7.2	Basic Assumptions	160
7.3	Bending Analysis of Laminated Beams	161
7.4	Examples	166
	7.4.1 Bending of a Laminated Beam under Uniform Load	166
	7.4.2 A Laminated Cantilever Beam under Tip Loading	167
7.5	Eigenvalue Problems of Laminated Beams	168
	References and Suggested Readings	169

Chapter Eight   Finite Element Analysis of Composite Structures   171

8.1	Introduction	171
8.2	The Finite Element Method	171
8.3	Advantages and Disadvantages of the Finite Element Method	172
8.4	Basic Steps in the Finite Element Method in Static Problem	173
8.5	The Basic Problems	175
	8.5.1 Mass Matrix	175
	8.5.2 Stiffness Matrix	175
	8.5.3 Damping Matrix	175
8.6	Solution Techniques	175
	8.6.1 Static Analysis	175
	8.6.2 Free Vibration Analysis	176
8.7	Eigenvalue Solution Techniques	176
	8.7.1 Simultaneous Iteration Technique	177
8.8	Transient Vibration Analysis	178
8.9	Stiffness Matrix	179

8.10	Direct Stiffness Method		180
8.11	Overall Stiffness Matrix		183
8.12	Isoparametric Element for the Analysis of Laminated Plates		186
	8.12.1	Constitutive Relations	187
	8.12.2	Plate Element Formulation	190
	8.12.3	Mass Matrix of the Plate	194
8.13	Formulation of the Composite Stiffener Element		194
8.14	Formulation of the Composite Beam Element		194
	8.14.1	Formulation	194
8.15	Isoparametric Stiffener Stiffness Matrix		199
	8.15.1	Stiffener Mass Matrix	206
8.16	Finite Element Analysis of the Laminated Composite Shell		206
8.17	FEM of Laminated Plates Based on Higher Order Plate Theory		213
8.18	FEM of Laminated Plates Based on Higher Order Theory: Another Approach		215
	8.18.1	Strain–Displacement Relationship	216
	8.18.2	Stress–Strain Relationship	217
	8.18.3	Derivation of the Stiffness Matrix	221
	8.18.4	Boundary Conditions	222
8.19	Numerical Examples		223
	8.19.1	Bare Composite Plates Under Transverse Load	223
	8.19.2	Bare Composite Plate in Free Vibration	223
	8.19.3	Composite Cylindrical Shell Under Transverse Load	224
	8.19.4	Free Vibration of a Composite Shallow Shell	225
	References and Suggested Readings		225

Chapter Nine Hygrothermal Effects in Laminates — **228**

9.1	Introduction	228
9.2	Effect of Hygrothermal Forces on Mechanical Behaviour	229
9.3	Micromechanics of Hygrothermal Properties	231
9.4	Unidirectional Lamina: Hygrothermal Strains	232
9.5	Free Thermal Strains	233
9.6	Stress–Strain Relationship of a Lamina when Free Thermal Strains are Included	234
9.7	Hygrothermoelastic Stress–Strain Relations	236
9.8	CTE and CME of Laminates	241
9.9	Determination of Stresses in a Laminate due to Hygrothermoelastic Forces	241
9.10	Residual Stresses	242
9.11	Warpage	243
9.12	Examples	244

xii *Mechanics of Composite Materials and Structures*

	References and Suggested Readings	252
	Exercise 9	252

Chapter Ten Failure Theories and Strength of a Unidirectional Lamina — 254

10.1	Introduction	254
10.2	Micromechanics of Failure of Unidirectional Lamina	254
	10.2.1 Longitudinal Tension	254
	10.2.2 Longitudinal Compression	256
	10.2.3 Transverse Tension	259
	10.2.4 Transverse Compression	259
	10.2.5 Inplane Shear	260
10.3	Anisotropic Strength and Failure Theories	260
	10.3.1 Maximum Stress Theory	261
	10.3.2 Maximum Strain Theory	262
	10.3.3 Deviatoric Strain Energy Theory (Tsai-Hill)	264
	10.3.4 Interactive Tensor Polynomial Theory (Tsai-Wu)	266
10.4	Examples	268
10.5	Importance of Shear Stress	276
10.6	Example on Shear Strength	276
10.7	Choice of Failure Criteria	277
10.8	Typical Strength Properties	278
	References and Suggested Readings	278
	Exercise 10	279

Chapter Eleven Analysis of Laminate Strength — 281

11.1	Introduction	281
11.2	Possible Modes of Failure	281
11.3	Stress Analysis at First-Ply Failure	281
11.4	Ultimate Laminate Failure or Analysis of Last-Ply Failure	283
	11.4.1 Total-Ply Failure Method	283
	11.4.2 Partial-Ply Failure Method	283
11.5	Examples	284
11.6	Interlaminar Stress	299
11.7	Prediction Methods	303
	References and Suggested Readings	304
	Exercise 11	305

Chapter Twelve Design of Fibre Reinforced Composite Structures — 307

12.1	Introduction	307
12.2	Composite Structural Design	308

		12.2.1	The Design Spiral	309
		12.2.2	Design Criteria	309
		12.2.3	Design Allowables	310
		12.2.4	Material Selection in Composite Design	311
		12.2.5	Selection of Configuration and Manufacturing Process	312
	12.3	Laminate Design		312
		12.3.1	Selection of Laminate	313
		12.3.2	Laminate Design Problem	313
		12.3.3	Laminate Design Procedure	314
	12.4	Mathematical Analysis of the Laminate		314
		12.4.1	Estimation of Shear Force	317
		12.4.2	Estimation of Deflection	319
		12.4.3	Mathematical Algorithm	320
	12.5	Design Examples		320
		12.5.1	Design of a Tension Member	321
		12.5.2	Laminate Design for Strength	322
		12.5.3	Laminate Design for Stiffness	323
		12.5.4	Composite Panels Subjected to Combined Inplane Loads	334
	12.6	Design of Single Skin Panels		338
	12.7	Design of Stiffened Structures		339
		12.7.1	Design of Composite Stiffeners	340
		12.7.2	Types of Composite Stiffeners	340
		12.7.3	Stiffener Design Parameters	340
		12.7.4	Design of a Longitudinally Stiffened Panel	343
		References and Suggested Readings		343

Chapter Thirteen Composite Joints — **345**

13.1	Introduction			345
13.2	Classes of Laminate Joints			345
13.3	Bonded Joints			345
		13.3.1	Stress Distribution	345
		13.3.2	Modes of Failure	349
		13.3.3	Merits and Demerits of Adhesive Bonded Joints	349
13.4	Mechanical Joints			349
		13.4.1	The Failure Modes of the Mechanical Joints	350
		13.4.2	Advantages and Disadvantages of Mechanical Joints	350
13.5	Preliminary Design of the Adhesively Bonded Composite Joints			350

13.6	Preliminary Design of Composite Bolted Joints		357
	13.6.1 Bearing Failure		357
	13.6.2 Tension Failure		358
13.7	Composite Multi-Bolt Joints		359
	13.7.1 Design of Composite Multi-Bolt Joints		359
13.8	Other Approaches		360
	References and Suggested Readings		361
	Answers to Exercise Problems		364
	Index		369

PREFACE

Fibre reinforced plastic composite (FRP) materials have a wide range of applications in various engineering artefacts such as offshore structures, maritime structures, ships, aerospace structures, civil engineering structures, machine components, chemical industrial applications, etc. The applications are either weight critical such as in aerospace and offshore platform topsides or performance critical such as the non-magnetic and non-corrosive composites in naval minehunters. Though the scope of intelligent exploitation of composites is ample, the paucity of adequate knowledge among many practicing engineers has until now been a major deterrent. As such a lot of effort has gone into enunciating newer and better approaches to impart knowledge on FRP composite materials, structural mechanics and structural analysis. The present book is an attempt to present an integrated and unified approach to FRP related topics.

The importance of the subject has been recognised in recent years by many universities, where undergraduate and postgraduate programmes on the subject are currently in vogue. A number of doctoral and research programmes are also been undertaken. Further, as part of the continuing education programme, many universities and commercial organisations are offering short-term courses on this subject.

A number of books already exist on FRP composite mechanics. They can be broadly classified under three headings. The first envisages the subject from point of view the material science covering the chemical and morphological features of composite mechanics. Books in the second category treat the subject primarily from micro-mechanics and lamination theory standpoint. Books in the third category deal with the analysis of FRP composite structural components.

This book transcends the traditional classification mentioned above. It presents the micromechanics, lamination theory and the analysis of composite structural components in a unified and integrated manner. The material of the book has been developed while teaching undergraduate and postgraduate students of the Department of Ocean Engineering and Naval Architecture of the Indian Institute of Technology, Kharagpur. Starting from basic concepts, the reader is introduced to with more advanced topics gradually. The first draft of the book was prepared by the author during his assignment as a Visiting Professor at the University of Southampton in 1998. The author gratefully acknowledges the help received by him from Professor R.A. Shenoi of the Department of Ship Science, University of Southampton. The manuscript of the book has been typed by Parimal Kumar Ray. The necessary drawings have been prepared by Chinmoy Mukherjee and Parimal Kumar Ray. The author is grateful to them. He wishes to put on record the help received from Lt. N. Kumar, a postgraduate student while dealing with Chapter 12 and thanks his wife Ratna for the patience and forbearance shown by her during the preparation of the manuscript.

Finally, the author expresses his sincere thanks and heart-felt gratitude to Mr. Madhu Reddy, Director of Universities Press (India) Private Limited and his competent staff for the excellent co-operation and generous help received from them.

The financial assistance granted by the Quality Improvement Programme Cell at the Indian Institute of Technology, Kharagpur for the preparation of the manuscript is gratefully acknowledged.

Kharagpur
June 2004

Madhujit Mukhopadhyay

CHAPTER 1

INTRODUCTION TO COMPOSITE MATERIALS

1.1 DEFINITIONS

A composite material is defined as a material system which consists of a mixture or a combination of two or more distinctly differing materials which are insoluble in each other and differ in form or chemical composition.

Thus, a composite material is labelled as any material consisting of two or more phases. Many combinations of materials may, therefore, be termed as composite materials, such as concrete, mortar, reinforced rubbers, conventional multiphase alloys, fibre reinforced plastics, fibre reinforced metals and similar fibre impregnated materials.

Two-phase composite materials are classified into two broad categories: particulate composites and fibre reinforced composites. Particulate composites are those in which particles having various shapes and sizes are dispersed within a matrix in a random fashion. As the distribution of particles is random and as the particles are of varying shapes and sizes, these composites are treated as quasi-homogeneous and quasi-isotropic. Examples of particulate composites are mica flakes reinforced with glass (non-metallic particles in a non-metallic matrix), aluminium particles in polyurethane rubber (metallic particles in a non-metallic matrix), lead particles in copper alloys (metallic particles in a metallic matrix) and silicon carbon particles in aluminium (non-metallic particles in a metallic matrix).

Particulate composites are used for electrical applications, welding, machine parts and other purposes. Particulate composites made of tungsten and molybdenum particles dispersed in silver and copper matrices are used for electrical contact applications as well as electrodes welding. Lead particles mixed with copper alloy and steel improve machinability. In machine parts where high surface hardness is required particulate matrix is formed by mixing tungsten carbide particles in a cobalt matrix. Titanium carbide in cobalt or nickel is very much suited for high temperature applications.

Fibre reinforced composite materials consist of fibres of significant strength and stiffness embedded in a matrix with distinct boundaries between them. Both fibres and matrix maintain their physical and chemical identities, yet their combination performs a function which cannot be done by each constituent acting singly. Fibres of fibre reinforced plastic (FRP) may be short or continuous. It appears obvious that FRP having continuous fibres is indeed more efficient.

Classification of FRP composite materials into four broad categories has been done according to the matrix used [1.1] They are polymer matrix composites, metal matrix composites, ceramic matrix composites and carbon/carbon composites [Table 1.1]. Polymer matrix composites are made of thermoplastic or thermoset resins reinforced with fibres such as glass, carbon or boron. Metal matrix composites consist of a matrix of metals or alloys reinforced with metal fibres such as boron or carbon. Ceramic matrix composites consist of ceramic matrices reinforced with ceramic fibres such as silicon carbide, alumina or silicon nitride. They are mainly effective for high temperature applications. Carbon/carbon composites consist of graphite carbon matrix reinforced with graphite fibres. In addition to the above, there are other types of composites as well. The flake composites consist of a matrix reinforced with flakes which may be of different types such as glass flakes, mica flakes and metal flakes. The distribution of the flakes throughout the matrix provide a considerable barrier to moisture, gas and chemical transport. It can suitably be used for obtaining high thermal and electrical resistance or conductivity.

Table 1.1 Classification of FRP composite materials [1.1]

Matrix type	Fibre	Matrix
Polymer	E-glass S-glass Carbon (graphite) Aramid (Kevlar) Boron	Epoxy Polyimide Polyester Thermoplastics Polysulfone
Metal	Boron Borsil Carbon (graphite) Silicon carbide Alumina	Aluminium Magnesium Titanium Copper
Ceramic	Silicon carbide Alumina Silicon nitride	Silicon carbide Alumina Glass-ceramic Silicon nitride
Carbon	Carbon	Carbon

In-filled or skeletal composites or continuous three-dimensional structural matrix is filled by a second material. Laminar composites consist of thin layers of different materials bonded together.

Of all the types of composites discussed above, the most important is the fibre reinforced composites or filamentary type composites – this is from the application point of view. This book will deal with fibre reinforced polymer matrix composite materials.

1.2 HISTORY OF FIBRE REINFORCED COMPOSITES

For millenniums, fibre had been used as reinforcement for making components of structural construction. There are Biblical references dating back to 2000 b. c. or earlier, to the straw reinforced mud bricks and composite bows found in Egypt and Mongolia. The development can be traced through to the 'daub and wattle' construction of buildings in Europe in the Middle Ages. The Japanese Samurai warriors used laminated metals in their swords in order to obtain the desirable material properties. In the nineteenth century, iron rods were used as reinforcements for masonry resulting in reinforced masonry construction. Asbestos fibres were used for reinforcement in phenolic resins in the early part of this century. The process of obtaining strong glass fibres was developed in the late 1930s and the development of the first commercial unsaturated polymer resins came a little later. The first glass fibre boat was built at the time of the Second World War in 1942. Reinforced plastics started to be used more or less at the same time in electrical components and aircrafts. Filament winding was invented in 1946 and incorporated into missile application in the 1950s. Advanced composites stem from the development of the first boron and then high strength carbon fibres in the 1950–1965 period. The first application of advanced composites to aircraft components was made in 1970. Kevlar or aramid fibres were developed in 1973 by Du Pont. From the 1970s, the area of application of composites has expanded in many directions. Among them are aerospace structures, automotive, sports equipment, biomedical products, high performance vessels and many other areas. The current emphasis is on the development of metal matrix, ceramic matrix, and carbon/carbon composites.

1.3 CONSTITUENT MATERIALS

The major constituents of a fibre reinforced composite material are reinforcing fibre, matrix, coupling agents, coatings and fillers. Fibres are the principal load carrying members while the matrix which surrounds it, keeps them in proper location and correct orientation [1,2]. Matrix acts as the medium by which the load is transferred through the fibres by means of shear stress. Matrix protects the fibre from environmental damages caused by elevated temperature and humidity. Coupling agents and coatings applied to the fibres improve their wettings with the matrix and also facilitate bonding across the fibre–matrix interface. The major purpose of using fillers in some polymeric matrices is to reduce cost and achieve a better dimensional stability.

1.3.1 Fibres

Materials in fibre form are stronger and stiffer than that used in a bulk form. There is a likely presence of flaws in bulk material which affects its strength while internal flaws are mostly absent in the case of fibres. Further, fibres have strong molecular or crystallographic alignment and are in the shape of very small crystals. Fibres have also a low density which is advantageous.

Fibre is the most important constituent of a fibre reinforced composite material. They also occupy the largest volume fraction of the composite. Reinforcing fibres as such can take up only its tensile load. But when they are used in fibre reinforced composites, the surrounding matrix enables the fibre to contribute to the major part of the tensile, compressive, flexural or shear strength and stiffness of FRP composites.

Tensile stress–strain curves of a few typical fibres are presented in Fig. 1.1. They are linear up to failure for all reinforcing fibres. Further, the strains at the failure of the fibres are exceedingly low. Fibres exhibit brittle mode at failure. They are, however, prone to damage while handling as well as during contact with other surfaces. Properties of some important fibres are presented in Table 1.2.

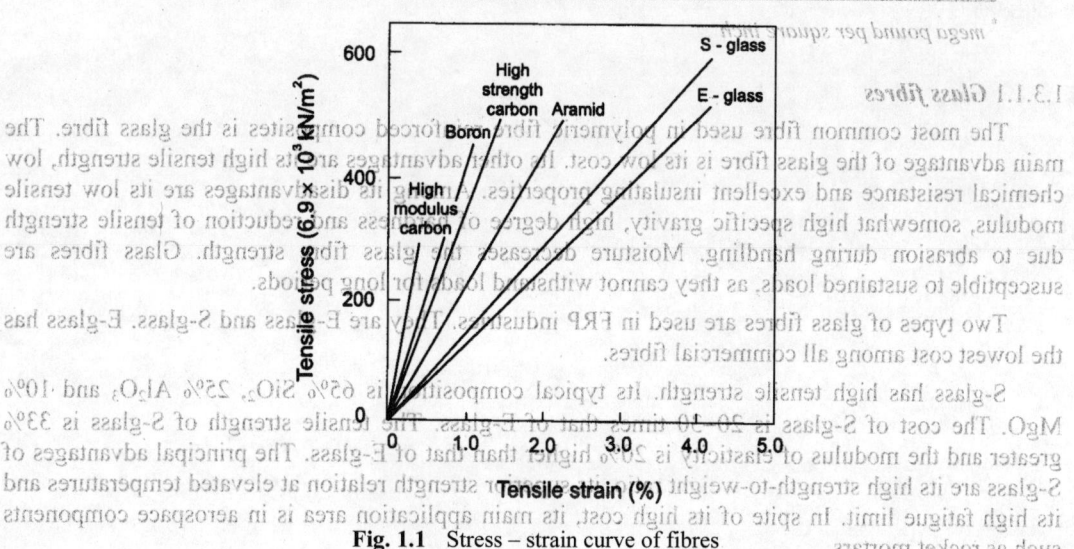

Fig. 1.1 Stress – strain curve of fibres

Fibres are used either as a single or as a combination of two types-chopped strand mat (CSM) or woven roving (WR). By chopping strands to short lengths (5–75 mm), they can be directly used in spray-up lamination. CSM consists of randomly-oriented fibres bound with an emulsion or powder binder. In boat construction, weights of CSM vary from 300 to 900 g/m^2.

A fabric is constructed of interlaced yarns, fibres or filaments. Typical glass–fibre fabrics are manufactured by interlacing warp (lengthwise) yarns and fill (crosswise) yarns on conventional weaving looms. By the weave of a fabric we can understand the way the warp yarns and fill yarns are interlaced. The popular weave patterns are plain, twill, leno and unidirectional. Plain weave is the oldest and most common textile weave. In this, one warp end is woven over one fill yarn and under the next and the process is repeated.

Woven roving material is of plain weave and is balanced. They are also available in biased form. Fibre orientation of WR may be ± 45° direction. They are also obtained with equal distribution of fibres in 0/90° and ± 45° directions. Weights of WR vary from 200 to 900 g/m^2 [1.3].

Table 1.2 Raw fibre properties[1.1–1.2].

Fibre	Typical diameter (μm)	Specific gravity	Tensile modulus GPa (Msi)*	Tensile strength GPa (Ksi)	Ultimate elongation %	Coefficient of thermal expansion 10^{-6}/°C
Glass						
E-glass	10	2.54	72.4 (10.5)	3.45 (500)	4.8	5.0
S-glass	10	2.49	86.9 (12.6)	4.30 (625)	5.7	2.9
Aramid						
Kevlar 49	11.9	1.45	131 (19)	3.62 (525)	2.8	− 2.2
Kevlar 149	11.9	1.47	179 (26)	3.45 (500)	1.9	
Carbon						
T-300	7	1.76	231 (33.5)	3.65 (530)	1.4	− 0.55
	8	1.80	395 (57)	2.48 (360)	0.7	
	10	2.15	758 (110)	2.45 (350)	0.32	
Boron	-	0.93	400 (60)	3.4 (500)	-	-

* *mega pound per square inch*

1.3.1.1 *Glass fibres*

The most common fibre used in polymeric fibre reinforced composites is the glass fibre. The main advantage of the glass fibre is its low cost. Its other advantages are its high tensile strength, low chemical resistance and excellent insulating properties. Among its disadvantages are its low tensile modulus, somewhat high specific gravity, high degree of hardness and reduction of tensile strength due to abrasion during handling. Moisture decreases the glass fibre strength. Glass fibres are susceptible to sustained loads, as they cannot withstand loads for long periods.

Two types of glass fibres are used in FRP industries. They are E-glass and S-glass. E-glass has the lowest cost among all commercial fibres.

S-glass has high tensile strength. Its typical composition is 65% SiO_2, 25% Al_2O_3 and 10% MgO. The cost of S-glass is 20–30 times that of E-glass. The tensile strength of S-glass is 33% greater and the modulus of elasticity is 20% higher than that of E-glass. The principal advantages of S-glass are its high strength-to-weight ratio, its superior strength relation at elevated temperatures and its high fatigue limit. In spite of its high cost, its main application area is in aerospace components such as rocket mortars.

1.3.1.2 Carbon fibres

Carbon fibres are characterised by a combination of high strength, high stiffness and light weight. Carbon fibres are produced by polymeric fibre percusors or pitch fibre percusors.

The advantages of carbon fibres are their very high tensile strength-to-weight ratio, high tensile modulus-to-weight ratio, very low coefficient of thermal expansion and high fatigue strength. The disadvantages are their low impact resistance and high electrical conductivity. Due to the high cost, the use of carbon fibres is justified only in weight critical structures, that is, mostly applied to aerospace industry.

Carbon fibres are categorised into two types: high strength and high modulus. The high modulus type is more expensive as it requires higher production temperature while the high strength variety is more popular.

Carbon fibres are commercially available within a wide range of tensile modulus: from 207 GPa (30×10^6 psi) to 802 GPa (125×10^6 psi). In general, low modulus fibres have many advantages over high modulus fibres such as having lower specific gravity, lower cost, higher tensile and compressive strengths and higher ultimate strains.

Until relatively recently, the fabrics for use with resin matrices have been two-dimensional crimped weaves with warp and fill yarns going over and under one another in a manner similar to classical forms of clothing materials. Modern advances have been made in a number of directions which include the following

1. Multi-dimensional weaving
2. 3-D weaving to provide greater transverse strength
3. Crimpless weaves with secondary yarns knitted to hold together collimated straight primary yarns in one or more unidirectional plies

These advancements are mainly related to carbon fibres.

1.3.1.3 Aramid fibres

The first significant group of polymeric reinforcement fibres is polyaramid fibres developed as Kevlar by Du Pont [1.4]. Kevlar aramid is made of carbon, hydrogen, oxygen and nitrogen and is essentially an aromatic organic compound. The advantages of aramid fibres are their low density, high tensile strength and low cost. Of all the available commercial fibres, it has the highest tensile strength-to-weight ratio. Glass-fibre composites weigh 65% more than composites made of aramid fibres of equivalent stiffness.

Characteristics of Kevlar 49 are its high strength and stiffness, light weight, vibration damping, resistance to damage, fatigue and stress ruptures.

These is another variety of aramid fibre available – Kevlar 29 – which is of low density and high strength. Kevlar 29 is used in ropes, cables and coated fabrics for inflatables.

The principal disadvantages of aramid fibres are their low compressive strength and the difficulty in cutting or machining. These fibres have complex anisotropic structure due to which local crumpling and fibrillation of individual fibre takes place. This further reduces the strength in compression and bending. For structures or structural components where compression and bending are predominant such as in a shell, aramid fibres can be used only when it is hybridized with glass or carbon fibres.

A more advanced variety of Kevlar fibre is Kevlar 149. Of all commercially available aramid fibres, it has the highest tensile modulus as it has 40% higher modulus than Kevlar 49. The strain at failure for Kevlar 149 is, however, lower than that of Kevlar 49.

Aramid fibres are costlier than E-glass, but are cheaper than carbon fibres.

1.3.1.4 *Boron fibres*

Boron fibres are characterized by their very high tensile modulus, the range of which is 379–414 GPa (55–60×10^6 psi). Boron fibres have relatively large diameters and due to this they are capable of withstanding large compressive stress and providing excellent resistance to buckling. Boron fibres are, however, costly and in fact are costlier than most varieties of carbon fibres. The application area of boron fibres at present is restricted to aerospace industries only.

Close to the outer surface of the boron layer, a state of biaxial compression exists, which makes the fibre less sensitive to mechanical damage. The adverse radioactivity of boron fibres with metals is reduced by chemical vapour deposition of silicon carbide on boron fibres, which produces borosic fibres.

1.3.1.5 *Ceramic fibres*

Ceramic fibres are mainly used in application areas dealing with elevated temperature. Examples of ceramic fibres are silicon carbide (SiC) and aluminium oxide (Al_2O_3). In metal matrices where boron fibres exhibit adverse radioactivity, both the above ceramic fibres are found suitable. Continuous ceramic fibre has an added advantage in that they have properties such as high strength, high elastic modulus with high temperature capabilities and are free from environmental attack.

1.3.2 **Polymeric Matrix**

Polymers are divided into two broad categories: thermoplastic and thermoset. Thermoplastic polymers are those which are heat softened, melted and reshaped as many times as desired. But a thermoset polymer cannot be melted or reshaped by the application of heat or pressure.

Depending on the particular thermoplastic material used, thermoplastic matrix components can, however, be used over a wide range of temperature – from 100°C to 300°C. The advantages of thermoplastic matrices are their improved fracture toughness over the thermoset matrix and their potential of much lower cost in the manufacturing of finished composites.

There are various reasons why thermoplastic polymers are not used for the manufacture of FRPs. Some of them are mentioned here. Perhaps the greatest drawback of the thermoplastic polymer is that it can be used only at ambient temperature. A significant problem is encountered while mixing fibrous material with a thermoplastic matrix due to the high viscosity of the latter at normal temperature. To make matters worse, thermoplastic polymers exhibit considerable strain at relatively low stresses. However, in the chemical industry, they are used in a range of products due to their property of chemical inertness, toughness and pleasing appearance.

Traditionally, thermoset polymers (also called resins) are widely used as a matrix material for fibre reinforced composites in structural composite components. The wetout from simple mixing of fibres and matrix is good. Thermoset polymers improve thermal stability and chemical resistance. The main disadvantages are their limited storage life at low temperature, the considerable time wastage using this matrix in fabrication in the mould and low value of strains to failure. The properties of both these types of matrices are given in Tables 1.3 and 1.4.

For the purpose of a simple classification, we may divide the thermosets into five categories
(i) Polyester resin, (ii) epoxy resin, (iii) vinyl ester resin, (iv) phenolic resin and (v) high performance resin.

Table 1.3 Typical properties of thermosetting resins

Properties	Thermosetting polymers			
	Polyester	Vinyl ester	Epoxy	Phenolic
Specific gravity	1.1 – 1.5	1.12	1.2	1.15
Tensile modulus (GPa)	1.1 – 4.5	3 – 4	2 – 6	3
Poisson's ratio	0.36	-	0.37	-
Tensile strength (Mpa)	40 – 90	65 – 90	35 – 130	50 – 75
Compressive strength (Mpa)	90 – 250	127	100 – 200	200
Elongation (%)	2.5	1 – 5	1 – 8.5	2
Coeff. of thermal expansion ($10^{-6}/°C$)	60 – 200	53	45 – 70	-
Water absorption (%)	0.1 – 0.3	-	0.1 – 0.4	-

1.3.2.1 *Polyester resins*

The most commonly used resin in glass reinforced plastic construction is the polyester resin and they have exhibited good performance. An unsaturated polyester resin is formed by the reaction of (i) a saturated difunctional acid, (ii) an unsaturated difunctional acid and (iii) a difunctional glycol. Compound (i) is usually an aromatic acid (e.g.,phthalic acid) whilst (ii) is usually aliphatic and component (iii) may be aromatic or aliphatic. These three components react together to form a polyester resin. The resin is dissolved in a vinyl monomer, usually sterene to form a syrupy liquid.

Table 1.4 Typical properties of some structural thermoplastic resins [1.3]

Material	SG*	Young's modulus GPa)	Tensile yield stress (MPa)	Tensile failure strain (%)	Heat distortion temp. (°C)	Comments
ABS (acrylonitrile butadiene styrene)	1.05	3	35	50	100	Used in some small craft, e.g., surfboards; poor weathering
PET (polyethylene terephthalate)	1.35	2.8	80	80	75	Used mainly in injection moulding; creep-susceptible
HDPE (high-density polyethylene)	0.95	1.0	30	600–1200	60	Low-cost, tough, water-resistant, creep and fatigue-susceptible
PA (polyamide, nylon 6/6)	1.15	2.2	75	60	75	Tough, fatigue-resistant; susceptible to moisture effects
PC (polycarbonate)	1.2	2.3	60	100	130	Good impact and fatigue resistant
PES (polyethersulphone)	1.35	2.8	84	60	203	Tough, temperature and fire resistant; used in aerospace components
PEI (polyetherimide)	1.3	3.0	105	60	200	as PES
PEEK (polyetherether ketone)	1.3	3.7	92	50	140	as PES

* specific gravity

The main advantages of polyester resins are their reasonable cost and ease with which they can be used.

1.3.2.2 *Epoxy resins*

Epoxy resins are mostly used in aerospace structures for high performance applications. It is also used in marine structures, rarely though, as cheaper varieties of resins other than epoxy are available.

The extensive use of epoxy resins in industry is due to: (1) the ease with which it can be processed, (2) excellent mechanical properties in composites and (3) high hot and wet strength properties (150°C). Performance of epoxies is superior to polyester resins due to their superior mechanical properties and better resistance to degradation by water and other solvents.

The chemistry of the epoxy resin components is such that it gives a better adhesion to reinforcing fibre than polyester resins.

1.3.2.3 *Vinyl ester resins*

Being a combination of the principles of both epoxy and polyester resin chemistry, vinyl ester resins have a close resemblance to polyester resins, but has a chemical similarity to epoxy resins. Vinyl ester resin is superior to polyester resin because it offers greater resistance to water. These resins provide superior chemical resistance and superior retention properties of strength and stiffness at elevated temperature.

In construction and marine industries, vinyl ester resins have been widely used in boat construction. The application of vinyl ester is limited mainly in the USA to small high performance hulls such as racing canoes and speed boats.

Vinyl ester resins are between polyester resins and epoxies from the cost point of view.

1.3.2.4 *Phenolic resins*

The main characteristics of phenolic resins are their excellent fire resistance properties. As such, they are now introduced in high temperature application areas. The recently developed cold-cure varieties of phenolic resins are used for contact moulding of structural laminates.

Phenolic resins have inferior mechanical properties to both polyester resins and epoxy resins, but have higher maximum operating temperature, much better flame retardant and smoke and toxic gas emission characteristics. Due to the above advantages, phenolic resins are the only matrix used in aircraft interior and other locations of public occupancy.

Phenolic resins are increasingly used in internal bulkheads, decks and furnishings in ships.

1.3.2.5 *High performance resins*

Attempts are on for the development of matrices with better properties at elevated temperatures. It has been observed that processing characteristics deteriorate with the increase of thermal stability.

The general chemical approach is to devise resins which incorporate aromatic, hetero-cyclic, and ladder polymer elements. The most highly developed systems are bismaleimide (BMI) and polimide (PI).

1.3.3 **Prepegs**

If fibre and matrix were available commercially as one entity, it avoids the procurement of fibre and matrix separately. Partly cured matrix resins act as a binder to a well laid out fibre system. These fibres are known as prepegs. They may be unidirectional or woven [1.5]. Due to its tacky texture, it is easy to handle and can best be used in moulding of complex geometrical shapes. The prepegs, however, have a limited shelf life. The manufacturers of these materials specify the condition at which they need to be stored in freezers.

1.3.4 **Fillers and Other Additives**

Fillers may be added to the polymeric matrix for one or more of the following reasons: (a) reduction of cost, (b) increase of modulus, (c) control of viscosity and (d) production of a smoother surface.

The most common filler in polyester and vinyl ester resins is calcium carbonate. It not only reduces the cost, but also lessens mould shrinkage. Examples of other fillers are clay, mica and glass microspheres. Although fillers increase the modulus of an unreinforced matrix, they also tend to reduce its strength and impact resistance.

The impact strength and crash resistance of brittle thermosetting polymers can be improved by mixing them with small amounts of elastic elastomeric toughners.

In addition to fillers and toughners, colourants, flame retardants and ultraviolet absorbers may be added to the matrix. Toughners are used for improving impact strength. Typical examples of toughners are silicon, rubber and butadiene styrene. The purpose of the colourant is to obtain the required colour. Typical examples are titanium dioxide, barium sulphate, magnesium carbonate and cadmium reds and yellows. Addition of flame retardants give fire retardant properties to the composites. Typical examples are red phosphorous, antimony trioxide, butyl acid phosphate and aluminium hydroxide. To prevent discolourisation of the composite from exposure to sunlight, ultraviolet absorbents are used. Typical examples are acetyl salicyclic acid and benzotriazoles hydroxphenyl.

1.4 LAMINA AND LAMINATE

A lamina or a ply is formed by a combination of a large number of fibres in a thin layer of matrix. Fibres in the lamina may be continuous or discontinuous, arranged in a specific direction or in a random orientation. A unidirectional lamina is one where the fibres in a lamina run parallel to one another in a particular direction. It is natural that discrete fibre composites will have lower strength and modulus than continuous fibre composites. However, with the random orientation of the fibre, it is possible to obtain nearly equal mechanical and physical properties in all directions in the plane of the lamina. The thickness of the lamina ranges from 0.1–1 mm. The standard thickness of a unidirectional ply is 0.125 mm whereas typical thickness of the woven ply is 0.25 mm.

The principal coordinate axes of an orthotropic lamina are shown in Fig. 1.2. The three principal axes are: axis 1, the longitudinal axis in the direction of fibre, axis 2, the transverse axis in the direction normal to the fibre and axis 3 in the direction normal to the plane of the lamina.

A laminate is formed by stacking several laminas (Fig. 1.3). It is the most common form of fibre reinforced composites. It is made of a desired thickness so as to enable it to support a given load and maintain a given deflection. Fibre orientation of each lamina and stacking sequence of various layers can be varied to obtain a wide range of physical and mechanical properties of composites.

1.5 GENERAL CHARACTERISTICS OF FRPS

Traditional structural materials like steel and aluminium are considered as isotropic. The properties of the fibre reinforced composites are, however, strongly dependent on the direction of measurement. For example, properties such as the tensile strength and the tensile modulus in a unidirectional FRP attain maximum values in the longitudinal direction of fibres. At any other angle of measurement, mechanical properties attain lower values, being minimum at perpendicular to the longitudinal direction of fibres. Similar angular dependence is observed for other physical and mechanical properties such as the coefficient of thermal expansion, thermal conductivity and strength. Bi– or multidirectional reinforcement gives a more balanced set of properties. Although the magnitude of these properties are lower than the longitudinal properties of unidirectional composites, they represent a considerable advantage over common structural materials on a unit weight basis.

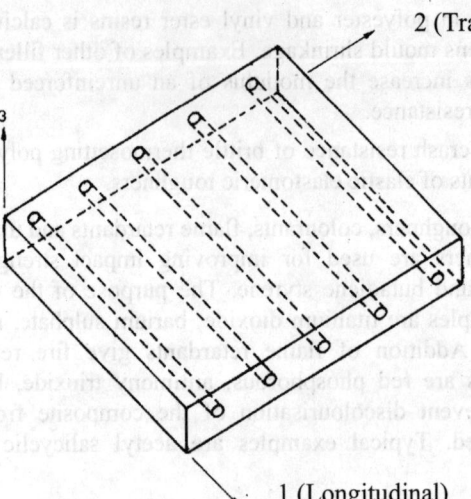

Fig. 1.2 Principal coordinate axes of a lamina

The design of metal structures is more or less straight forward, but not so for FRPs which exhibit different properties in different directions. But due to this anisotropy, properties of the FRP can be tailored to meet the design requirement. This flexibility enables the designer to selectively reinforce a structure in the direction of major stresses, increase its stiffness in a preferred direction and produce structures with zero coefficient of thermal expansion.

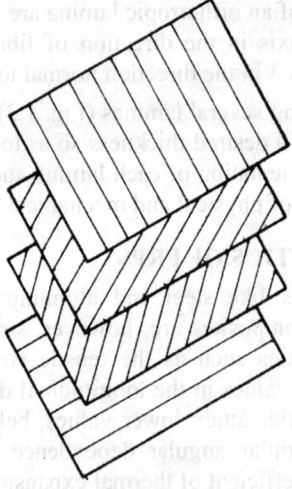

Fig 1.3 Laminate construction

1.6 WHY FRPS ?

The development of advanced fibre reinforced composite materials has been considered as the biggest technical revolution after the jet engine [1.10]. Fibre reinforced composites possess high strength and stiffness. Some of these materials perform equally well or better than many traditional metallic materials. In addition, fatigue strength-to-weight ratios as well as fatigue damage tolerance of many composite laminates are excellent.

Coefficients of thermal expansion for many fibre reinforced composites are much lower than those of metals. As such composite structures exhibit a better dimensional stability over a wide range of temperature variation. However, differences in thermal expansion between metals and composite materials may create undue thermal stresses when they are used in conjunction, for example, in an attachment.

Fibre reinforced composites possess high internal damping. This leads to a better vibrational energy absorption within the material, and results in reduced transmission of noise, vibration and harshness (NVH).

In unfavourable environments, metals are usually susceptible to corrosion. The non-corroding behaviour of fibre reinforced composites is an added advantage. Cracks and flaws in metals grow during the service life of the structure. Though they are easy to detect, their repair work may not be simple. Damage of composite structures is usually internal in nature and can be detected only by sophisticated non-destructive testing. Protective coatings are applied on the surface of composite structures to take care of likely damages on the surface. Many polymer matrix composites absorb moisture from the surroundings which creates dimensional changes as well as adverse internal stresses in the material.

A significant advantage of the FRP construction is its low maintenance cost.

The cost of raw materials of composite structures such as fibres and auxiliary materials are high in fabrication and assembly operations. Though composite structures are much lighter, their tooling costs are also less and they usually possess less number of parts. In conventional structural materials, though the cost of raw materials is less, more often the cost involved with tooling, machinery and assembly is high, thus offsetting the initial advantage.

Structures made of fibre reinforced composite are lighter than those made of conventional materials. In space vehicles, reduction in weight is linked to fuel savings. Spacecraft may have weight savings as much as 40 percent if fibre reinforced composite structures are used.

1.7 MICROMECHANICS AND MACROMECHANICS

Micromechanics deals with the deformation and stress in the basic constituents of a structure. It deals with local failures such as matrix failure, fibre failure and interface/interphase failure. As such the constituent materials are examined on a microscopic scale without recourse to their internal structure.

Macromechanics deals with the behaviour of the composite material presumed as homogeneous and the effects of the composite material are detected only as average apparent properties of composites. In macromechanics, properties along the length and perpendicular to the fibre direction are considered.

1.8 PROPERTIES OF TYPICAL COMPOSITE MATERIALS

Various types of composite materials are possible by choosing different constituents mixed in different proportions. Mechanical, thermal and electrical properties of some typical composite laminates are given in Table 1.5. It can be seen from Table 1.5 that composites have higher specific modulus and higher specific strength than steel.

Fibre properties dictate the stiffness of unidirectional composites. Figure 1.4 indicates the stress–strain curves of a typical unidirectional composite. The stress–strain curve for aluminium is presented in the figure for the sake of comparison. For higher strength of unidirectional composites, the ultimate strain is generally lower at failure. For some composites, the increase of stiffness has resulted in lowering the strength (curves 4, 5, 6, 7). The stress–strain curve for unidirectional composites in general is linear.

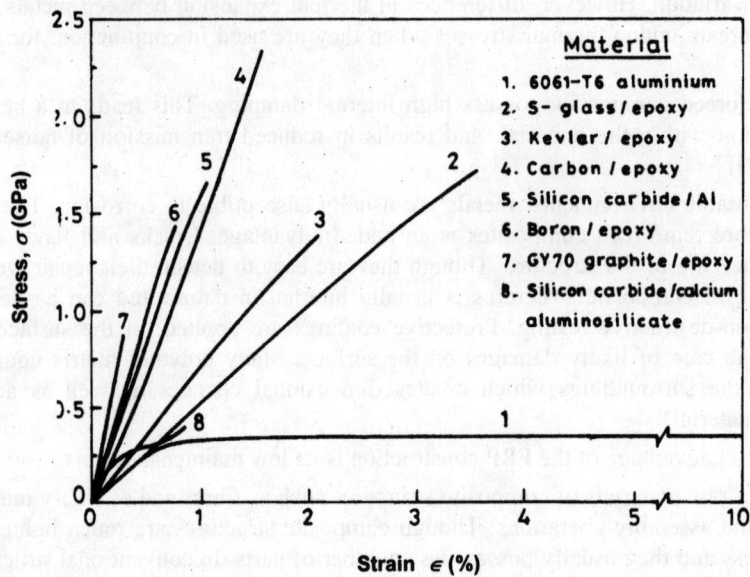

Fig. 1.4 Stress-strain curves of typical unidirectional composites in the fibre direction

1.9 APPLICATION OF COMPOSITES

Commercial and industrial applications of fibre reinforced composites are diverse and varied. Some of these applications are ships and submarines, aircrafts and spacecrafts, trucks and rail vehicles, automobiles, robots, civil engineering structures and prosthetic devices. The main application areas may be broadly classified as follows

1. Marine field
2. Aircraft and space
3. Automotive
4. Sporting goods

1.9.1 Marine Field

Use of composites in the marine field is growing steadily since the early 1950s. Initial applications of FRPs were limited to small crafts such as lifeboats and pleasure boats. Now, structures of several hundred tonnes are regularly produced and used. Potential applications in the marine field range from small components such as radar domes, masts, and piping to large-scale structures, submersibles and offshore structure modules [1.3].

Glass reinforced plastics (GRP) are extensively used in the construction of boat hulls including yachts, lifeboats, dinghies, canoes, speed boats, fishing boats and passenger launches. The popularity of GRP with boat builders lies in its competitive low cost (in comparison to wooden hulls), a trouble-free performance, low maintenance cost and aesthetics.

Table 1.5 Properties of some typical unidirectional composite lamina[1.7, 1.20, 1.27]

Material	Fibre volume fraction	Specific gravity	Young's modulus (GPa)	Shear modulus (GPa)	Tensile strength (MPa)	Compressive strength (MPa)	Shear strength (MPa)	Specific Young's modulus	Specific tensile strength	Longitudinal thermal expansion coefficient [10^{-6}/°C]	Longitudinal moisture expansion coefficient
E-glass polyester (CSM)	0.18	1.5	8	3	100	140	75	5.3	67	30	0
E-glass polyester	0.43	1.8	30	3.5	750	600	-	16.7	417	-	0
E-glass epoxy	0.55	2.10	39	3.8	1080	620	89	15	514	7	0
S-glass epoxy	0.50	2.00	43	4.5	1280	690	69	21.5	640	5	0
Carbon epoxy (high modulus)	0.62	1.70	300	20	700	650		176	412	- 0.9	-
Carbon epoxy (high strength)	0.62	1.60	140	15	1500	0		87	937	- 2	0.01
Kevlar epoxy (Kevlar 149/epoxy)	0.60	1.38	87	2.2	1280	335	49	63	927	- 0.1	0
Graphite epoxy	0.57	1.59	294	4.9	589	491	49	140	370		0
Boron epoxy	0.50	2.03	201	5.4	1380	1600	62	99	680	6.1	0
Mild steel	-	7.8	207	8.0	325	340	190	26.5	42	12	0

GRP has been successfully used in military and commercial hovercrafts. Feasibility studies have indicated substantial savings in hull weight by using hybrid glass/carbon laminates in place of steel and aluminium for the construction of hydrofoils and fast patrol boats [1.8]. Construction of a GRP vessel proves to be costly, but their maintenance works out to be cheaper.

Ultra-high performance sailing craft and power boats have advanced composite construction [1.9].

FRP vessels may be constructed for special purposes, such as transportation of corrosive cryogenic bulk cargo, requirement of a non-magnetic hull or requirement of considerable savings in weight for better performance.

Naval applications of FRP include mine countermeasure vessels (MCMV), landing craft, fast patrol boats and submersibles. GRP has found a favourable application in MCMVs where a hull with negligible magnetic signature is required in order to avoid the activation of magnetic mines. Additional developments in naval application include weapon enclosures, gun enclosures, rudders, dry dock shelters, missiles, blast shields, ladders, deck drains, rails, radomes, masts and stacks.

FRP is used in submarines for flooded nose fairings using planes and non-pressure hull decks. A specific requirement for an underwater vessel is that of high specific compressive strength. Care should be exercised while designing to check failure against buckling, under fatigue and impact loads and against creep. GRP submersibles have been successfully used in offshore operations.

Other marine applications of FRP include submarine casings and appendages, superstructure of ships, warship radomes, sonar domes, ship's piping and ventilation systems, oil and water storage tanks, floats and buoys for fishing and mine sweeping purpose.

Hull–superstructure interaction can be avoided by using a low modulus material like the GRP. The elastic modulus of GRP is less than 10% that of steel, while the strength of the GRP is comparable with that of steel.

GRP sheathing is used to protect wooden hulls from bores, leakage and rot. Sheathing is a cover made by one or two plies of CSM attached by polyester or epoxy resin.

1.9.2 Aircraft and Space

Optimally an aircraft requires a reduction in weight to attain greater speed and increased payload and fibre-reinforced composite have been found to be ideal for this purpose. No doubt for one of the most important application areas of FRP is in the field of civil and commercial aircrafts. Carbon fibres either alone or in the hybridized condition is used for a large number of aircraft components. Carbon and Kevlar have become the major material used in many wing, fuselage and empennage components. They are also used in secondary structures such as elevator facesheets, horizontal stabilizers, upper rudder et al., of many commercial aircrafts.

FRP with epoxy as the resin is used for the manufacture of helicopter blades. One of the main reasons why FRP is used for rotor blades is the ability of the material to tailor the dynamic frequency of the blade to its operating parameters. FRPs are more suitable for blade application than metals, as with this material, blades of any shape can be manufactured without any additional cost, an advantage which does not hold good for metal blades whose shapes are limited to those which can be extended, machined or rolled.

A missile structure, when made of FRP is light, and has an increased range of action and payload. A missile structure made of FRP reduces the weight of the structure considerably. In ICBMs and other missile systems, graphite composites are used for its high stiffness, strength and minimum weight.

There is a wide variation of temperature in space and as such the dimensional stability of spacecraft components to maintain precise alignment of communication and sensor systems is a major requirement. Graphite and Kevlar fibres are well suited for space applications because of their high specific strength and modulus and low coefficient of thermal expansion. Strength and stiffness of composites are major considerations for the aircraft whereas stiffness and low coefficient of thermal expansion are the major requirements for space applications.

Some of the application areas of FRP to spacecraft are antennas, booms, support trusses and struts. Carbon–epoxy composite tubes are used in constructing truss structures for Low Earth Orbit (LEO) satellites and interplanetary satellites.

1.9.3 Automotive Field

FRPs have been used in many parts of the car.

The exterior part of the car such as hood or door panels requires sufficient stiffness. The other requirement is that it should offer maximum resistance to dent formation (damage tolerance). Resins like polyeurathanes enable the damage tolerance to be limited to acceptable values. Further, a good surface finish is highly desirable.

Application of FRPs include the chassis components as well, such as corvette rear leaf spring and unileaf E-glass reinforced epoxy.

In racing cars, parts of the engines are made of graphite–epoxy. Connecting rods which are subjected to fatigue are now made of composites for better performance. Other parts of the engine where composites are used are push rods, rocker arms, pistons, cylinder leads and engine blocks. The advantage of using advanced composites in engines is higher speed with the production of more power and the simultaneous reduction of engine weight.

Use of FRP components in automotive industries is much less than the aircraft industries. For aircraft components, the basic technique used for manufacture is the hand lay-up, whereas in automotive industries more sophisticated techniques of fabrication such as compression moulding, resin transfer moulding, filament winding and pultrusion are used.

1.9.4 Sporting Goods

Many sporting goods are made of FRPs nowadays. One of the major advantages of using FRP is the reduction of weight.

Tennis rackets or snow skis are made as a sandwich structure – FRP with carbon or boron fibre as the skin and the core formed by soft and light urethane foam which enables the structure to have a weight reduction without any decrease in stiffness.

FRPs enable damping of vibrations. Therefore, shock resulting from the impact of the ball on the tennis racket which is transmitted to the arm of the player will dampen out at a quicker rate.

Other application areas of fibre reinforced polymers in sports are fishing rods, bicycle frames, archery bows, sail boats and kayaks, oars, paddles, canoe hulls, racket balls, rackets, javelins, helmets, golf club staff, hockey sticks, athletic shoe soles and heels, surfboards and many other items.

REFERENCES AND SUGGESTED READINGS

1.1 Isaac M. Daniel and Ori Ishai, *Engineering Mechanics of Composite Materials*, Oxford University Press, 1994.

1.2 P. K. Mallick, *Fiber Reinforced Composites*, 2nd Edition, Marcel Dekker Inc., USA, 1993.

1.3 C. S. Smith, *Design of Marine Structures in Composite Materials*, Elsevier Applied Science, England, 1990.

1.4 D. Pamington (Ed.), *Carbon and High Performance Fibres Directory*, 4th Edition, Pammac Directories Ltd., U. K., 1988.

1.5 M. H. Datoo, *Mechanics of Fibrous Composites*, Elsevier Applied Science, London, 1991.

1.6 G. Luben, (Ed.), *Handbook of Composites*, Van Nastrand Inc, New York, 1986.

1.7 A. F. Johnson, *Engineering Design Properties of GRP*, Publication No. 215/1, British Plastic Federation, 1978.

1.8 *Engineering Materials Handbook*, Vol. I - Composites, American Society of Metals, 1987.

1.9 C. S. Smith and A. H. Monks, Design of High Performance Hulls in the Fibre Reinforced Plastics, *Proceedings of the Symposium on Small Fast Warships and Security Vessels*, RINA, London, May, 1982.

1.10 Jack Vinson, *The Behaviour of Sandwich Structures of Isotropic and Composite Materials*, Technomic Pub. Co., 1999

1.11 R. A. Shenoi and J. F. Wellicome, *Composite Materials in Marine Structures*, Vol. I & II, Cambridge University Press, 1993.

1.12 R. F. Gibson, *Principles of Composite Materials Mechanics*, McGraw Hill, 1993.

1.13 J-M. Berholet, *Composite Materials: Mechanical Behaviour and Structural Analyses*, Springer-Verlag, 1998.

1.14 A. L. Kalamkarov and A. G. Kolpakov, *Analyses, Design and Optimisation of Composite Structures*, John Wiley & Sons, 1997.

1.15 R. M. Jones, *Mechanics of Composite Materials*, 2nd Edition., McGraw Hill, 1999.

1.16 J. R. Vinson and R.L. Sierakowski, *The Behaviour of Structures Composed of Composite Materials*, Kluwer Academics, 1986.

1.17 R. M. Christiansen, *Mechanics of Composite Materials*, R. E. Krieger Pub Co., 1991.

1.18 B. D. Agarwal and L. J. Broutman, *Analysis and Performance of Fibre Composites*, John Wiley and Sons, 1990.

1.19 J. Hult and F. G. Rammerstorfer, *Engineering Mechanics of Fibre Reinforced Polymers and Composite Structures*, Springer-Verlag, 1995.

1.20 M. M. Schwartz, *Composite Materials*, Vol. 1 & 2, Prentice Hall, 1997.

1.21 E. J. Barbero, *Introduction to Composite Materials Design*, Taylor and Francis, 1998.

1.22 D. Hull and T. W. Clyne, *An Introduction to Composite Materials*, 2nd Edition., Cambridge University Press, 1996.

1.23 O. O. Ocahoa and J. N. Reddy, *Finite Element Analysis of Composite Laminates*, Kluwer Academics, 1996.

1.24 L. Hollaway, *Polymer Composites for Civil and Structural Engineering*, Blackie Academic & Professional London, 1993.

1.25 M. W. Hyer, *Stress Analysis of Fiber-Reinforced Composite Materials*, McGraw-Hill, International Editions, Massachussets, 1998.

1.26 D. Hull, *An Introduction to Composites Materials*, Cambridge University Press, U.K., 1981.

1.27 B. W. Rosen, *Mechanics of Composite Strengthening in Fibre Composite Materials,* American Society for Metals, Ohio, USA, 1965

CHAPTER 2

PROCESSING OF FRP COMPOSITES

An extensive range of well-established processing methods is available for FRP composites. These vary from simple labour intensive methods suitable for one-off products to automated methods for producing large numbers of complex components. The method of processing selected by a manufacturer depends on factors such as shape, cost, number of components and required performance.

2.1 CONTACT MOULDING

2.1.1 Mould Preparation

By far the most common method of fabrication for large structures such as ship hulls is contact moulding in an open female mould using cold curing polyester resin and E-glass reinforcement. The first step in the fabrication process is the mould preparation. For small to medium size structures, moulds are usually fabricated in GRP, in which case a male plug, commonly of wooden construction finished in GRP, is first assembled whose external shape defines the structure to be built. Very large moulds for ship construction may be of steel or aluminium construction lined with an epoxy paste or similar filler to allow fairing out of welded distortions. Mould preparation is usually completed by wax polishing and application of polyvinyl alcohol (PVA) or an equivalent release agent. Lamination is usually started by application of a pigmented gel coat of good quality resin, deposited in the mould by brush or spray (to a thickness between 0.3–0.5 mm), the main purpose of which is to provide a smooth external surface. Lamination is then continued, before the gel coat has fully cured, using one of the following two methods– spray-up or hand lay-up.

2.1.2 Spray-up

Glass fibre rovings, chopped to a length of 25–50 mm, are sprayed simultaneously with polyester resin, the latter being mixed with a catalyst and accelerator at the spray gun. The glass–resin mixture is consolidated by manual rolling, providing a laminate with a fibre weight fraction of 0.25–0.3. Much of the labour involved in hand lay-up is eliminated by this fabrication process, which tends itself to automated, production line manufacture of large numbers of small, low performance hulls. Control of thickness, however, is difficult and the quality of laminate is generally lower than can be obtained by hand lay-up.

2.1.3 Hand lay-up

Resin mixed with a catalyst is deposited liberally on the gel coat or on a previous ply of impregnated reinforcement by a roller-dispenser, brush or spray gun. Each ply of reinforcement, in the form of CSM with a real weight of 300–600 g/m^2 or woven rovings with a real weight in the range of 400–800 g/m^2, is dispensed from a roll, typically 1–1.5 m wide, and is wetted out and consolidated by rolling or brushing into the wet resin. In WR adjacent strips of reinforcement within a ply may be lapped or butted; in either case the strips of reinforcement forming the subsequent plies must be staggered to avoid a continuous line of weakness in the material. The resulting laminate usually has a fibre weight fraction between 0.45 and 0.55.

This requires little capital equipment but is labour intensive. It is particularly suited for a limited number of a particular structure. The main disadvantages of the method are the low reinforcement content and the difficulty in removing all the trapped air; hence the mechanical properties are not as good as in other processes.

2.2 COMPRESSION MOULDING METHODS

2.2.1 Matched Die Moulding

This method is widely used for long production runs for components ranging in size from small domestic items to doors and cab panels for large commercial vehicles. The material to be shaped is pressed between heated matched dies, as illustrated in Fig. 2.1. The pressure required depends on the flow characteristics of the feed material and may be as high as 50 MPa but is usually less than 10 MPa. The feed material flows into the contours of the mould and when the temperature is high enough, rapidly cures. The time for the complete moulding process depends on the feed material, on the dimension of the components and on whether pre-heating of the feed has been employed to shorten the time. The time required typically range from several seconds to several minutes. Good mould detail and dimensional accuracy are possible although the cost of a complex tool steel die has to be considered.

Two forms of feed which are particularly suited to matched die moulding are sheet moulding compounds (SMC) and dough moulding compounds (DMC). SMC is a prepared sheet of resin–fibre blend which contains all the necessary additives such as curing agent, release agent, and pigment. It reduces the number of components to be stored, is clean to use and results in a good consistency in the finished component. As all the constituents are pre-mixed, SMC has a shelf life of around three to six months at room temperature. DMC is also a blend of all the necessary constituents, but only short fibres are used. The resulting fibrous mixture has the consistency of a dough or putty and can be readily made into accurately measured quantities for the feeding process. The shelf life of DMC is less than that of SMC.

2.2.2 Forming Methods Employing Gas Pressure

These forming methods are sometimes known as bag moulding processes and can be categorised under three broad headings.

The first of these is vacuum bag moulding in which, unlike the case of matched die moulding, only one mould is required. This process, [see Fig. 2.2(a)], may be regarded as an extension of the contact moulding process. It involves placing over the mould a flexible membrane, separated from the uncured laminate by a film of PVA, polythene or equivalent material, sealing the edges and evacuating the air under the membrane so that the laminate is subjected to a pressure of up to 1 bar. Curing may be accelerated by placing the component in an oven or employing a heated mould.

Autoclave moulding is a modification of vacuum forming that uses pressures in excess of atmospheric pressure (e.g., 5–15 bar) to produce high density, reproducible products for critical applications such as those needed in the aerospace industry. The mould is situated in an autoclave (pressurised oven), [see Fig. 2.2(b)], which has facilities for heating and pressurising by a gas, usually nitrogen.

The pressure bag works on a similar principle in that a pressure in excess of atmospheric pressure is used for shaping but it is cheaper as it does not require an autoclave. A flexible bag is placed over the lay-up on the mould. Inflation of the bag by compressed air, forces the lay-up into the mould as shown in Fig. 2.2(c).

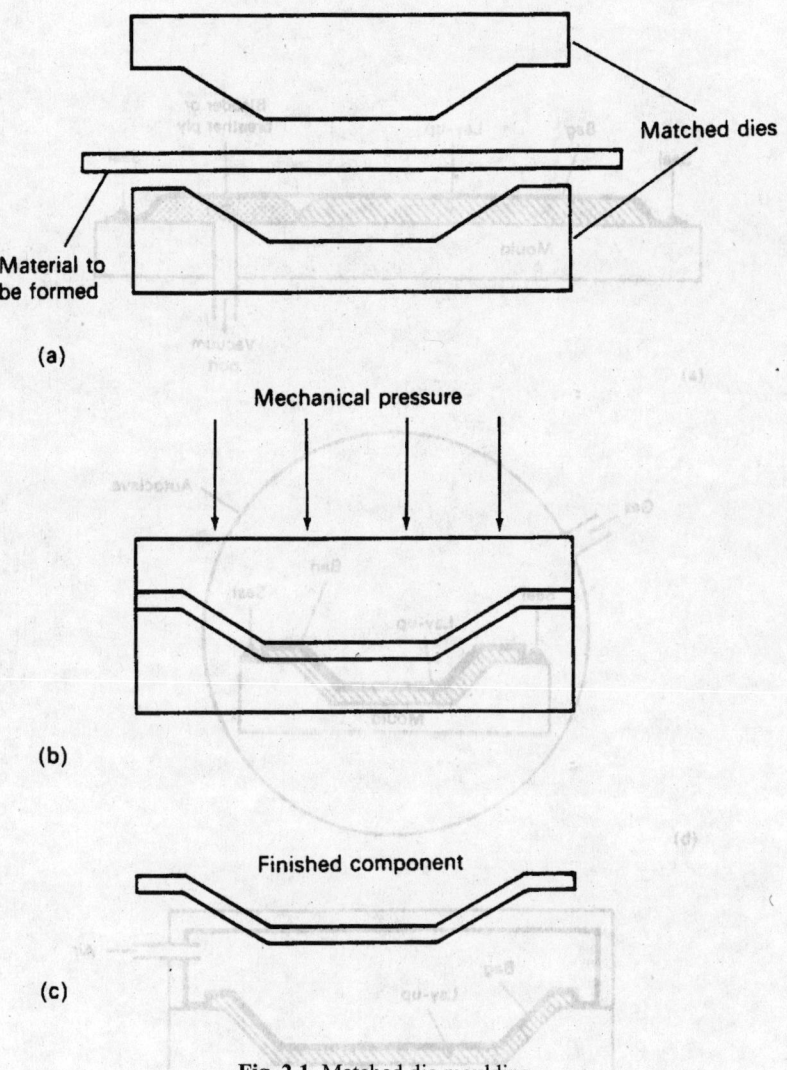

Fig. 2.1 Matched die moulding

2.2.3 Low Pressure, Closed Mould system

The methods considered in this section consist of placing the reinforcement in a closed mould and then inserting the resin material into the mould to infiltrate the reinforcement.

In resin transfer moulding (RTM), the low viscosity resin is injected into the closed mould using low pressure and is subsequently cured. A consequence of the use of low pressures is that inexpensive moulds, made for example from GRP, have sufficient strength. Such moulds facilitate the manufacture of complex shapes and large components without the need for high cost tooling.

However, as the mould material does not have good high temperature properties, curing has to be carried out slowly, to restrict any temperature rise which could damage the mould. In fact, the production cycle is long. For large components it may even take days, as the infiltration stage is also slow owing to the low pressures involved.

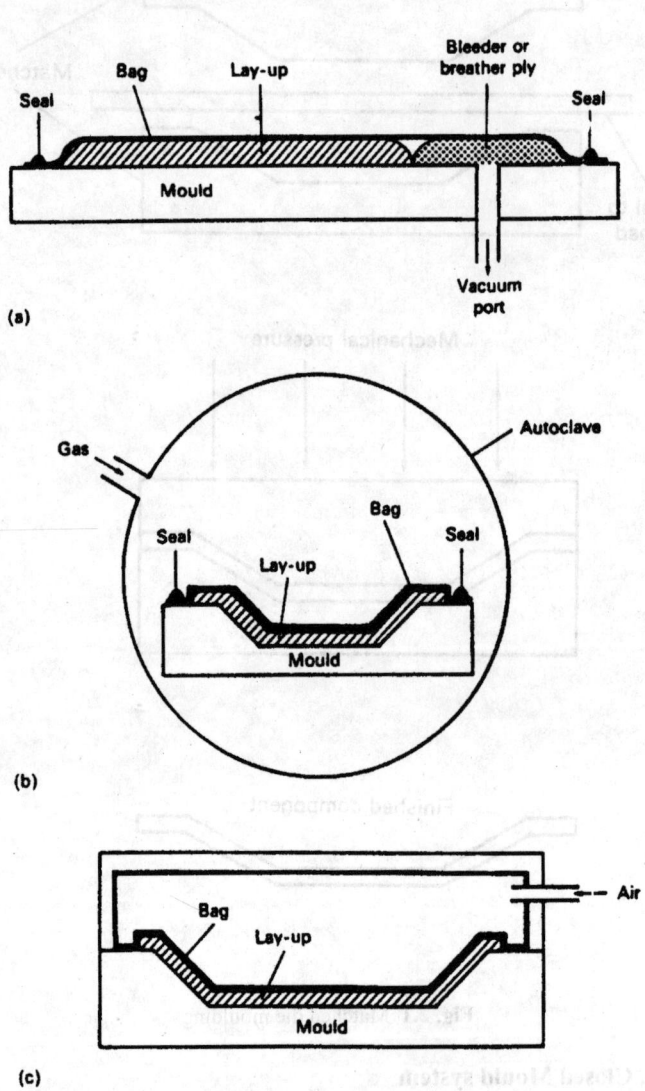

Fig. 2.2 Open mould forming methods employing gas pressure:
(a) Vacuum forming; (b) Autoclave moulding; (c) Pressure bagging

The low pressures required for RTM may be obtained by extracting the air from the mould and allowing atmospheric pressure, or even lower pressure, to force the resin into the mould. This variant of RTM is called vacuum-assisted resin injection moulding (VARIM).

Instead of using pre-catalysed resin with a slow cure, it is possible to mix two fast reacting components to make the resin just prior to injection into the mould containing the pre-form. The components are mixed at high pressures in an impingement mixing chamber and then injected into a mould where the pressure is usually less than 1 MPa (Fig. 2.3). This is followed by a rapid curing so that the cycle time for this process, which is known as reinforced reaction injection moulding (RRIM), is far less than that for VARIM and is typically 1–2 minutes.

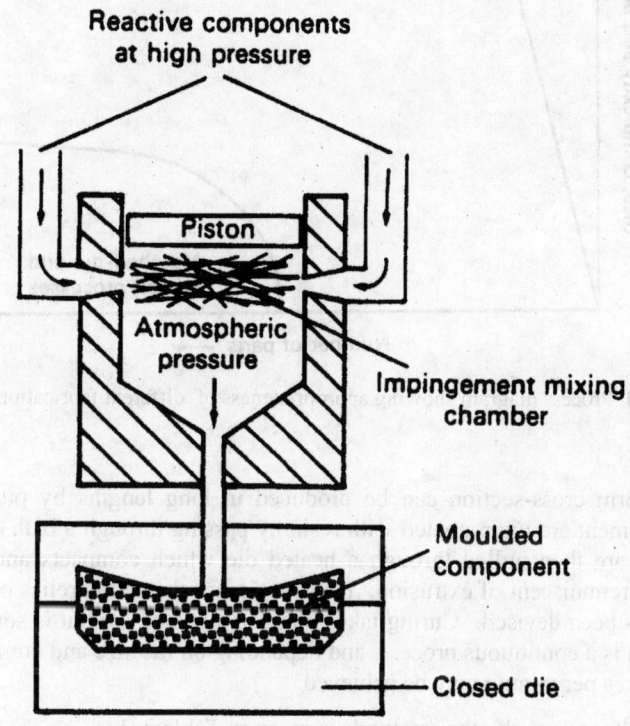

Fig. 2.3 Diagram of reinforced reaction injection moulding

It is important to appreciate the relative merits of the different processing methods and to know under what circumstances a particular method is likely to be selected for manufacture. It is therefore appropriate to recap some of the main features of the methods discussed so far. Hand lay-up can be used to produce complex and/or large structures and components in small quantities. The properties obtained are variables depending on the ratio of constituents used. Capital costs here are low, but it is labour intensive and slow. Therefore these methods are used in region A of Fig. 2.4. The equipment for matched die moulding methods is expensive, but components can be produced rapidly. These, and related methods, are especially suited for the production of large number of components the complexity, of which is limited by the need to use steel dies (region C). RTM processes lie between the two extremes (region B); they are employed for relatively small runs on simple components and for longer runs on more complex components.

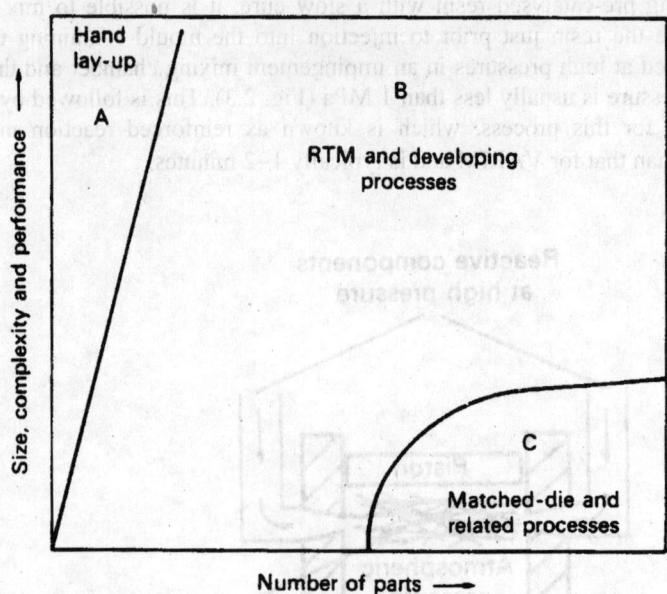

Fig. 2.4 Process diagram showing appropriateness of different fabrication methods

2.2.4 Pultrusion

Rods of uniform cross-section can be produced in long lengths by pultrusion. Continuous rovings of reinforcement are impregnated with resin by passing through a bath of resin (Fig.2.5). The impregnated fibres are then pulled through a heated die which compacts and shapes the required profile in a manner reminiscent of extrusion. However, since the action relies on a pulling action, the name pultrusion has been devised. Curing takes place in a heated die but is sometimes completed in an oven. Pultrusion is a continuous process, and depending on the size and complexity of the section, rates of several metres per minute may be achieved.

A comparison of pultrusion with other methods is given in Table 2. 1.

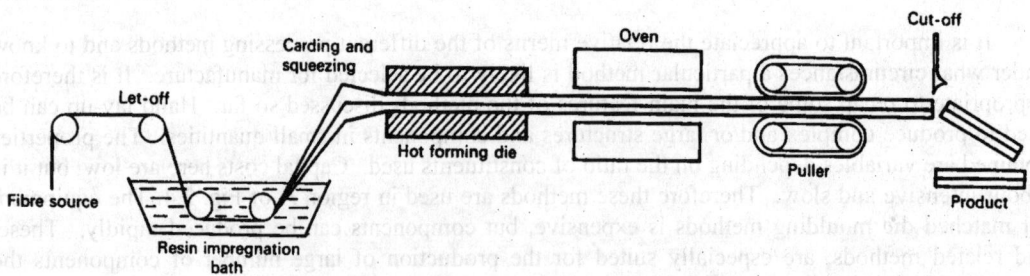

Fig. 2.5 Schematic diagram of the pultrusion process

Table 2.1 Comparison of costs and efficiencies of FRP production processes

Process	Typical cycle time	Equipment capital (Rs. 100,000)	Mould capital (Rs. 100,000)	Product value per cycle (Rs.)	Product value per hour (Rs.)	Process efficiency*
Compression	3 min	37.5	3.5–14	0.75–3.75	14–75	365–1428
Autoclave	8h	105	0.75	7.5–75	1–9	8.3–79
Filament	4h	14–75	0.75	7.5–75	2–18	119–250
VARIM	10–60 min	3.5–7	0.2–0.75	0.75–7.5	4.5–7.5	120–1000
Pressure Bag	1h	3.5	0.08–0.35	0.75–3	0.75–3	200–800
Spray	3h	3.5	0.08–0.75	3.75–19	1.5–6	400–1600
Hand Lay-up	5h	0	0.08–0.75	3.755–19	1.2	10000
Pultrusion	0.5–3 m/min	37.5–75	1.5–7.5	2.25	68–405	2884–5400

Process Efficiency = (Product value per hour × 10^{10}) / Total capital

2.3 FILAMENT WINDING

Structures in the form of bodies of revolution, including cylindrical and spherical shells and cylinders with hemispherical or torispherical end closures may be fabricated economically and to high performance standards by filament winding. Fabrication is carried out by winding reinforcing fibres, in the form of a continuous roving, which may be impregnated with resin just prior to winding (wet-winding) or may be pre-impregnated with partially cured resin (dry-winding), on to a mandrel which defines the internal geometry (Fig. 2.6). Winding may be helical, in which case the mandrel rotates continuously while the fibre-feed arm shuttles back and forth at speeds regulated to provide the required winding angles. Or the winding may be polar, in which case the usual process comprises the rotation of the fibre-feed arm in a longitudinal plane around a stationary mandrel. Mandrels, which must be able to withstand compression induced by the winding tension and possibly also high curing temperatures, may be steel tubes in the case of tubular windings.

Filament wound components include underwater pipelines and ships pipework, oil and water storage tanks, air bottles, buoys, radomes, torpedo hulls, helicopter blades, etc.

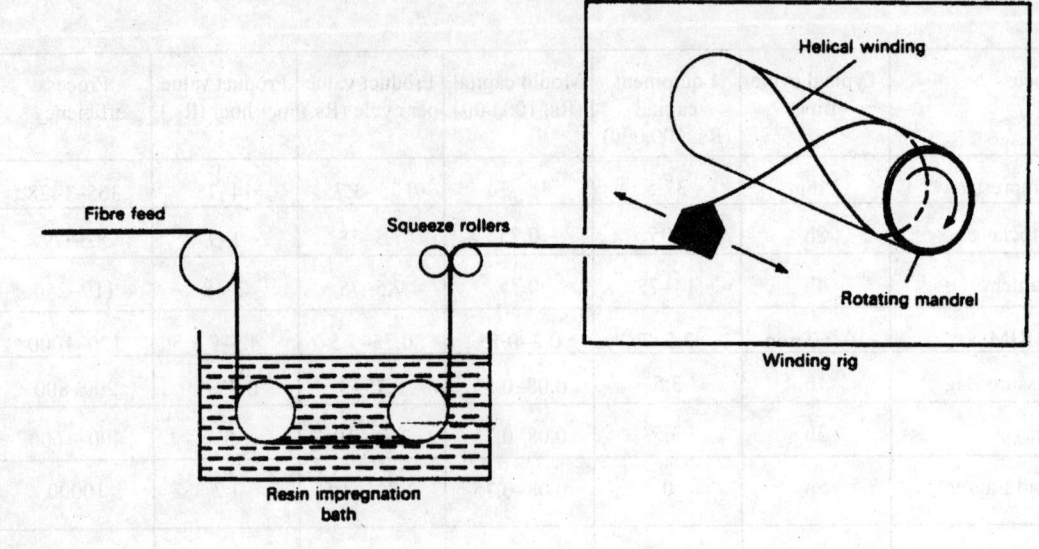

Fig. 2.6 Schematic diagram of a filament winding process

CHAPTER 3

MICROMECHANICAL ANALYSIS OF COMPOSITE STRENGTH AND STIFFNESS

3.1 INTRODUCTION

Elastic properties of fibre reinforced plastic materials are considered as a function of properties of fibres and resins. Properties of the fibre - reinforced composite on the basis of its constituent elements can be done by a variety of methods [3.1].

1. Strength of materials approach
2. Self-consistent model
3. Variational approach
4. Numerical methods
5. Empirical equations

Earlier attempts made for the evaluation of mechanical properties are based on strength of materials approach. The derivation is based on simple assumptions, such as the existence of uniform stress or uniform strain in its constituents. It has resulted in a fibre–matrix interaction [3.2] or a rule of mixtures formulation [3.3], where the constituents – the fibre and the matrix, are assumed to function together in series or parallel. Though the estimation of the longitudinal properties on this basis are found to be adequately accurate, the transverse and shear properties of the composite are underestimated by this approach.

The theory of elasticity solution has been obtained for the elastic components by modelling a FRP to consist of a fibre surrounded by a cylindrical matrix [3.4, 3.5]. This self-consistent model neglects the interaction between fibres and as such underestimates the composite properties for higher fibre volume ratios.

Application of energy theorems to the composite has shown that the assumption of series or parallel connection yields upper and lower bounds of properties. Subsequent work dealt primarily with the requirement of the energy methods to achieve closer spacing of bounds and to provide solutions for a range of possible fibre–matrix geometries [3.6, 3.7, 3.8].

The difficulty in applying numerical methods such as the finite difference method, the finite element method etc., is that they do not yield a closed bound solution [3.9]. The result may at best be available in graphical form.

Empiricism has been introduced for the prediction of properties of composites. Some empirical factor based on experiments is introduced between the upper and lower bound solutions for more exact prediction of properties.

In the following, a single layer of fibre - reinforced composite material is considered – the fibre and the matrix are placed side by side alternately [Fig. 3.1].

3.2 VOLUME AND WEIGHT FRACTIONS

One of the primary factors that determine the properties of composites is the relative proportion of the fibre and the matrix. The relative proportion can be expressed in terms of weight or volume fractions. Weight fractions are easier to obtain during manufacture or by an experimental method after manufacture. Volume fractions on the other hand are more convenient for theoretical calculations. Hence, it is desirable to determine expressions for conversion between weight and volume fractions.

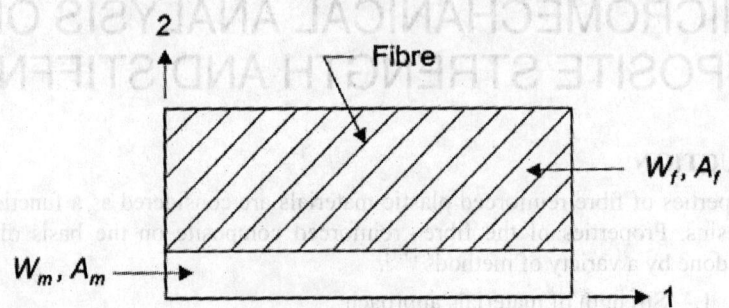

Fig. 3.1 Idealisation of the unidirectional composite

Consider a case of weights/volumes of fibres, matrix and net composite being w_f/v_f, w_m/v_m and w_c/v_c respectively (subscripts c, f and m stand for composite, fibre and matrix respectively). Let this volume and weight fractions be given by V and W respectively.

$$V_f = \frac{v_f}{v_c}, \quad V_m = \frac{v_m}{v_c} \tag{3.1}$$

We know,

$$v_f + v_m = v_c \tag{3.2}$$

Dividing both sides of (3.2) by v_c and using the relationship of (3.1) yields

$$V_f + V_m = 1 \tag{3.3}$$

Similarly,

$$W_f = \frac{w_f}{w_c} \quad \text{and} \quad W_m = \frac{w_m}{w_c} \tag{3.4}$$

Again,

$$w_f + w_m = w_c \tag{3.5}$$

Following a similar procedure as above

$$W_f + W_m = 1 \tag{3.6}$$

If ρ indicates density, then from (3.5), we get

$$\rho_f v_f + \rho_m v_m = \rho_c v_c \tag{3.7}$$

Dividing both sides of (3.7) by v_c and using (3.1) yields

$$\rho_c = \rho_f V_f + \rho_m V_m \tag{3.8}$$

Similarly, replacing volumes in terms of weights and densities, (3.2), yields

$$\frac{w_c}{\rho_c} = \frac{w_f}{\rho_f} + \frac{w_m}{\rho_m} \qquad (3.9)$$

Combining (3.9) with (3.4), yields

$$\rho_c = \frac{1}{\dfrac{W_f}{\rho_f} + \dfrac{W_m}{\rho_m}} \qquad (3.10)$$

Thus, the density of a composite material is expressed in (3.10) in terms of its weight fractions. Further, noting (3.4) and substituting the weight in terms of densities, yields

$$W_f = \frac{w_f}{w_c} = \frac{\rho_f v_f}{\rho_c v_c} = \frac{\rho_f}{\rho_c} V_f \qquad (3.11a)$$

Similarly,

$$W_m = \frac{\rho_m}{\rho_c} V_m \qquad (3.11b)$$

The above equations can be generalised if the number of constituents are more than two.

$$\rho_c = \sum (\rho_i V_i), \quad \rho_c = \frac{1}{\sum \left(\dfrac{W_i}{\rho_i}\right)}, \quad W_i = \frac{\rho_i}{\rho_c} V_i \qquad (3.12)$$

Due to the presence of voids or air bubbles, theoretical and experimental values will not match. If the theoretical and experimental densities are ρ_{ct} and ρ_{ce} respectively and if V_v is the volume fraction of voids, then

$$\rho_{ce} \cdot v_c = \rho_{ct} (v_c - v_v) \qquad (3.13)$$

or

$$V_v = \frac{\rho_{ct} - \rho_{ce}}{\rho_{ct}} \qquad (3.14)$$

3.3 ASSUMPTIONS AND LIMITATIONS

Before developing methods by which structural properties of a composite may be assessed, it must be recognised that the actual nature of fibre reinforced composite lamina is extremely complex and in order to produce micromechanical solutions, certain simplifying assumptions are made.

The basic assumptions made are as follows

1. Both the fibre and the matrix are homogeneous and isotropic.
2. The fibre, the matrix and the resulting composite exhibit linear elastic behaviour.
3. Perfect bond exists between fibres and matrices so that no slippage occurs at the interface.
4. Fibres are uniform, regularly spaced and perfectly aligned.
5. The matrix is free of voids.
6. The lamina is in a stress-free state (i.e., no residual stresses are present)

3.4 LONGITUDINAL STRENGTH AND STIFFNESS

In addition to the assumptions mentioned above, a further assumption made is that the applied loads are parallel to the fibre direction.

As there is no slippage between the fibre and the matrix, longitudinal strains experienced by the fibre, the matrix and the composite are equal.

$$\varepsilon_c = \varepsilon_f = \varepsilon_m = \varepsilon_1 \tag{3.15}$$

Referring to Fig. 3.2, the load carried by composite, P_1 is shared between fibres P_f and matrix P_m.

$$P_1 = P_f + P_m \tag{3.16}$$

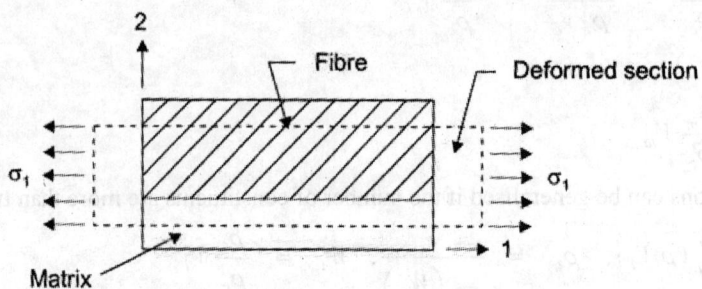

Fig. 3.2 Longitudinally stressed lamina

The loads in turn can be written in terms of stresses and the corresponding cross-sectional areas. Therefore, (3.16) can be written as

$$\sigma_1 A_c = \sigma_{1f} A_f + \sigma_m A_m$$

or, $$\sigma_1 = \sigma_{1f} \cdot \frac{A_f}{A_c} + \sigma_m \cdot \frac{A_m}{A_c} \tag{3.17}$$

where σ_1 is the average stress acting on the cross-section A_c, σ_{1f} and σ_m are stresses in the fibre and the matrix respectively.

For composites with parallel fibres, the volume fractions are equal to area fractions. Therefore,

$$\frac{A_f}{A_c} = V_f, \quad \frac{A_m}{A_c} = V_m \tag{3.18}$$

Substituting (3.18) in (3.17) yields

$$\sigma_1 = \sigma_{1f} V_f + \sigma_m V_m \tag{3.19}$$

Noting (3.5) and expressing stresses in terms of the strain and elastic modulus of (3.19) yields

$$E_1 \varepsilon_1 = E_{1f} \varepsilon_1 V_f + E_m \varepsilon_1 V_m$$

or, $$E_1 = E_{1f} V_f + E_m V_m \tag{3.20}$$

where E_1 is the longitudinal modulus of the composite (E indicates the corresponding modulus). Equations (3.19) and (3.20) are known as the rule of mixtures.

Equation (3.20) shows that composite longitudinal modulus is intermediate between the fibre and matrix moduli [Fig. 3.3].

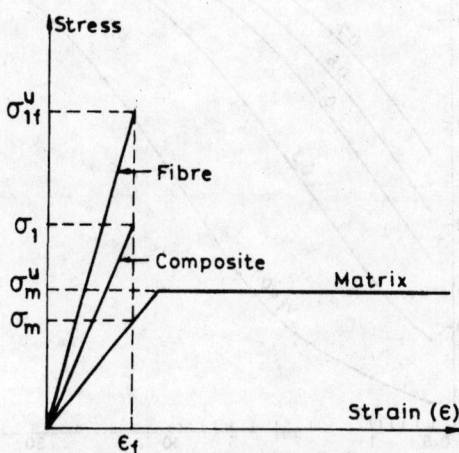

Fig. 3.3 Stress–strain relations of the composite and its components
(fibre dominated strength)

The fraction of the load carried by fibres in a unidirectional lamina is given by

$$\frac{P_{1f}}{P_1} = \frac{\sigma_{1f} V_f}{\sigma_{1f} V_f + \sigma_m (1-V_f)} = \frac{E_{1f} V_f}{E_{1f} V_f + E_m (1-V_f)} \tag{3.21}$$

The fraction of the load carried by the fibre is plotted as a function of the ratio of the moduli of the fibre to that of the matrix and V_f [Fig. 3.4]. In polymeric matrix composites, $E_{1f}/E_m > 10$. Thus for $V_f > 0.25$, fibres carry more than 75 percent of the composite load. Increasing fibre volume fraction can increase the fibre load fraction. However, there is a limit beyond which fibres cannot be packed in a matrix – the practical limit is 80% of the volume fraction; theoretically however, cylindrical fibres can be packed to almost 90% of the volume fraction.

A slightly modified version suggested by Whitney and Riley [3.5] takes account of the variation of the fibre alignment or straightness by introducing a factor K (with $0 \leq K \leq 1$) such that

$$E_1 = KE_{1f} - (KE_{1f} - E_m)V_m \tag{3.22}$$

With $K=1$, (3.22) reverts back to (3.20).

Equations (3.19) and (3.20) are generalised for multiphase materials as follows

$$\sigma_1 = \sum \sigma_i V_i, \quad E_1 = \sum E_i V_i \tag{3.23}$$

Equation (3.20) can also be written as

$$E_1 = (E_{1f} - E_m)V_f + E_m \tag{3.24}$$

Since E_m is much smaller than E_{1f}, (3.24) clearly shows how dependent E_1 is upon the volume of the fibre present. This relationship is graphically shown in Fig. 3.5.

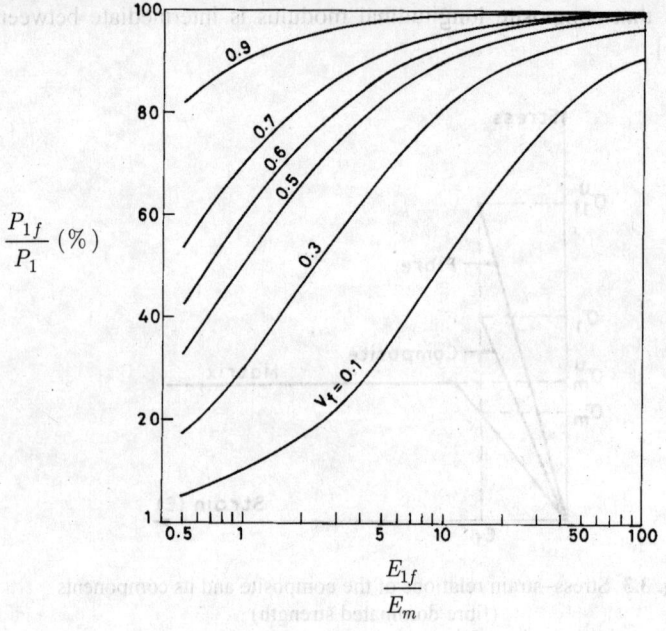

Fig. 3.4 Variation of load carried by the fibre with volume fraction of fibre for varying modular ratio

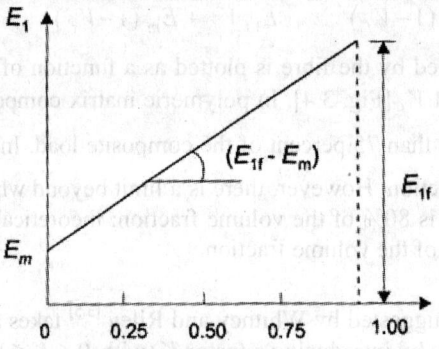

Fig. 3.5 Variation of E_1 with fibre content

3.5 TRANSVERSE MODULUS

A unidirectional composite with parallel fibres is stressed in the transverse direction, that is, the load is applied perpendicular to the parallel fibres (Fig. 3.6). Each layer has the same area on which the load acts, and thus, each layer will carry the same load and experience the same stress, that is,

$$\sigma_2 = \sigma_{2f} = \sigma_m \tag{3.25}$$

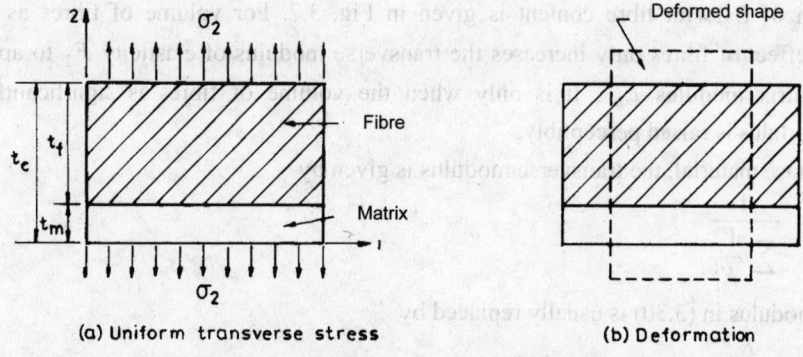

Fig. 3.6 Lamina under transverse normal stress

The composite transverse elongation δ_2 is the sum of the fibre and matrix elongation δ_f and δ_m respectively.

$$\delta_2 = \delta_f + \delta_m \tag{3.26}$$

Because each layer is assumed to be uniform in thickness, it follows that the cumulative thickness of the fibre and matrix layers will be proportional to their respective volume fractions.

Expressing elongation of (3.26) in terms of strains and thickness, yields

$$\varepsilon_2 t_c = \varepsilon_{2f} t_f + \varepsilon_m t_m \tag{3.27}$$

where t_f and t_m are the thicknesses of the fibre and the matrix of the composite having thickness t_c.

Dividing both sides by t_c and noting that the thickness is proportional to the volume fraction, yields

$$\varepsilon_2 = \varepsilon_{2f} \frac{t_f}{t_c} + \varepsilon_m \frac{t_m}{t_c}$$

or,

$$\varepsilon_2 = \varepsilon_{2f} V_f + \varepsilon_m V_m \tag{3.28}$$

Replacing strain terms by their corresponding modulus and stresses, yields

$$\frac{\sigma_2}{E_2} = \frac{\sigma_{2f}}{E_{2f}} V_f + \frac{\sigma_m}{E_m} V_m \tag{3.29}$$

where E_{2f} is the transverse modulus of the fibre.

Noting the relations of (3.25), (3.29) becomes

$$\frac{1}{E_2} = \frac{V_f}{E_{2f}} + \frac{V_m}{E_m} \tag{3.30}$$

Equation (3.30) can be written as

$$\frac{1}{E_2} = \frac{V_f}{E_{2f}} + \frac{1 - V_f}{E_m} \tag{3.31}$$

The variation of E_2 with fibre content is given in Fig. 3.7. For volume of fibres as large as 50 percent, the effect of fibres only increases the transverse modulus of elasticity E_2 to approximately twice the matrix modulus E_m. It is only when the volume of fibres is significantly high that transverse modulus is raised perceptibly.

For a multiphase material, the transverse modulus is given by

$$E_2 = \frac{1}{\sum \frac{V_i}{E_i}} \tag{3.32}$$

The matrix modulus in (3.30) is usually replaced by

$$E'_m = \frac{E_m}{1 - v_m^2} \tag{3.33}$$

where v_m is the Poisson's ratio of the matrix. This takes care of the constraint imposed by the fibres in the fibre direction on the matrix. Thus, (3.30) becomes

$$\frac{1}{E_2} = \frac{V_f}{E_{2f}} + \frac{V_m}{E'_m} \tag{3.34}$$

The transverse modulus is a matrix dominated property. The nature of stresses in the matrix surrounding the fibres is pretty complex. Equation (3.34) tends to underestimate the transverse modulus.

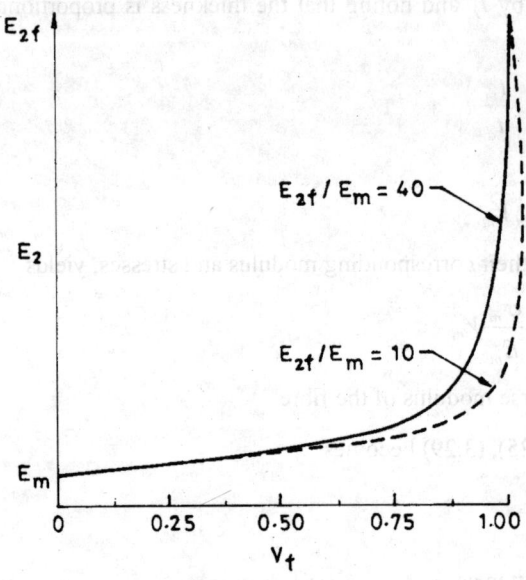

Fig. 3.7 Variation of E_2 with fibre content

Halpin and Tsai [3.16] have suggested a semi-empirical relationship for the evaluation of E_2.

$$E_2 = E_m \frac{1 + \xi_1 \eta_1 V_f}{1 - \eta_1 V_f} \qquad (3.35)$$

where

$$\eta_1 = \frac{E_{2f} - E_m}{E_{2f} + \xi_1 E_m} \qquad (3.36)$$

where ξ_1 is the reinforcing efficiency factor for transverse loading. Equation (3.35) gave good correlation with experimental results for ξ_1 between 1 and 2. If ξ_1 is evaluated from reliable experimental results of E_2, then using this value of ξ_1, E_2 can be determined for the same composite for a wide range of fibre-volume ratios.

There is an element of inconsistency in the assumptions made for the derivation of (3.34). The stress–strain relations of fibre and matrix are

$$\varepsilon_{2f} = \frac{\sigma_2}{E_{2f}}, \qquad \varepsilon_m = \frac{\sigma_2}{E_m} \qquad (3.37)$$

As E_{2f} and E_m are different, there is a transverse strain mismatch at the boundary between the fibre and the matrix. Further, the transverse stresses in the fibre or the matrix are not likely to be the same. A rigorous solution for the transverse strength can be obtained only when the displacements across the boundary between the fibre and matrix are totally matched and for obtaining such a solution, equations from the theory of elasticity are to be used. The error inherent with such inconsistencies can be found only from experimental results.

Another observation can be made for this solution. As the Poisson's ratios of the fibre and the matrix are different, longitudinal stresses are induced in the fibre and the matrix with the balance force acting as shearing stresses at the fibre–matrix boundary.

3.6 INPLANE SHEAR MODULUS

Derivation of G_{12}, the inplane shear modulus is analogous to that of E_2. The composite element is subjected to uniform shearing and complementary shearing stresses along the boundaries, as shown in Fig. 3.8a.

Shearing stresses in the fibre and the matrix are equal. Thus,

$$\tau_c = \tau_f = \tau_m \qquad (3.38)$$

Referring to Fig. 3.8, total shear deformation of the composite, Δ_c, is the sum of the shear deformation of the fibre and the matrix, Δ_f and Δ_m.

$$\Delta_c = \Delta_f + \Delta_m \qquad (3.39)$$

The shear deformation in each constituent can be expressed as the product of the shear strain and the corresponding thickness. Therefore,

$$\Delta_c = \gamma_c t_c, \qquad \Delta_f = \gamma_f t_f, \qquad \Delta_m = \gamma_m t_m \qquad (3.40)$$

Combining equations (3.39) and (3.40) yields

$$\gamma_c t_c = \gamma_f t_f + \gamma_m t_m$$

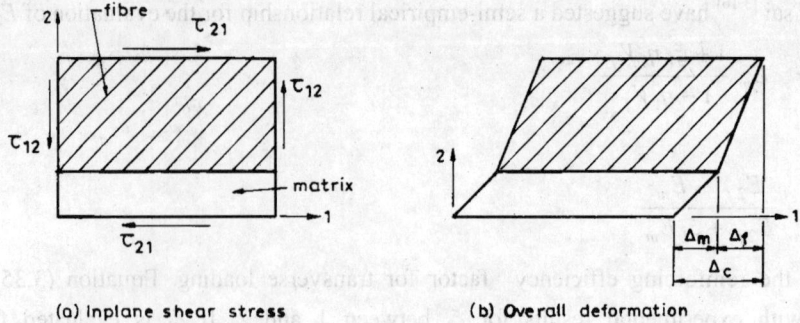

Fig. 3.8 Element under inplane shear stress

or, $\quad \gamma_c = \gamma_f \dfrac{t_f}{t_c} + \gamma_m \dfrac{t_m}{t_c}$ (3.41)

The cumulative thickness of the fibres and matrix is proportional to the respective volume fractions. Therefore, (3.41) can be written as

$$\gamma_c = \gamma_f V_f + \gamma_m V_m \qquad (3.42)$$

Assuming linear behaviour of shear stress and shear strain, (3.42) can be written as

$$\dfrac{\tau_c}{G_{12}} = \dfrac{\tau_f}{G_{12f}} V_f + \dfrac{\tau_m}{G_m} V_m \qquad (3.43)$$

where G_{12} is the inplane shear modulus of the composite, G_{12f} and G_m are the shear modulus of the fibre and matrix respectively.

Noting the equality of stresses given by (3.38), (3.43) reduces to

$$\dfrac{1}{G_{12}} = \dfrac{V_f}{G_{12f}} + \dfrac{V_m}{G_m} \qquad (3.44)$$

Variation of G_{12} with the fibre content is shown in Fig. 3.9. The approach however underestimates the inplane shear modulus. G_{12} is also a matrix dominated property.

Halpin and Tsai have proposed a semi-empirical formula for inplane shear modulus as

$$G_{12} = G_m \dfrac{1 + \xi_2 \eta_2 V_f}{1 - \eta_2 V_f} \qquad (3.45)$$

where,

$$\eta_2 = \dfrac{G_{12f} - G_m}{G_{12f} + \xi_2 G_m} \qquad (3.46)$$

and ξ_2 is the reinforcing efficiency factor for the inplane shear.

For $\xi_2 = 1$, experimental results compare best with (3.44) when it becomes

$$G_{12} = G_m \dfrac{(G_{12f} + G_m) + V_f (G_{12f} - G_m)}{(G_{12f} + G_m) - V_f (G_{12f} - G_m)} \qquad (3.47)$$

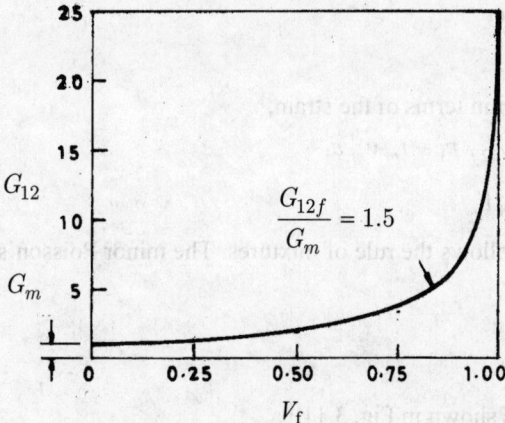

Fig. 3.9 Variation of G_{12} with the fibre content

3.7 POISSON'S RATIO

For inplane loading of a composite, two Poisson's ratios need to be considered. v_{12} is referred to as the major Poisson's ratio which relates the longitudinal strain to the transverse strain. The other one v_{21} is termed as minor Poisson's ratio which relates the transverse strain to the longitudinal strain.

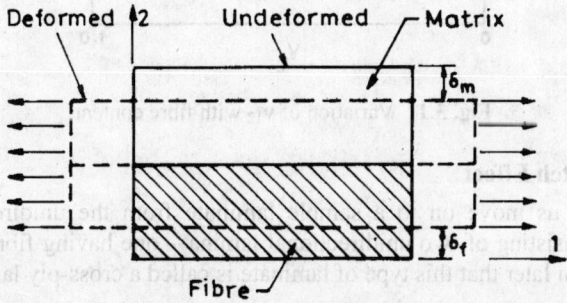

Fig. 3.10 Model for prediction of Poisson's ratio

The same model as that used for the determination of transverse modulus is used for the prediction of major Poisson's ratio. The inplane load in this case is applied parallel to the fibre (Fig. 3.10). By definition

$$v_{12} = -\frac{\varepsilon_2}{\varepsilon_1} \tag{3.48}$$

where ε_2 is the lateral strain of the element in the 2-direction.

Transverse strains in the composite fibres and matrix can be written in terms of longitudinal strains and Poisson's ratio. Thus,

$$(\varepsilon_2)_f = -v_{12f}\varepsilon_1, \qquad (\varepsilon_2)_m = -v_m \varepsilon_1 \tag{3.49}$$

Referring to Fig. 3.10, we get

$$\delta_c = \delta_f + \delta_m \tag{3.50}$$

Substituting the deformation in terms of the strain,

$$-t_c \nu_{12} \varepsilon_1 = -t_f \nu_{12f} \varepsilon_1 - t_m \nu_m \varepsilon_1 \tag{3.51}$$

or, $\quad \nu_{12} = \nu_{12f} V_f + \nu_m V_m \tag{3.52}$

The major Poisson's ratio follows the rule of mixtures. The minor Poisson's ratio can be derived from the classical equation as

$$\frac{\nu_{12}}{E_1} = \frac{\nu_{21}}{E_2} \tag{3.53}$$

The variation ν_{12} with V_f is shown in Fig. 3.11.

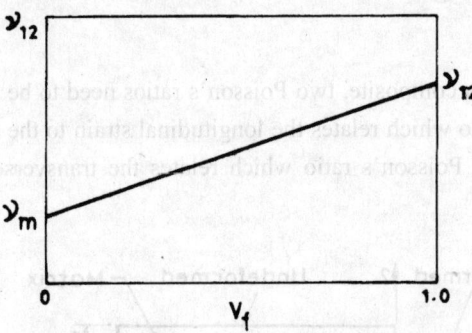

Fig. 3.11 Variation of ν_{12} with fibre content

3.7.1 Poisson's Mismatch Effect

For a change let us move on to a sample laminate from the unidirectional lamina. Let us consider a laminate consisting of two unidirectional lamina – one having fibre orientation 0° and the other 90°. We shall learn later that this type of laminate is called a cross-ply laminate.

Suppose an inplane load is acting at 0°. Under the load, 0 degree ply will extend parallel to the load and will contract perpendicular to the load. This contraction is however restricted by 90° plies. As such tensile stresses are generated at angles to the fibres tending to split the plies. The resulting strain depends on the difference in Poisson's ratio for loading parallel and perpendicular to the fibres and on the thickness and moduli of the plies of the laminate. These strains are larger in the GRP laminates than in the carbon FRP laminates and often lead to 0° splitting.

3.8 PROBLEMS ON MICROMECHANICAL ANALYSIS

Problem 3.1

The E-glass fibres in a polyester resin is 35% by weight.

Given $\rho_f = 2.50 \text{ gm/ml}$ and $\rho_m = 1 \text{ gm/ml}$

calculate V_f and ρ_c for the lamina.

Method 1

$W_f = 0.35$, $W_m = 1 - 0.35 = 0.65$

$\rho_c = \dfrac{1}{\dfrac{0.35}{2.50} + \dfrac{0.65}{1}} = 1.266 \text{ gm/ml}$

$W_f = \dfrac{w_f}{w_c} = \dfrac{\rho_f v_f}{\rho_c v_c} = \dfrac{\rho_f}{\rho_c} V_f$

$V_f = \dfrac{\rho_c}{\rho_f} W_f = \dfrac{1.266}{2.50} \cdot 0.35 = 0.1772$

$V_m = 1 - 0.1772 = 0.8228$

Method 2

Consider a composite of weight = 1 gm

Weight of fibres in it = 0.35 gm

Weight of matrix = 1 − 0.35 = 0.65 gm

Volume of fibre $= \dfrac{0.35}{2.50} = 0.14$ ml

Volume of matrix $= \dfrac{0.65}{1} = 0.65$ ml

Volume of 1 gm of composite = 0.14 + 0.65 = 0.79 ml

Fibre volume fraction $= \dfrac{0.14}{0.79} = 0.1772$

Matrix volume fraction $= 1 - V_f = 1 - 0.1772 = 0.8228$

Composite density is $\rho_c = \dfrac{1}{0.79} = 1.266$ gm/ml

Example 3.2

Assume that the fibres in a composite lamina are arranged in a hexagonal array as shown in the Fig. 3.12. Determine the maximum fibre volume fraction that can be picked in this arrangement?

Referring to Fig. 3.12 (b),

No. of fibres in a unit cell $= \dfrac{60}{360} \times 3 = \dfrac{1}{2}$

Fibre cross-sectional area in a unit cell $= \dfrac{1}{2}\left(\dfrac{\pi d^2}{4}\right) = \dfrac{\pi d^2}{8}$

Unit cell area $= \dfrac{1}{2} \dfrac{\sqrt{3}}{2} a.a = \dfrac{\sqrt{3}}{4} a^2$

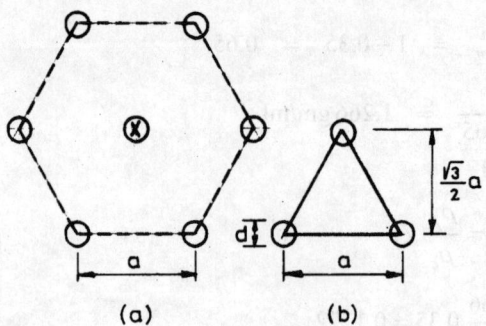

Fig. 3.12 Example 3.2

Fibre volume fraction, $V_f = \dfrac{\pi d^2}{8} \times \dfrac{4}{\sqrt{3}\, a^2} = 0.907 \dfrac{d^2}{a^2}$

or, $a = \sqrt{0.907}\, \dfrac{d}{V_f^{1/2}} = 0.952 \dfrac{d}{V_f^{1/2}}$

Interfibre spacing (R) between the central fibre and each fibre is given by

$$R = a - d = \left(\dfrac{\sqrt{0.907}}{V_f^{1/2}} - 1\right) d$$

For maximum volume fraction $R = 0$.
Therefore,

$$V_{f_{max}} = 0.907$$

Example 3.3

For an E-glass polyester sample, the following are the data for a resin burnt–off test

Weight of an empty crucible = 10.1528 gm

Weight of crucible + sample before burnt-off = 10.5219 gm

Weight of crucible + sample after burnt-off = 10.3221 gm

Calculate the fibre weight fraction, fibre volume fraction and density of the composite sample. Assume $\rho_f = 2.50$ gm/ml and $\rho_m = 1.20$ gm/ml.

Do you expect the calculated value higher or lower than the actual value?

Total weight of the composite = 10.5219 − 10.1528 = 0.3691 gm

Weight of the fibre = 10.3221 − 10.1528 = 0.1693 gm

Weight of the matrix = 0.3691 − 0.1693 = 0.1998 gm

Therefore,

Fibre weight fraction, $W_f = \dfrac{0.1693}{0.3691} = 0.4587$

Matrix weight fraction, $W_m = \dfrac{0.1998}{0.3691} = 0.5413$

Density of the composite is $\rho_c = \dfrac{1}{\dfrac{0.4587}{2.50} + \dfrac{0.5413}{1.20}} = 1.5759$

Fibre volume fraction is $V_f = \dfrac{\rho_c}{\rho_f} W_f = \dfrac{1.5759}{2.50} \times 0.4587 = 0.2891$

Example 3.4

A glass/epoxy specimen weighing 0.98 gm was burnt and the weight of the remaining fibres was found to be 0.49 gm. Densities of glass and epoxy are 2.4 gm/ml and 1.20 gm/ml respectively. Determine the density of composites in the absence of voids. If the actual density of the composite was measured to be 1.50 gm/ml, what is the void fraction?

$w_c = 0.98 \text{ gm}, \quad w_m = 0.98 - 0.49 = 0.49 \text{ gm}, \quad w_f = 0.49 \text{ gm}$

$W_f = \dfrac{w_f}{w_c} = \dfrac{0.49}{0.98} = 0.5$

$W_m = \dfrac{w_m}{w_c} = \dfrac{0.49}{0.98} = 0.5$

$\rho_c = \dfrac{1}{\dfrac{W_f}{\rho_f} + \dfrac{W_m}{\rho_m}}$

$= \dfrac{1}{\dfrac{0.5}{2.4} + \dfrac{0.5}{1.20}}$

$= 1.6 \text{ gm/ml}$

Volume fraction of the void is given by

$V_v = \dfrac{\rho_{ct} - \rho_{ce}}{\rho_{ct}} = \dfrac{1.6 - 1.50}{1.6} = 0.0625 \text{ gm/ml}$

Example 3.5

Evaluate the transverse modulus E_2 of a composite lamina with the following properties

$E_{2f} = 14.8 \text{ GPa}, \quad E_m = 3.45 \text{ GPa}, \quad v_m = 0.36, \quad V_f = 0.65$

Determine E_2 by both the strength of materials approach and the Halpin–Tsai relationship using $\xi_1 = 1$

Strength of materials approach

$E'_m = \dfrac{E_m}{1 - v_m^2} = \dfrac{3.45}{1 - (0.36)^2} = 3.964$

$V_f = 0.65, \quad V_m = 1 - V_f = 1 - 0.65 = 0.35$

Therefore,
$$E_2 = \frac{14.8 \times 3.964}{0.65 \times 3.964 + 0.35 \times 14.8} = 7.56 \text{ GPa}$$

Using the Halpin–Tsai relation:
$$\eta_1 = \frac{E_{2f} - E_m}{E_{2f} + \xi_1 E_m} = \frac{14.8 - 3.45}{14.8 + 1 \times 3.45} = 0.622$$

Halpin–Tsai relation gives
$$E_2 = E_m \frac{1 + \xi_1 \eta_1 V_f}{1 - \eta_1 V_f}$$
$$= 3.45 \frac{1 + 1 \times 0.622 \times 0.65}{1 - 0.622 \times 0.65}$$
$$= 8.133 \text{ GPa}$$

Example 3.6

Calculate the longitudinal modulus and tensile strength of a unidirectional composite containing 60 percent by volume of carbon fibres ($E_{1f} = 294$ GPa and $\sigma_{1fu} = 5.6$ GPa) in a toughened epoxy matrix ($E_m = 3.6$ GPa, $\sigma_{mu} = 105$ MPa). Compare these values with the experimentally determined values of $E_1 = 162$ GPa, $\sigma_{1u} = 2.94$ GPa . What fraction of the load is carried by fibres in the composite?

$$V_f = 0.6, \quad V_m = 1 - 0.6 = 0.4$$
$$E_1 = E_{1f} V_f + E_m V_m$$
$$= 294 \times 0.6 + 3.6 \times 0.4$$
$$= 177.84 \text{ GPa}$$

The experimentally determined value of E_1 of 162 GPa is less than the theoretical value obtained by the rule of mixtures.

Similarly,
$$\sigma_{1u} = \sigma_{1fu} V_f + \sigma_{mu} V_m$$
$$= 5.6 \times 0.6 + 0.105 \times 0.4$$
$$= 3.402 \text{ GPa}$$

The theoretical value of the tensile strength is higher than the experimental value.

The fraction of the load carried by the fibres is given by
$$\frac{P_{1f}}{P_c} = \frac{\sigma_{1fu} \cdot V_f}{\sigma_{1u}} = \frac{5.6 \times 0.6}{3.402} = 0.9876$$

Example 3.7

Determine the inplane shear modulus G_{12} of a glass /epoxy composite with properties
$$G_{12f} = 28 \text{ GPa}, \quad G_m = 1300 \text{ MPa}, \quad V_f = 0.6$$

using the strength of materials approach and the Halpin–Tsai relationship with $\xi_2 = 1$.

$$V_m = 1 - 0.6 = 0.4$$

$$G_{12} = \frac{G_{12f} \cdot G_m}{G_m V_f + G_{12f} V_m}$$

$$G_{12} = \frac{28 \times 1.3}{0.6 \times 28 + 0.4 \times 1.3} = 2.10 \text{ GPa}$$

Using the Halpin–Tsai formula,

$$\eta_2 = \frac{G_{12f} - G_m}{G_{12f} + \xi_2 G_m}$$

$$= \frac{28 - 1.3}{28 + 1 \times 1.3} = 0.9113$$

$$G_{12} = G_m \frac{1 + \xi_2 \eta_2 V_f}{1 - \eta_2 V_f}$$

$$= 1.3 \times \frac{1 + 0.9113 \times 1 \times 0.6}{1 - 0.9113 \times 0.6}$$

$$= 4.44 \text{ GPa}$$

Example 3.8

For the problem given in Fig. 3.13, the strength of materials approach can be used to determine the apparent Young's modulus for the general problem of a composite material with the inclusion of an arbitrary shape. $A_2(x)$ is the area of the distribution of inclusion. If $E = \frac{F}{\delta}$, then determine the expression for the modulus.

Referring to Fig. 3.13,

Total area $A = 1 \times 1 = 1$

Area of intersection $= A_2$

Area of remaining material $= A_1 = 1 - A_2$

Now, the total force can be expressed as a product of the stresses in the body and corresponding area

$$F = \sigma_1 A_1 + \sigma_2 A_2$$
$$= E_1 \varepsilon A_1 + E_2 \varepsilon A_2$$

This E_2 is not to be confused with the transverse modulus.
Expressing A_1 in terms of A_2 as above

$$\varepsilon = \frac{F}{E_1 + (E_2 - E_1) A_2}$$

Now,

$$\delta = \int d\delta = \int_0^1 \varepsilon\, dx$$

$$= \int_0^1 \frac{F}{E_1 + (E_2 - E_1) A_2}\, dx$$

$$\delta = \frac{F}{E}$$

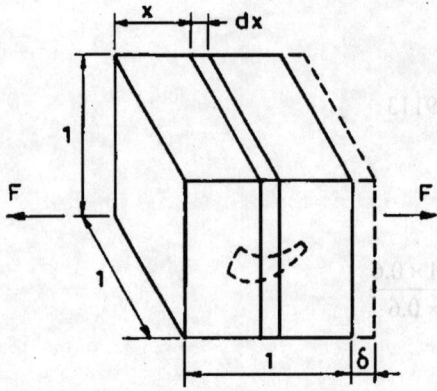

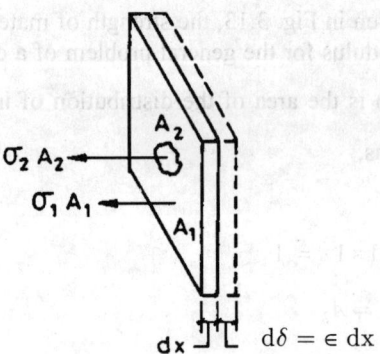

Fig. 3.13 Example 3.8

Example 3.9

The longitudinal modulus of a glass reinforced plastic lamina is to be doubled by substituting some of the glass fibres with carbon fibres. The total fibre volume remains unchanged at 0.5. Calculate the fraction of carbon fibres.

Given $E_c = 300\,\text{GPa}$, $E_g = 70\,\text{GPa}$, $E_m = 5\,\text{GPa}$

The longitudinal modulus of GRP is given by its rule of mixtures as

$$E_1 = E_g V_g + E_m V_m$$

V_g is the volume fraction of glass fibres = 0.5, $V_m = 1 - 0.5 = 0.5$

$$E_1 = 0.5 \times 70 + 0.5 \times 5 = 37.5 \text{ GPa}$$

Now, E_1 is to be doubled by addition of carbon fibres. The new value of E_1 is

$$E_1 = 2 \times 37.5 = 75 \text{ GPa}$$

Let V_c be the volume fraction of carbon fibres.

Therefore, the volume fraction of glass fibre is

$$V_g = 0.5 - V_c$$

Therefore,

$$V_c \times 300 + V_g \times 70 + V_m \times 5 = 75$$

or, $V_c \times 300 + (0.5 - V_c) \times 70 + 0.5 \times 5 = 75$

or, $V_c = 0.163$

Therefore, the required volume fraction of carbon is 0.163.

Example 3.10

The density of a semi-crystalline polymer matrix is expressed as

$$\rho_m = \rho_{mc} V_{mc} + \rho_{ma} V_{ma}$$

where ρ_{mc} and ρ_{ma} are the densities of the crystalline and the amorphous phase respectively and V_{mc} and V_{ma} are the corresponding volume fractions.

The density of the fibre of the composite is 1.8 g/ml and the density of the composite is 1.6 g/ml. Given $V_f = 0.6$, $\rho_{mc} = 1.4$ g/ml and $\rho_{ma} = 1.25$ g/ml.

Determine the volume and weight fractions of the crystalline and amorphous phase of the matrix.

$$V_m = 1 - V_f = 1 - 0.6 = 0.4$$

We know

$$\rho_c = \rho_f V_f + \rho_m V_m$$

$$1.6 = 1.8 \times 0.6 + \rho_m 0.4$$

or, $\rho_m = 1.3$

Now, in the matrix, let the volume fraction of the crystalline part be V_{mc}.

Therefore, the volume fraction of the amorphous part, $V_{ma} = 1 - V_{mc}$

$$1.3 = 1.4 V_{mc} + 1.25 (1 - V_{mc})$$

or, $V_{mc} = 0.33$

Therefore, within the matrix, $V_{ma} = 1 - 0.33 = 0.67$

The volume fraction of the crystalline part of the composite is,

$$\bar{V}_{mc} = 0.33 \times 0.4 = 0.132$$

The volume fraction of the amorphous part of the composite is,

$$\bar{V}_{ma} = 0.67 \times 0.4 = 0.268$$

The weight fraction of the crystalline part of the matrix

$$= \frac{\rho_{mc}}{\rho_c} \bar{V}_{mc} = \frac{1.4}{1.6} 0.132 = 0.1155$$

The weight fraction of the amorphous part of the matrix

$$= \frac{\rho_{ma}}{\rho_c} \bar{V}_{ma} = \frac{1.25}{1.6} 0.268 = 0.2094$$

REFERENCES AND SUGGESTED READINGS

3.1 Z. Hashin, Analysis of composite materials – A survey, *J. Appl. Mech.*, V.50, 1983, pp. 481-505.

3.2 B. W. Schaffer, Stress–strain relations for reinforced plastics parallel and normal to their internal filaments, *AIAA J.*, V.1, No. 2, 1984, pp. 348–352.

3.3 R. H. Sonneborn, *Fibre Glass Reinforced Plastics*, Van Nostrand Reinhold, New York, 1966.

3.4 R. Hill, Theory of mechanical properties of fibre-strengthened materials: III. Self-consistent models, *J. of Mechanics and Physics of Solids*, Vol.4, 1966, pp.1537–1542.

3.5 J. M. Whitney and M. B. Riley, Elastic properties of fibre reinforced composite materials, *AIAA J.*, Vol.4, 1966, pp. 1537–1542.

3.6 R. A. Shenoi and J. F. Wellicome, *Composite Materials in Maritime Structures*, Vol.1, *Fundamental Aspects*, Cambridge University Press, U. K., 1993.

3.7 R. M. Jones, *Mechanics of Composite Materials*, 2nd Edition McGraw Hill, New York, 1999.

3.8 B. D. Agarwal and L. J. Broutman, *Analysis and Performance of Fibre Composites*, 2nd Edition, John Wiley, New York, 1990.

3.9 D. F. Adams and D. R. Doner, Transverse normal loading of a unidirectional composite, *Journal of Composite Materials*, Vol. 1, 1967, pp. 152–164.

3.10 J. C. Halpin and S. W. Tsai, Effects on environmental factors on composite materials, Div Force Technical Report AFML-TR-67-423, Wright Aeronautical Labs, Dayton, Ohio, 1967.

EXERCISE 3

3.1 Determine E_2 by both the strength of materials approach and the Halpin–Tsai relation for a composite with the following properties

E_{2f} = 360 GPa, E_m = 65 GPa, v_m = 0.3, V_f = 0.35

Assume $\xi_2 = 2$

3.2 A unidirectional fibre composite contains 60% by volume of HMS-4 carbon fibres in an epoxy matrix. The tensile strength of the fibre is 2.48 GPa and the elastic modulus is 345 GPa. For the matrix E_m = 3.45 GPa and σ_m = 138 MPa. Calculate the longitudinal tensile strength and the longitudinal elastic modulus of the composite.

3.3 A unidirectional fibre reinforced composite contains 55 % by volume of fibres. Given E_{1f} = 300 GPa, σ_{1f} = 5.6 GPa, E_m = 3.5 GPa, σ_m = 100 MPa, ε_{mu} = 3%. Calculate the longitudinal tensile modulus, tensile strength and failure strength.

3.4 In a unidirectional lamina, fibres of round cross-section are arranged in the form of a square array (Fig. 3.13). Calculate the theoretical fibre volume fraction in the composite lamina. What is the maximum fibre volume fraction that can be arranged in this fashion?

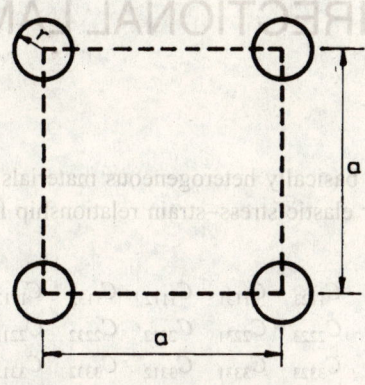

Problem 3.4

3.5 What will you conclude if for determination of E_2, equal strains in the fibre and matrix are assumed instead of equal stresses?

3.6 The weight of the matrix is measured to be 35% of the weight of the composite. What is the fibre volume fraction? The specific gravities of glass and epoxy are 2.58 and 1.22 respectively.

3.7 The material in a composite beam is a particular type of carbon fibre reinforced epoxy. The density of the carbon fibre is 1.8 gm/ml, longitudinal modulus of the fibre is 228 GPa and V_f = 0.6. To reduce the cost, carbon fibres are needed to be replaced by an equal volume percentage of E-glass fibre (density of E-glass = 2.54 gm/ml and elastic modulus = 72.4 GPa), the depth of the beam is increased three-fold.

Assuming that the cost of the carbon fibres, E-glass fibres and epoxies are Rs. 1225 per kg, Rs. 122.50 per kg and Rs. 182 per kg respectively, determine the percentage material cost saved. Specific gravity of epoxy is 1.25.

CHAPTER 4

ELASTIC PROPERTIES OF THE UNIDIRECTIONAL LAMINA

4.1 INTRODUCTION

Fibre reinforced plastics are basically heterogeneous materials. They are treated as anisotropic materials. The most general linear elastic stress–strain relationship for anisotropic materials is given by

$$\begin{Bmatrix} \sigma_{11} \\ \sigma_{22} \\ \sigma_{33} \\ \sigma_{23} \\ \sigma_{31} \\ \sigma_{12} \\ \sigma_{32} \\ \sigma_{13} \\ \sigma_{21} \end{Bmatrix} = \begin{bmatrix} C_{1111} & C_{1122} & C_{1133} & C_{1123} & C_{1131} & C_{1112} & C_{1132} & C_{1113} & C_{1121} \\ C_{2211} & C_{2222} & C_{2233} & C_{2223} & C_{2231} & C_{2212} & C_{2232} & C_{2213} & C_{2221} \\ C_{3311} & C_{3322} & C_{3333} & C_{3323} & C_{3331} & C_{3312} & C_{3332} & C_{3313} & C_{3321} \\ C_{2311} & C_{2322} & C_{2333} & C_{2323} & C_{2331} & C_{2312} & C_{2332} & C_{2313} & C_{2321} \\ C_{3111} & C_{3122} & C_{3133} & C_{3123} & C_{3131} & C_{3112} & C_{3132} & C_{3113} & C_{3121} \\ C_{1211} & C_{1222} & C_{1233} & C_{1223} & C_{1231} & C_{1212} & C_{1232} & C_{1213} & C_{1221} \\ C_{3211} & C_{3222} & C_{3233} & C_{3223} & C_{3231} & C_{3212} & C_{3232} & C_{3213} & C_{3221} \\ C_{1311} & C_{1322} & C_{1333} & C_{1323} & C_{1331} & C_{1312} & C_{1332} & C_{1313} & C_{1321} \\ C_{2111} & C_{2122} & C_{2133} & C_{2123} & C_{2131} & C_{2112} & C_{2132} & C_{2113} & C_{2121} \end{bmatrix} \begin{Bmatrix} \varepsilon_{11} \\ \varepsilon_{22} \\ \varepsilon_{33} \\ \varepsilon_{23} \\ \varepsilon_{31} \\ \varepsilon_{12} \\ \varepsilon_{32} \\ \varepsilon_{13} \\ \varepsilon_{21} \end{Bmatrix} \quad (4.1)$$

Equation (4.1) in tensor notations can be written as

$$\sigma_{ij} = C_{ijkl} \, \varepsilon_{kl}$$
$$(i, j, k, l = 1, 2, 3) \quad (4.2)$$

where C_{ijkl} are called stiffness components. We have employed rectangular cartesian coordinates with usual cartesian tensor notations involving the summation of repeated indices.

The stress–strain relationship can also be written as

$$\varepsilon_{ij} = S_{ijkl} \, \sigma_{kl} \quad (4.3)$$

where S_{ijkl} are the compliance components.

Matrix $[C_{ijkl}]$ is the inverse of the matrix $[S_{ijkl}]$ and vice versa.

Thus a general anisotropic material requires 81 elastic constants.

4.2 STRESS–STRAIN RELATIONSHIP

Stress and strain tensors are required to be symmetric, that is,

$$\sigma_{ij} = \sigma_{ji}, \, \varepsilon_{ij} = \varepsilon_{ji} \quad (4.4)$$

The symmetry of stress and strain tensors reduces the number of independent elastic constants to 36. Contracted notations are used in mechanics of composites. The stress tensors are written as follows

$\sigma_{11} = \sigma_1, \sigma_{22} = \sigma_2, \sigma_{33} = \sigma_3, \sigma_{23} = \sigma_4, \sigma_{31} = \sigma_5, \sigma_{12} = \sigma_6$ \hfill (4.5)

In contracted notations, the stress–strain relationship for an anisotropic body is (Fig. 4.1)

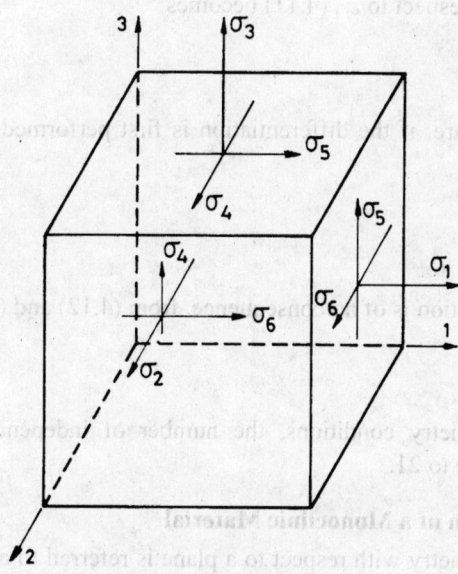

Fig. 4.1 Stress components

$$\begin{Bmatrix} \sigma_1 \\ \sigma_2 \\ \sigma_3 \\ \sigma_4 \\ \sigma_5 \\ \sigma_6 \end{Bmatrix} = \begin{bmatrix} C_{11} & C_{12} & C_{13} & C_{14} & C_{15} & C_{16} \\ C_{21} & C_{22} & C_{23} & C_{24} & C_{25} & C_{26} \\ C_{31} & C_{32} & C_{33} & C_{34} & C_{35} & C_{36} \\ C_{41} & C_{42} & C_{43} & C_{44} & C_{45} & C_{46} \\ C_{51} & C_{52} & C_{53} & C_{54} & C_{55} & C_{56} \\ C_{61} & C_{62} & C_{63} & C_{64} & C_{65} & C_{66} \end{bmatrix} \begin{Bmatrix} \varepsilon_1 \\ \varepsilon_2 \\ \varepsilon_3 \\ \varepsilon_4 \\ \varepsilon_5 \\ \varepsilon_6 \end{Bmatrix}$$ (4.6)

or, $\quad \sigma_i = C_{ij}\, \varepsilon_j$ \hfill (4.7)

$(i, j = 1, 2, 3, 4, 5, 6)$

Similarly,
$$\varepsilon_i = S_{ij}\, \sigma_j$$ (4.8)

The strain energy functions reduce the number of independent components still further.
The work done per unit volume is given by

$$W = \tfrac{1}{2}\, \sigma_i\, \varepsilon_i \quad (i = 1, \ldots\ldots, 6)$$ (4.9)

Substituting σ_i from (4.7) into (4.9), yields

$$W = \tfrac{1}{2}\, C_{ij}\, \varepsilon_i\, \varepsilon_j$$ (4.10)

Stress can be expressed as a derivative of strain.

$$\sigma_i = \frac{\partial W}{\partial \varepsilon_i} = C_{ij}\,\varepsilon_j \qquad (4.11)$$

Differentiating further with respect to ε_j, (4.11) becomes

$$C_{ij} = \frac{\partial^2 W}{\partial \varepsilon_i \partial \varepsilon_j} \qquad (4.12)$$

Based on the above procedure, if the differentiation is first performed with respect to ε_j and then with respect to ε_i, we get

$$C_{ji} = \frac{\partial^2 W}{\partial \varepsilon_j \partial \varepsilon_i} \qquad (4.13)$$

Since the order of differentiation is of no consequence, from (4.12) and (4.13), we get

$$C_{ij} = C_{ji} \qquad (4.14)$$

Similarly,

$$S_{ij} = S_{ji} \qquad (4.15)$$

Taking advantage of symmetry conditions, the number of independent elastic constants in the stress–strain relations reduce to 21.

4.2.1 Stress–Strain Relation of a Monoclinic Material

Materials having symmetry with respect to a plane is referred to as monoclinic materials. It has been shown that for monoclinic materials, C_{ij} has 13 independent components [4.1, 4.2].

$$[C_{ij}] = \begin{bmatrix} C_{11} & C_{12} & C_{13} & 0 & 0 & C_{16} \\ & C_{22} & C_{23} & 0 & 0 & C_{26} \\ & & C_{33} & 0 & 0 & C_{36} \\ & & & C_{44} & C_{45} & 0 \\ & \text{symmetrical} & & & C_{55} & 0 \\ & & & & & C_{66} \end{bmatrix} \qquad (4.16)$$

The coordinate 3 is normal to the plane of symmetry.

4.2.2 Stress–Strain Relation of a Specially Orthotropic Material

An orthotropic material has three mutually perpendicular planes of material symmetry. It is referred to as specially orthotropic material when the reference system of coordinates is selected along the principal planes of material symmetry. The number of independent elastic components for such a case reduces to 9.

$$[C_{ij}] = \begin{bmatrix} C_{11} & C_{12} & C_{13} & 0 & 0 & 0 \\ & C_{22} & C_{23} & 0 & 0 & 0 \\ & & C_{33} & 0 & 0 & 0 \\ & & & C_{44} & 0 & 0 \\ & \text{symmetrical} & & & C_{55} & 0 \\ & & & & & C_{66} \end{bmatrix} \qquad (4.17)$$

The compliance matrix $[S_{ij}]$ can be obtained from the inverse of $[C_{ij}]$

4.2.3 Stress–strain Relation for a Transversely Isotropic Material

A plane of isotropy is a plane where at every point in the planes, mechanical properties are the same in all directions [Fig. 4.2]. In transversely isotropic material, one of the planes for the orthotropic case is taken to be the plane of isotropy.

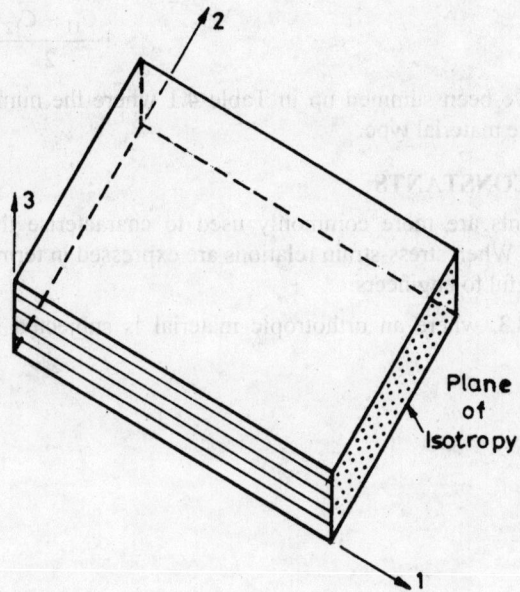

Fig. 4.2 Transversely isotropic material

In Fig. 4.2, the plane of isotropy consists of that containing axes 2 and 3. As such in the stress–strain relation, axes 2 and 3 can be interchanged. The stiffness matrix of a transversely isotropic material thus becomes

$$[C_{ij}] = \begin{bmatrix} C_{11} & C_{12} & C_{12} & 0 & 0 & 0 \\ & C_{22} & C_{23} & 0 & 0 & 0 \\ & & C_{22} & 0 & 0 & 0 \\ & symmetrical & & \frac{1}{2}(C_{22} - C_{23}) & 0 & 0 \\ & & & & C_{66} & 0 \\ & & & & & C_{66} \end{bmatrix} \quad (4.18)$$

The number of independent elastic constants reduce to 5 for this case – $C_{11}, C_{12}, C_{22}, C_{23}$ and C_{66}.

4.2.4 Stress–Strain Relation of an Isotropic Material

An isotropic material has the same elastic properties in all directions. As such an infinite number of material symmetry exists through a point. In this case, number of elastic constants reduce to two.

$$[C_{ij}] = \begin{bmatrix} C_{11} & C_{12} & C_{12} & 0 & 0 & 0 \\ & C_{11} & C_{12} & 0 & 0 & 0 \\ & & C_{11} & 0 & 0 & 0 \\ & & & \dfrac{C_{11}-C_{12}}{2} & 0 & 0 \\ & symmetrical & & & \dfrac{C_{11}-C_{12}}{2} & 0 \\ & & & & & \dfrac{C_{11}-C_{12}}{2} \end{bmatrix} \quad (4.19)$$

All the above results have been summed up in Table 4.1 where the number of independent elastic constants are related to the material type.

4.3 ENGINEERING CONSTANTS

Engineering constants are more commonly used to characterise the behaviour of FRP than stiffness and compliance. When stress-strain relations are expressed in terms of engineering constants, it becomes more meaningful to engineers.

Referring to Fig. 4.3, where an orthotropic material is subjected to uniaxial tension σ_1, the stress–strain relations are

$$\varepsilon_1 = \frac{\sigma_1}{E_1}$$

$$\varepsilon_2 = -\frac{\nu_{12}}{E_1}\sigma_1$$

$$\varepsilon_3 = -\frac{\nu_{13}}{E_1}\sigma_1 \quad (4.20)$$

$$\varepsilon_4 = \varepsilon_5 = \varepsilon_6 = 0$$

Table 4.1 Materials and independent elastic constants [4.1]

Material	No. of independent elastic constants
1. General anisotropic material	81
2. Anisotropic material with symmetric stress and strain components	36
3. Anisotropic material with energy considerations	21
4. Monoclinic material	13
5. Specially orthotropic material	9
6. Orthotropic material with transverse isotropy	5
7. Isotropic material	2

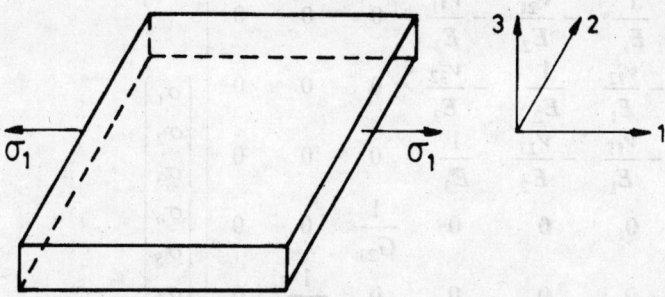

Fig. 4.3 Orthotropic material subjected to longitudinal tension

where E_1 is the modulus of elasticity in the 1-axis direction. ε_1, ε_2 and ε_3 are the axial strains along the 1, 2 and 3 - axis system.

ε_4, ε_5 and ε_6 are the shearing strains and ν_{12}, and ν_{13} are Poisson's ratio.

Referring to Fig. 4.4, for an orthotropic material subjected to pure inplane shear σ_6, the stress–strain relations are

$$\varepsilon_1 = \varepsilon_2 = \varepsilon_3 = \varepsilon_4 = \varepsilon_5 = 0 \qquad (4.21)$$

$$\varepsilon_6 = \frac{\sigma_6}{G_{12}}$$

where G_{12} is the inplane shear modulus of the material.

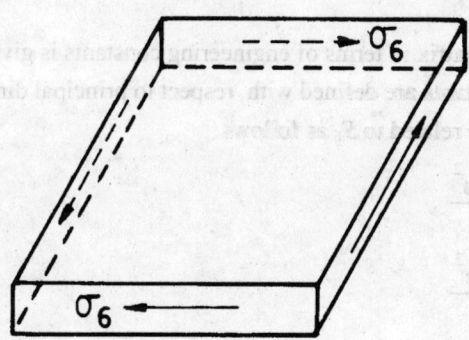

Fig. 4.4 Orthotropic material subjected to inplane shear

Likewise, the stress–strain relations for other stress components acting independently can be determined.

Thus, if all stress components σ_1, ...σ_6 act on an orthotropic material, then the stress–strain relationship assumes the following form

$$\begin{Bmatrix} \varepsilon_1 \\ \varepsilon_2 \\ \varepsilon_3 \\ \varepsilon_4 \\ \varepsilon_5 \\ \varepsilon_6 \end{Bmatrix} = \begin{bmatrix} \frac{1}{E_1} & -\frac{\nu_{21}}{E_2} & -\frac{\nu_{31}}{E_3} & 0 & 0 & 0 \\ -\frac{\nu_{12}}{E_1} & \frac{1}{E_2} & -\frac{\nu_{32}}{E_3} & 0 & 0 & 0 \\ -\frac{\nu_{13}}{E_1} & -\frac{\nu_{23}}{E_2} & \frac{1}{E_3} & 0 & 0 & 0 \\ 0 & 0 & 0 & \frac{1}{G_{23}} & 0 & 0 \\ 0 & 0 & 0 & 0 & \frac{1}{G_{31}} & 0 \\ 0 & 0 & 0 & 0 & 0 & \frac{1}{G_{12}} \end{bmatrix} \begin{Bmatrix} \sigma_1 \\ \sigma_2 \\ \sigma_3 \\ \sigma_4 \\ \sigma_5 \\ \sigma_6 \end{Bmatrix} \quad (4.22)$$

As the compliance matrix is symmetric, we get

$$\frac{\nu_{21}}{E_2} = \frac{\nu_{12}}{E_1}$$

$$\frac{\nu_{13}}{E_1} = \frac{\nu_{31}}{E_3} \quad (4.23)$$

$$\frac{\nu_{23}}{E_2} = \frac{\nu_{32}}{E_3}$$

In general,

$$\frac{\nu_{ij}}{E_i} = \frac{\nu_{ji}}{E_j} \quad (i, j = 1, 2, 3) \quad (4.24)$$

Therefore, the compliance matrix in terms of engineering constants is given by (4.22).
Nine main engineering constants are defined with respect to principal directions of the material.
Stiffness coefficients C_{ij} are related to S_{ij} as follows

$$C_{11} = \frac{S_{22}S_{33} - S_{23}^2}{S}$$

$$C_{22} = \frac{S_{11}S_{33} - S_{13}^2}{S}$$

$$C_{33} = \frac{S_{11}S_{22} - S_{12}^2}{S} \quad (4.25)$$

$$C_{12} = \frac{S_{13}S_{23} - S_{12}S_{33}}{S}$$

$$C_{23} = \frac{S_{11}S_{13} - S_{23}S_{11}}{S}$$

$$C_{13} = \frac{S_{12}S_{23} - S_{13}S_{22}}{S}$$

$$C_{44} = \frac{1}{S_{44}}, \quad C_{55} = \frac{1}{S_{55}}, \quad C_{66} = \frac{1}{S_{66}}$$

where

$$S = \begin{vmatrix} S_{11} & S_{12} & S_{13} \\ S_{21} & S_{22} & S_{23} \\ S_{31} & S_{32} & S_{33} \end{vmatrix} \tag{4.26}$$

$$= S_{11}S_{22}S_{33} - S_{11}S_{23}^2 - S_{22}S_{13}^2 - S_{33}S_{12}^2 + 2S_{12}S_{23}S_{31}$$

From the relations of (4.22), we can write

$$C_{11} = \frac{1 - v_{23}v_{32}}{E_2 E_3 \Delta}$$

$$C_{22} = \frac{1 - v_{13}v_{31}}{E_1 E_3 \Delta}$$

$$C_{33} = \frac{1 - v_{12}v_{21}}{E_1 E_2 \Delta}$$

$$C_{12} = \frac{v_{12} + v_{31}v_{23}}{E_2 E_3 \Delta} = \frac{v_{12} + v_{13}v_{32}}{E_1 E_3 \Delta} \tag{4.27}$$

$$C_{23} = \frac{v_{32} + v_{12}v_{31}}{E_1 E_3 \Delta} = \frac{v_{23} + v_{21}v_{13}}{E_1 E_2 \Delta}$$

$$C_{13} = \frac{v_{13} + v_{12}v_{23}}{E_1 E_2 \Delta} = \frac{v_{31} + v_{21}v_{32}}{E_2 E_3 \Delta}$$

$$C_{44} = G_{23}, \quad C_{55} = G_{13}, \quad C_{66} = G_{12}$$

where

$$\Delta = \frac{1}{E_1 E_2 E_3} \begin{vmatrix} 1 & -v_{21} & -v_{31} \\ -v_{12} & 1 & -v_{32} \\ -v_{13} & -v_{23} & 1 \end{vmatrix} \tag{4.28}$$

For a transversely isotropic material with the 2-3 plane as the plane of isotropy, the following relationship holds good

$$\begin{aligned} E_2 &= E_3 \\ G_{12} &= G_{13} \\ v_{12} &= v_{13} \end{aligned} \tag{4.29}$$

4.4 STRESS–STRAIN RELATIONS OF A THIN LAMINA

For structural applications, fibre reinforced composite materials are used in the form of thin laminates. Let us investigate the case of a single lamina in plane stress. In such a case all stress components in the out-of-plane direction (3-direction) are zero, that is,

$$\sigma_3 = \sigma_4 = \sigma_5 = 0 \tag{4.30}$$

Substituting the stress values of (4.30) into (4.17), the stress-strain equation becomes

$$\begin{Bmatrix} \sigma_1 \\ \sigma_2 \\ 0 \\ 0 \\ 0 \\ \sigma_6 \end{Bmatrix} = \begin{bmatrix} C_{11} & C_{12} & C_{13} & 0 & 0 & 0 \\ C_{21} & C_{22} & C_{23} & 0 & 0 & 0 \\ C_{31} & C_{32} & C_{33} & 0 & 0 & 0 \\ 0 & 0 & 0 & C_{44} & 0 & 0 \\ 0 & 0 & 0 & 0 & C_{55} & 0 \\ 0 & 0 & 0 & 0 & 0 & C_{66} \end{bmatrix} \begin{Bmatrix} \varepsilon_1 \\ \varepsilon_2 \\ \varepsilon_3 \\ \varepsilon_4 \\ \varepsilon_5 \\ \varepsilon_6 \end{Bmatrix} \tag{4.31}$$

which in an expanded form gives

$$\begin{aligned}
\sigma_1 &= C_{11}\varepsilon_1 + C_{12}\varepsilon_2 + C_{13}\varepsilon_3 \\
\sigma_2 &= C_{12}\varepsilon_1 + C_{22}\varepsilon_2 + C_{23}\varepsilon_3 \\
0 &= C_{13}\varepsilon_1 + C_{23}\varepsilon_2 + C_{33}\varepsilon_3 \\
\varepsilon_4 &= 0 \\
\varepsilon_5 &= 0 \\
\sigma_6 &= C_{66}\varepsilon_6
\end{aligned} \tag{4.32}$$

Solving (4.32), yields

$$\begin{aligned}
\sigma_1 &= \left(C_{11} - \frac{C_{13}C_{13}}{C_{33}} \right)\varepsilon_1 + \left(C_{12} - \frac{C_{13}C_{23}}{C_{33}} \right)\varepsilon_2 \\
&= Q_{11}\varepsilon_1 + Q_{12}\varepsilon_2 \\
\sigma_2 &= \left(C_{12} - \frac{C_{23}C_{13}}{C_{33}} \right)\varepsilon_1 + \left(C_{22} - \frac{C_{23}C_{23}}{C_{33}} \right)\varepsilon_2 \\
&= Q_{12}\varepsilon_1 + Q_{22}\varepsilon_2 \\
\sigma_6 &= C_{66}\varepsilon_6 = Q_{66}\varepsilon_6
\end{aligned} \tag{4.33}$$

Equation (4.33) written in matrix form becomes

$$\begin{Bmatrix} \sigma_1 \\ \sigma_2 \\ \sigma_6 \end{Bmatrix} = \begin{bmatrix} Q_{11} & Q_{12} & 0 \\ Q_{12} & Q_{22} & 0 \\ 0 & 0 & Q_{66} \end{bmatrix} \begin{Bmatrix} \varepsilon_1 \\ \varepsilon_2 \\ \varepsilon_6 \end{Bmatrix} \tag{4.34}$$

or, $\{\sigma\}_{1,2} = [Q]_{1,2}\{\varepsilon\}_{1,2}$ \hfill (4.35)

Here, Q_{ij} is known as the reduced stiffness component, and is given by

$$Q_{ij} = C_{ij} - \frac{C_{i3}C_{j3}}{C_{33}} \quad (i, j = 1, 2, 6) \tag{4.36}$$

The stress–strain relationship can similarly be written in terms of compliances

$$\begin{Bmatrix} \varepsilon_1 \\ \varepsilon_2 \\ \varepsilon_6 \end{Bmatrix} = \begin{bmatrix} S_{11} & S_{12} & 0 \\ S_{12} & S_{22} & 0 \\ 0 & 0 & S_{66} \end{bmatrix} \begin{Bmatrix} \sigma_1 \\ \sigma_2 \\ \sigma_6 \end{Bmatrix} \tag{4.37}$$

or, in compact form, (4.37) becomes

$$\{\varepsilon\}_{1,2} = [S]_{1,2}\{\sigma\}_{1,2} \tag{4.38}$$

When the principal material axis systems are chosen as the directions in which the stress components are to be determined, then (4.34) and (4.37) are valid. For this case, the stress–strain relationship is valid through four independent engineering constants.

The stiffness in terms of engineering constants along the principal material axis are

$$Q_{11} = \frac{E_1}{1-\nu_{12}\cdot\nu_{21}}$$

$$Q_{22} = \frac{E_2}{1-\nu_{12}\cdot\nu_{21}} \tag{4.39}$$

$$Q_{12} = \frac{\nu_{21}E_1}{1-\nu_{12}\nu_{21}} = \frac{\nu_{12}E_2}{1-\nu_{12}\nu_{21}}$$

$$Q_{66} = G_{12}$$

Compliances in terms of engineering constants are

$$S_{11} = \frac{1}{E_1}$$

$$S_{22} = \frac{1}{E_2} \tag{4.40}$$

$$S_{12} = -\frac{\nu_{12}}{E_1} = -\frac{\nu_{21}}{E_2}$$

$$S_{66} = \frac{1}{G_{12}}$$

Thus, four independent engineering constants are E_1, E_2, G_{12} and ν_{12} (or ν_{21}). It may be noted that ν_{12} and ν_{21} are related as indicated in (4.22).

Relationship between Q_{ij} and S_{ij} are as follows

$$S_{11} = \frac{Q_{22}}{Q_{11}Q_{22} - Q_{12}^2}$$

$$S_{12} = \frac{-Q_{12}}{Q_{11}Q_{22} - Q_{12}^2}$$

$$S_{22} = \frac{Q_{11}}{Q_{11}Q_{22} - Q_{12}^2} \tag{4.41}$$

$$S_{66} = \frac{1}{Q_{66}}$$

Conversely,

$$Q_{11} = \frac{S_{22}}{S_{11}S_{22} - S_{12}^2}$$

$$Q_{12} = \frac{-S_{12}}{S_{11} S_{22} - S_{12}^2} \qquad (4.42)$$

$$Q_{22} = \frac{S_{11}}{S_{11} S_{22} - S_{12}^2}$$

$$Q_{66} = \frac{1}{S_{66}}$$

4.5 EXAMPLES : SERIES A

Example 4A.1

An isotropic lamina has $E = 100$ kN / mm^2 and $\nu = 0.25$.

Determine the reduced stiffness matrix and reduced compliance matrix.

The shear modulus G for the elastic material is

$$G = \frac{E}{2(1+\nu)} = \frac{100}{2 \times 1.25} = 40 \text{ kN/mm}^2$$

Let us first determine the reduced stiffness coefficients

$$Q_{11} = Q_{22} = \frac{E}{1-\nu^2} = \frac{100}{1-(0.25)^2} = 106.67 \text{ kN/mm}^2$$

$$Q_{12} = \nu Q_{11} = 0.25 \times 106.67 = 26.67 \text{ kN/mm}^2$$

$$Q_{66} = 40 \text{ kN/mm}^2$$

Therefore,

$$[Q] = \begin{bmatrix} 106.67 & 26.67 & 0 \\ 26.67 & 106.67 & 0 \\ 0 & 0 & 40 \end{bmatrix}$$

By inverting $[Q]$, one can obtain $[S]$. Another option of obtaining $[S]$ is to calculate all the elements of $[S]$ separately as have been done for $[Q]$. Calculations of $[S]$ have been done on the latter approach as given below

$$S_{11} = S_{22} = \frac{1}{100} = 10 \times 10^{-3} \text{ (kN/mm}^2)^{-1}$$

$$S_{12} = S_{21} = -\frac{0.25}{100} = -2.5 \times 10^{-3} \text{ (kN/mm}^2)^{-1}$$

$$S_{66} = \frac{1}{40} = 25 \times 10^{-3} \text{ (kN/mm}^2)^{-1}$$

Therefore,

$$[S] = \begin{bmatrix} 10 & -2.5 & 0 \\ -2.5 & 10 & 0 \\ 0 & 0 & 25 \end{bmatrix} \times 10^{-3} \text{ (kN/mm}^2)^{-1}$$

Example 4A.2

For an orthotropic lamina, engineering constants along the principal material axes are $E_1 = 150$ GPa, $E_2 = 20$ GPa, $G_{12} = 5$ GPa, $\nu_{12} = 0.2$.
Determine the reduced stiffness matrix $[Q]$.

$$\nu_{21} = \nu_{12}\frac{E_2}{E_1} = 0.2 \times \frac{20}{150} = 0.0267$$

$$1 - \nu_{12}\nu_{21} = 1 - 0.2 \times 0.0267 = 0.9947$$

$$Q_{11} = \frac{E_1}{1 - \nu_{12}\nu_{21}} = \frac{150}{0.9947} = 150.81 \text{ GPa}$$

$$Q_{12} = \nu_{21}Q_{11} = 0.0267 \times 150.81 = 4.027 \text{ GPa}$$

$$Q_{22} = \frac{E_2}{1 - \nu_{12}\nu_{21}} = \frac{20}{0.9947} = 20.11 \text{ GPa}$$

$$Q_{66} = G_{12} = 5 \text{ GPa}$$

Therefore,

$$[Q] = \begin{bmatrix} 150.81 & 4.027 & 0 \\ 4.027 & 20.11 & 0 \\ 0 & 0 & 5 \end{bmatrix} \text{ GPa}$$

Example 4A.3

Reduced stiffness matrix of an orthotropic lamina is given by

$$[Q] = \begin{bmatrix} 150.81 & 4.027 & 0 \\ 4.027 & 20.11 & 0 \\ 0 & 0 & 5 \end{bmatrix} \text{ GPa}$$

Determine E_1, E_2, G_{12} and ν_{12} of the orthotropic lamina

$$Q_{66} = G_{12} = 5$$

$$Q_{11} = 150.81$$

$$Q_{12} = \nu_{21}Q_{11}$$

or, $\nu_{21} = \dfrac{4.027}{150.81} = 0.0267$

$$\nu_{12} = \frac{4.027}{20.11} = 0.2$$

$$1 - \nu_{12}\nu_{21} = 1 - 0.2 \times 0.0267 = 0.9947$$

Therefore,

$$Q_{11} = \frac{E_1}{1 - \nu_{12}\nu_{21}} = 150.81$$

or, $E_1 = 150.81 \times 0.9947 = 150$ GPa

Similarly,

$$E_2 = 20.11 \times 0.9947 = 20 \text{ GPa}$$

4.6 TRANSFORMATION OF STRESS AND STRAIN

The stiffness of a composite changes with the change of ply orientation. A particular axis system is chosen for conveniently solving the problem - the axis system is known as the loading axis or the reference axis. For fibre reinforced composites, another axis system which is parallel and perpendicular to the fibre orientation is convenient for the calculation of material properties. As such the transformation of stresses and strains from one axis system to another is needed.

4.6.1 Transformation of Stress

The principal material axis system is indicated by the 1-2 axis and the reference axis system is shown by the *x-y* axis (Fig. 4.5). Figure 4.5(a) indicates an on-axis system, that is, where the principal material axis is coincident with the reference axis. Figure 4.5(b) depicts an off-axis system. Here, the reference axis system for a unidirectional composite is different from the material axis system. Counter-clockwise rotation of θ is taken as positive [4.3].

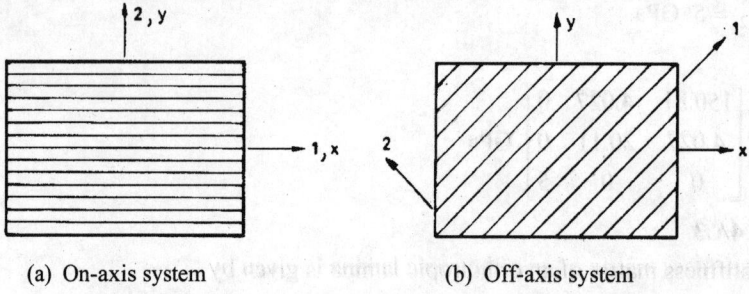

(a) On-axis system   (b) Off-axis system

Fig. 4.5 Axis system in an unidirectional stressed lamina

Unidirectional stressed lamina in an off-axis system is shown in Fig. 4.6. Stresses on planes coincident with the material axis system is shown in Fig. 4.7. The wedge is considered parallel and perpendicular to the fibre orientation.

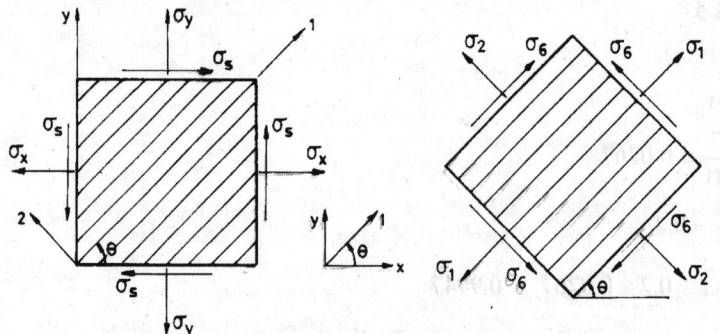

Fig. 4.6 Unidirectional stressed lamina

Referring to Fig. 4.7(a), the equilibrium of all horizontal and vertical forces of the wedge with unit area on the inclined plane yield the following equations

$$m\sigma_1 - n\sigma_6 = m\sigma_x + n\sigma_s \qquad (4.43)$$
$$n\sigma_1 + m\sigma_6 = n\sigma_y + m\sigma_s \qquad (4.44)$$

where

$$m = \cos\theta \text{ and } n = \sin\theta$$

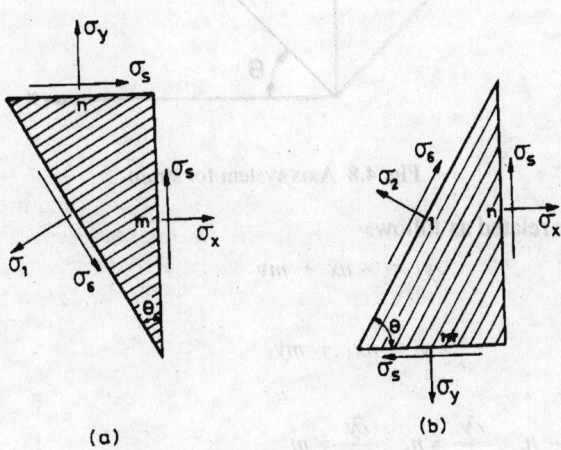

Fig. 4.7 Stresses on inclined plane

Solving (4.43) and (4.44), yields

$$\sigma_1 = m^2\sigma_x + n^2\sigma_y + 2mn\,\sigma_s \qquad (4.45)$$
$$\sigma_6 = -mn\,\sigma_x + mn\,\sigma_y + (m^2 - n^2)\,\sigma_s \qquad (4.46)$$

Similarly, referring to Fig. 4.7(b) and following similar steps as above, we get

$$\sigma_2 = n^2\sigma_x + m^2\sigma_y - 2mn\,\sigma_s \qquad (4.47)$$

Equations (4.45) to (4.47) written in matrix form becomes

$$\begin{Bmatrix}\sigma_1\\\sigma_2\\\sigma_6\end{Bmatrix} = \begin{bmatrix}m^2 & n^2 & 2mn\\n^2 & m^2 & -2mn\\-mn & mn & m^2-n^2\end{bmatrix}\begin{Bmatrix}\sigma_x\\\sigma_y\\\sigma_s\end{Bmatrix} \qquad (4.48)$$

or, $\{\sigma\}_{1,2} = [T]\{\sigma\}_{x,y}$ (4.49)

$[T]$ is known as the transformation matrix.

4.6.2 Transformation of Strain

Strain basically involves the geometry of the structure. The principal material axis is 1-2 and the reference axis is the xy axis (Fig. 4.8). Let u_1 and v_1 are the strains in the principal material axis direction 1 and 2 and u and v in the reference axis system.

The following strain–displacement relationships in the 1-2 axis system can be written as

$$\varepsilon_1 = \frac{\partial u_1}{\partial x_1}, \quad \varepsilon_2 = \frac{\partial v_1}{\partial y_1}, \quad \varepsilon_6 = \frac{\partial u_1}{\partial y_1} + \frac{\partial v_1}{\partial x_1} \qquad (4.50)$$

(x_1 and y_1 are corresponding to the axis system 1 and 2)

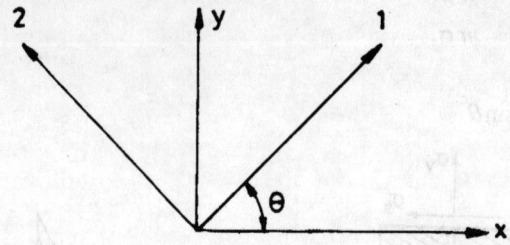

Fig. 4.8 Axis system for strain

Both the axis system are related as follows

$$x_1 = mx + ny \qquad\qquad y_1 = -nx + my \qquad (4.51)$$

Conversely,

$$x = mx_1 - ny_1 \qquad\qquad y = nx_1 + my_1 \qquad (4.52)$$

Therefore,

$$\frac{\partial x}{\partial x_1} = m, \quad \frac{\partial x}{\partial y_1} = -n, \quad \frac{\partial y}{\partial x_1} = n, \quad \frac{\partial y}{\partial y_1} = m \qquad (4.53)$$

Similarly, displacements in the two coordinate systems can be related as follows

$$u_1 = mu + nv, \qquad\qquad v_1 = -nu + mv \qquad (4.54)$$

and

$$u = mu_1 - nv_1, \qquad\qquad v = nu_1 + mv_1$$

Thus, the strain along the 1-axis is

$$\varepsilon_1 = \frac{\partial u_1}{\partial x} \cdot \frac{\partial x}{\partial x_1} + \frac{\partial u_1}{\partial y} \cdot \frac{\partial y}{\partial x_1}$$

or,

$$\varepsilon_1 = m \frac{\partial u_1}{\partial x} + n \frac{\partial u_1}{\partial y} \qquad (4.55)$$

Substituting u_1 from (4.54) with (4.55) yields

$$\varepsilon_1 = m \frac{\partial (mu + nv)}{\partial x} + n \frac{\partial (mu + nv)}{\partial y}$$

$$\varepsilon_1 = m \left[m \frac{\partial u}{\partial x} + n \frac{\partial v}{\partial x} \right] + n \left[m \frac{\partial u}{\partial y} + n \frac{\partial v}{\partial y} \right] \qquad (4.56)$$

By substituting appropriate strain terms in (4.56), yields

$$\varepsilon_1 = m^2 \varepsilon_x + n^2 \varepsilon_y + mn \varepsilon_s \qquad (4.57)$$

Adopting similar steps as above, the following equations result

$$\varepsilon_2 = n^2 \varepsilon_x + m^2 \varepsilon_y - mn \varepsilon_s \qquad (4.58)$$

$$\varepsilon_6 = -2mn \varepsilon_x + 2mn \varepsilon_y + (m^2 - n^2) \varepsilon_s \qquad (4.59)$$

Equations (4.57) to (4.59) written in matrix form becomes

$$\begin{Bmatrix} \varepsilon_1 \\ \varepsilon_2 \\ \dfrac{\varepsilon_6}{2} \end{Bmatrix} = \begin{bmatrix} m^2 & n^2 & 2mn \\ n^2 & m^2 & -2mn \\ -mn & mn & m^2-n^2 \end{bmatrix} \begin{Bmatrix} \varepsilon_x \\ \varepsilon_y \\ \dfrac{\varepsilon_s}{2} \end{Bmatrix} \qquad (4.60)$$

or, $\{\varepsilon\}_{1,2} = [T]\{\varepsilon\}_{x,y}$ \qquad (4.61)

By multiplying both sides of (4.48) and (4.60) by $[T]^{-1}$, we obtain for the stressed lamina (Fig. 4.7)

$$\begin{Bmatrix} \sigma_x \\ \sigma_y \\ \sigma_s \end{Bmatrix} = [T]^{-1} \begin{Bmatrix} \sigma_1 \\ \sigma_2 \\ \sigma_6 \end{Bmatrix} \qquad (4.62)$$

and

$$\begin{Bmatrix} \varepsilon_x \\ \varepsilon_y \\ \dfrac{\varepsilon_s}{2} \end{Bmatrix} = [T]^{-1} \begin{Bmatrix} \varepsilon_1 \\ \varepsilon_2 \\ \dfrac{\varepsilon_6}{2} \end{Bmatrix} \qquad (4.63)$$

where

$$[T]^{-1} = [T(-\theta)] = \begin{bmatrix} m^2 & n^2 & -2mn \\ n^2 & m^2 & 2mn \\ mn & -mn & m^2-n^2 \end{bmatrix} \qquad (4.64)$$

It may be noted that the laws of stress and strain transformation are independent of material properties, that is, they are the same for both isotropic and anisotropic materials.

4.7 TRANSFORMATION OF ELASTIC CONSTANTS

When a lamina is loaded in the reference axis xy, the relationship between stresses in the reference xy-axis and that in the principal material axis is given by (4.62).

Combining (4.62) and (4.34) result in

$$\begin{Bmatrix} \sigma_x \\ \sigma_y \\ \sigma_s \end{Bmatrix} = \begin{bmatrix} m^2 & n^2 & -2mn \\ n^2 & m^2 & 2mn \\ mn & -mn & m^2-n^2 \end{bmatrix} \begin{bmatrix} Q_{11} & Q_{12} & 0 \\ Q_{12} & Q_{22} & 0 \\ 0 & 0 & 2Q_{66} \end{bmatrix} \begin{Bmatrix} \varepsilon_1 \\ \varepsilon_2 \\ \dfrac{\varepsilon_6}{2} \end{Bmatrix} \qquad (4.65)$$

Substituting strains in the 1-2 axis in terms of xy-axis from (4.60) into (4.65), yields

$$\begin{Bmatrix} \sigma_x \\ \sigma_y \\ \sigma_s \end{Bmatrix} = \begin{bmatrix} m^2 & n^2 & -2mn \\ n^2 & m^2 & 2mn \\ mn & -mn & m^2-n^2 \end{bmatrix} \begin{bmatrix} Q_{11} & Q_{12} & 0 \\ Q_{12} & Q_{22} & 0 \\ 0 & 0 & 2Q_{66} \end{bmatrix} \begin{bmatrix} m^2 & n^2 & 2mn \\ n^2 & m^2 & -2mn \\ -mn & mn & m^2-n^2 \end{bmatrix} \begin{Bmatrix} \varepsilon_x \\ \varepsilon_y \\ \dfrac{\varepsilon_s}{2} \end{Bmatrix} \qquad (4.66)$$

Equation (4.66) is written as

$$\begin{Bmatrix}\sigma_x \\ \sigma_y \\ \sigma_s\end{Bmatrix} = \begin{bmatrix} Q_{xx} & Q_{xy} & 2Q_{xs} \\ Q_{yx} & Q_{yy} & 2Q_{ys} \\ Q_{sx} & Q_{sy} & 2Q_{ss} \end{bmatrix} \begin{Bmatrix}\varepsilon_x \\ \varepsilon_y \\ \varepsilon_s\end{Bmatrix} \quad (4.67)$$

The relationship between reduced stiffnesses are as follows

$$\begin{aligned}
Q_{xx} &= m^4 Q_{11} + n^4 Q_{22} + 2m^2n^2 Q_{12} + 4m^2n^2 Q_{66} \\
Q_{yy} &= n^4 Q_{11} + m^4 Q_{22} + 2m^2n^2 Q_{12} + 4m^2n^2 Q_{66} \\
Q_{xy} &= m^2n^2 Q_{11} + m^2n^2 Q_{22} + (m^4 + n^4) Q_{12} - 4m^2n^2 Q_{66} \\
Q_{xs} &= m^3n\, Q_{11} - mn^3\, Q_{22} + (mn^3 - m^3n) Q_{12} + 2(mn^3 - m^3n) Q_{66} \\
Q_{ys} &= mn^3\, Q_{11} - m^3n\, Q_{22} + (m^3n - mn^3) Q_{12} + 2(m^3n - mn^3) Q_{66} \\
Q_{ss} &= m^2n^2 Q_{11} + m^2n^2 Q_{22} - 2m^2n^2 Q_{12} + (m^2 - n^2)^2 Q_{66}
\end{aligned} \quad (4.68)$$

Equation (4.67) has elements of significant interest. It may be noted that there is coupling between normal stresses and shearing deformation as also between shearing stresses and normal strains. This coupling effect is absent when the loading is in the principal material axis direction.

4.8 TRANSFORMED REDUCED COMPLIANCE

The transformation of stress–strain relations in terms of compliance matrices are as follows

$$\begin{Bmatrix}\varepsilon_x \\ \varepsilon_y \\ \varepsilon_s\end{Bmatrix} = \begin{bmatrix} S_{xx} & S_{xy} & S_{xs} \\ S_{xy} & S_{yy} & S_{ys} \\ S_{xs} & S_{ys} & S_{ss} \end{bmatrix} \begin{Bmatrix}\sigma_x \\ \sigma_y \\ \sigma_s\end{Bmatrix} \quad (4.69)$$

For a specially orthotropic lamina, the stress–strain relationship is defined by four independent elastic constants in the principal material axis direction. They are E_1, E_2, G_{12} and ν_{12} (or ν_{21}). We change the notation of G_{12} to E_6 and G_{xy} to E_s. The stress–strain relationship in the principal material directions is given by

$$\begin{Bmatrix}\varepsilon_1 \\ \varepsilon_2 \\ \varepsilon_6\end{Bmatrix} = \begin{bmatrix} S_{11} & S_{12} & 0 \\ S_{12} & S_{22} & 0 \\ 0 & 0 & S_{66} \end{bmatrix}\begin{Bmatrix}\sigma_1 \\ \sigma_2 \\ \sigma_6\end{Bmatrix}$$

$$= \begin{bmatrix} \dfrac{1}{E_1} & -\dfrac{\nu_{21}}{E_2} & 0 \\ -\dfrac{\nu_{12}}{E_1} & \dfrac{1}{E_2} & 0 \\ 0 & 0 & \dfrac{1}{E_6} \end{bmatrix}\begin{Bmatrix}\sigma_1 \\ \sigma_2 \\ \sigma_6\end{Bmatrix} \quad (4.70)$$

The above stress–strain relation when transferred to any arbitrary yx axis system reduces to (4.69) in terms of compliances.

Suppose in (4.69) if σ_x is the only non-zero stress component, which means that $\sigma_y = \sigma_s = 0$, then

$$\varepsilon_x = \frac{\sigma_x}{E_x}$$

$$\varepsilon_y = -\frac{\nu_{xy}}{E_x}\sigma_x \qquad (4.71)$$

$$\varepsilon_s = \frac{\eta_{xs}}{E_x}\sigma_x$$

where ν_{xy} is the ratio of the transverse strain ε_y to the axial strain ε_x and η_{xs} the shear coupling coefficients is the ratio of shear strain ε_s to the axial strain ε_x.

Similarly, the following relations can be obtained

When only σ_y is operative   When only σ_s is operative

$$\varepsilon_x = -\frac{\nu_{yx}}{E_y}\sigma_y \qquad \varepsilon_x = \frac{\eta_{sx}}{E_s}\sigma_s$$

$$\varepsilon_y = \frac{\sigma_y}{E_y} \qquad \varepsilon_y = \frac{\eta_{sy}}{E_s}\sigma_s \qquad (4.72)$$

$$\varepsilon_s = \frac{\eta_{ys}}{E_y}\sigma_y \qquad \varepsilon_s = \frac{\sigma_s}{E_s}$$

Therefore, the superimposition of the three loadings results in the following equation.

$$\begin{Bmatrix}\varepsilon_x \\ \varepsilon_y \\ \varepsilon_s\end{Bmatrix} = \begin{bmatrix} \frac{1}{E_x} & -\frac{\nu_{yx}}{E_y} & \frac{\eta_{sx}}{E_s} \\ -\frac{\nu_{xy}}{E_x} & \frac{1}{E_y} & \frac{\eta_{sy}}{E_s} \\ \frac{\eta_{xs}}{E_x} & \frac{\eta_{ys}}{E_y} & \frac{1}{E_s} \end{bmatrix}\begin{Bmatrix}\sigma_x \\ \sigma_y \\ \sigma_s\end{Bmatrix} \qquad (4.73)$$

The symmetry of the compliance matrix yields

$$\frac{\nu_{xy}}{E_x} = \frac{\nu_{yx}}{E_y} \quad \text{or,} \quad \frac{\nu_{xy}}{\nu_{yx}} = \frac{E_x}{E_y}$$

$$\frac{\eta_{xs}}{E_x} = \frac{\eta_{sx}}{E_s} \quad \text{or,} \quad \frac{\eta_{xs}}{\eta_{sx}} = \frac{E_x}{E_s} \qquad (4.74)$$

$$\frac{\eta_{ys}}{E_y} = \frac{\eta_{sy}}{E_s} \quad \text{or,} \quad \frac{\eta_{ys}}{\eta_{sy}} = \frac{E_y}{E_s}$$

Comparing (4.73) with (4.69) yields

$$S_{xx} = \frac{1}{E_x}, \quad S_{yy} = \frac{1}{E_y}, \quad S_{ss} = \frac{1}{E_s}$$

$$S_{xy} = S_{yx} = -\frac{\nu_{xy}}{E_x} = -\frac{\nu_{yx}}{E_y} \qquad (4.75)$$

$$S_{xs} = S_{sx} = \frac{\eta_{xs}}{E_x} = \frac{\eta_{sx}}{E_s}$$

$$S_{ys} = S_{sy} = \frac{\eta_{ys}}{E_y} = \frac{\eta_{sy}}{E_s}$$

Following steps similar to those mentioned above while dealing with the reduced stiffness/compliance terms, it can be shown that

$$\begin{aligned}
S_{xx} &= m^4 S_{11} + n^4 S_{22} + 2m^2 n^2 S_{12} + m^2 n^2 S_{66} \\
S_{yy} &= n^4 S_{11} + m^4 S_{22} + 2m^2 n^2 S_{12} + m^2 n^2 S_{66} \\
S_{xy} &= m^2 n^2 S_{11} + m^2 n^2 S_{22} + (m^4 + n^4) S_{12} - m^2 n^2 S_{66} \\
S_{xs} &= 2m^3 n\, S_{11} - 2mn^3 S_{22} + 2(mn^3 - m^3 n) S_{12} + (mn^3 - m^3 n) S_{66} \\
S_{ys} &= 2mn^3 S_{11} - 2m^3 n\, S_{22} + 2(m^3 n - mn^3) S_{12} + (m^3 n - mn^3) S_{66} \\
S_{ss} &= 4m^2 n^2 S_{11} + 4m^2 n^2 S_{22} - 8m^2 n^2 S_{12} + (m^2 - n^2)^2 S_{66}
\end{aligned} \qquad (4.76)$$

4.9 TRANSFORMATION RELATIONS OF ENGINEERING CONSTANTS

[4.4]

Substituting compliances in terms of engineering constants as given by (4.76) into (4.70) yields

$$\frac{1}{E_x} = \frac{m^2}{E_1}(m^2 - n^2 \nu_{12}) + \frac{n^2}{E_2}(n^2 - m^2 \nu_{21}) + \frac{m^2 n^2}{E_6}$$

$$\frac{1}{E_y} = \frac{n^2}{E_1}(n^2 - m^2 \nu_{12}) + \frac{m^2}{E_2}(m^2 - n^2 \nu_{21}) + \frac{m^2 n^2}{E_6}$$

$$\frac{1}{E_s} = \frac{4m^2 n^2}{E_1}(1 + \nu_{12}) + \frac{4m^2 n^2}{E_2}(1 + \nu_{21}) + \frac{(m^2 - n^2)^2}{E_6} \qquad (4.77a)$$

$$\frac{\nu_{xy}}{E_x} = \frac{\nu_{yx}}{E_y} = \frac{m^2}{E_1}(m^2 \nu_{12} - n^2) + \frac{n^2}{E_2}(n^2 \nu_{21} - m^2) + \frac{m^2 n^2}{E_6}$$

$$\frac{\eta_{xs}}{E_x} = \frac{\eta_{sx}}{E_s} = \frac{2mn}{E_1}(m^2 - n^2 \nu_{12}) - \frac{2mn}{E_2}(n^2 - m^2 \nu_{21}) + \frac{mn^3 - m^3 n}{E_6}$$

$$\frac{\eta_{ys}}{E_y} = \frac{\eta_{sy}}{E_s} = \frac{2mn}{E_1}(n^2 - m^2 \nu_{12}) - \frac{2mn}{E_2}(m^2 - n^2 \nu_{21}) + \frac{m^3 n - mn^3}{E_6}$$

Equations (4.77a) can also be modified as follows

$$E_x = \frac{E_1}{m^4 + \left(\dfrac{E_1}{E_6} - 2\nu_{12}\right) n^2 m^2 + \dfrac{E_1}{E_2} n^4}$$

$$E_y = \frac{E_2}{m^4 + \left(\dfrac{E_2}{E_6} - 2\nu_{21}\right) n^2 m^2 + \dfrac{E_2}{E_1} n^4}$$

$$E_s = \frac{E_6}{n^4 + m^4 + 2\left(2\frac{E_6}{E_1}(1+2\nu_{12}) + 2\frac{E_6}{E_2} - 1\right)n^2m^2} \qquad (4.77b)$$

$$\nu_{yx} = \frac{\nu_{21}(n^4+m^4) - \left(1 + \frac{E_2}{E_1} - \frac{E_2}{E_6}\right)n^2m^2}{m^4 + \left(\frac{E_2}{E_6} - 2\nu_{21}\right)n^2m^2 + \frac{E_2}{E_1}n^2}$$

$$n_{xs} = \frac{2mn(m^2 - n^2\nu_{12}) - 2mn\frac{E_1}{E_2}(n^2 - m^2\nu_{21}) + \frac{E_1}{E_6}(mn^3 - m^3n)}{m^4 + \left(\frac{E_1}{E_6} - 2\nu_{12}\right)n^2m^2 + \frac{E_1}{E_2}n^4}$$

$$n_{ys} = \frac{\frac{2mn E_2}{E_1}(n^2 - m^2\nu_{12}) - 2mn(m^2 - n^2\nu_{21}) + \frac{E_2}{E_6}(m^3n - mn^3)}{m^4 + \left(\frac{E_2}{E_6} - 2\nu_{21}\right)m^2n^2 + \frac{E_2}{E_1}n^4}$$

4.10 EXAMPLES : SERIES B

Example 4B.1

For a particular FRP material, the following are the material properties in the principal material directions : $E_1 = 145$ GPa, $E_2 = 10.45$ GPa, $E_6 = 6.9$ GPa, $\nu_{12} = 0.28$.
Calculate $(\nu_{xy})_{45}$

$$\frac{\nu_{12}}{E_1} = \frac{\nu_{21}}{E_2}$$

or, $\nu_{21} = \frac{\nu_{12} \cdot E_2}{E_1} = \frac{0.28 \times 10.45}{145} = 0.02$

$m = \cos 45° = 0.707 = \sin 45° = n$

From (4.77a), we get

$$\frac{1}{E_x} = \frac{m^2}{E_1}(m^2 - n^2\nu_{12}) + \frac{n^2}{E_2}(n^2 - m^2\nu_{21}) + \frac{m^2n^2}{E_6}$$

$$= \frac{(0.707)^2}{145}(0.707^2 - 0.707^2 \times 0.28)$$

$$+ \frac{0.707^2}{10.45}(0.707^2 - 0.707^2 \times 0.02) + \frac{(0.707)^2(0.707)^2}{6.9}$$

$$= 0.0609$$

or, $E_x = 16.42$ GPa

Again, from (4.77a), we get

$$\frac{v_{xy}}{E_x} = \frac{m^2}{E_1}(m^2 v_{12} - n^2) + \frac{n^2}{E_2}(n^2 v_{12} - m^2) + \frac{m^2 n^2}{E_6}$$

or,
$$v_{xy} = 16.42 \left[\frac{(0.707)^2}{145}(0.707^2 \times 0.28 - 0.707^2) \right.$$
$$\left. + \frac{0.707^2}{10.45}(0.707^2 \times 0.02 - 0.707^2) + \frac{0.707^2 \times 0.707^2}{6.9} \right]$$

$$= 0.189$$

Example 4B.2

Two laminae are joined as shown in Fig. 4.9. Both are unidirectional lamina–fibres of one inclined at 0° and the other 45° to the loading axis. How is the specimen going to deform in the transverse direction? Properties of both lamina are same and they are

$E_1 = 145$ GPa, $E_2 = 10.45$ GPa, $E_6 = 6.9$ GPa and $v_{12} = 0.28$

We know

$$\frac{v_{12}}{E_1} = \frac{v_{21}}{E_2}$$

or, $v_{21} = 0.28 \times \dfrac{10.45}{145} = 0.02$

From (4.22), we get for the left lamina (1)

$$\varepsilon_1^{(1)} = \frac{\sigma_x}{E_1} = \frac{\sigma_x}{145}$$

$$\varepsilon_2^{(1)} = -\frac{v_{12}}{E_1}\sigma_x = -\frac{0.28}{145}\sigma_x = -01.93 \times 10^{-3}\sigma_x$$

In this case the principal material axis and the loading axis are coincident.
For the right 45° lamina (2), we get

$$\frac{1}{E_x} = \frac{m^2}{E_1}(m^2 - n^2 v_{12}) + \frac{n^2}{E_2}(n^2 - m^2 v_{21}) + \frac{m^2 n^2}{E_6}$$

$m = \cos 45° = 0.707$

$n = \sin 45° = 0.707$

From Example 4B.1, we know that

$E_x = 16.42$ GPa

Similarly, $v_{xy} = 0.189$

Therefore, for lamina (2), we get

$$\varepsilon_y^{(2)} = -\frac{v_{xy}}{E_x}\sigma_x = -\frac{0.189}{16.42}\sigma_x = -0.0115\,\sigma_x = -11.5 \times 10^{-3}\sigma_x$$

Fig. 4.9 Example 4.B.2

Therefore, comparing ε_y for both lamina, it is evident that lamina (2) will contract more than lamina (1). The possible deformed shape is shown in Fig. 4.10.

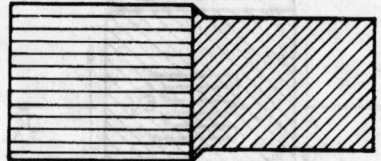

Fig. 4.10 Deformed shape

Example 4B.3

Determine Poisson's ratio v_{xy} at an angle $\theta = 30°$ with the fibre direction for a material with the following properties.

$E_1/E_2 = 3$, $E_6/E_2 = 0.5$ and $v_{12} = 0.25$

$E_1 = 3 E_2$, $E_6 = 0.5 E_2$

$$\frac{v_{12}}{E_1} = \frac{v_{21}}{E_2}, \quad \text{or,} \quad v_{21} = \frac{E_2}{E_1} v_{12} = \frac{0.25}{3} = 0.083$$

$$\frac{1}{E_x} = \frac{m^2}{E_1}(m^2 - n^2 v_{12}) + \frac{n^2}{E_2}(n^2 - m^2 v_{21}) + \frac{m^2 n^2}{E_6}$$

$m = \cos 30° = 0.866$

$n = \sin 30° = 0.15$

$$\frac{1}{E_x} = \frac{3}{4 \times 3 E_2}\left(\frac{3}{4} - \frac{1}{4} \times 0.25\right) + \frac{1}{4 E_2}\left(\frac{1}{4} - \frac{3}{4} \times 0.083\right) + \frac{3}{4} \times \frac{1}{4} \times \frac{2}{E_2}$$

$$= \frac{1}{E_2}[0.172 + 0.0469 + 0.375]$$

$$= \frac{0.5939}{E_2}$$

Now,

$$\frac{v_{xy}}{E_x} = \frac{m^2}{E_1}(m^2 v_{12} - n^2) + \frac{n^2}{E_2}(n^2 v_{21} - m^2) + \frac{m^2 n^2}{E_6}$$

or, $\quad 0.5939 \dfrac{v_{xy}}{E_2} = \dfrac{3}{4 \times 3E_2}\left(\dfrac{3}{4} \times 0.25 - \dfrac{1}{4}\right) + \dfrac{1}{4E_2}\left(\dfrac{1}{4} \times 0.083 - \dfrac{3}{4}\right)$

or, $\quad v_{xy} = 0.298$

Example 4B.4

An off-axis unidirectional lamina is loaded as shown in Fig. 4.11. Strain in the x-direction is measured with a strain gauge.

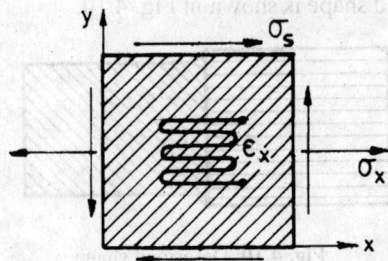

Fig. 4.11 Example 4B.4

a) Find an expression for shear coupling coefficient η_{sx} in terms of E_x, E_s, σ_x, σ_s and measured strain ε_x.

b) Determine ε_s.

$\varepsilon_x = 2 \times 10^{-3}$ m/m

$E_x = 58.7$ GPa, $E_s = 9.7$ GPa, $\sigma_x = 193$ MPa, $\sigma_s = 48.3$ MPa

$$\frac{\eta_{sx}}{E_s} = \frac{2mn}{E_1}(m^2 - n^2 v_{12}) - \frac{2mn}{E_2}(n^2 - m^2 v_{21}) + \frac{mn^3 - m^3 n}{E_6}$$

a) For this lamina, we can write

$$\varepsilon_x = \frac{\sigma_x}{E_x} + \frac{\eta_{sx}}{E_s}\sigma_s$$

or, $\quad \dfrac{\eta_{sx}}{E_s} = \dfrac{1}{\sigma_s}\left(\varepsilon_x - \dfrac{\sigma_x}{E_x}\right)$

or, $\quad \eta_{sx} = \dfrac{E_s}{E_x \sigma_s}(E_x \varepsilon_x - \sigma_x)$

b) $\quad \varepsilon_s = \dfrac{\eta_{xs}}{E_x}\sigma_x + \dfrac{\sigma_s}{E_s}$

We have

$$\eta_{sx} = \frac{9.7}{58.7 \times 48.3 \times 10^{-3}} [58.7 \times 2 \times 10^{-3} - 193 \times 10^{-3}]$$

$$= -0.259$$

Now,

$$\frac{\eta_{xs}}{E_x} = \frac{\eta_{sx}}{E_s}$$

Therefore,

$$\eta_{xs} = \frac{E_x}{E_s}\eta_{sx} = \frac{58.7}{9.7}(-0.259) = -1.565$$

$$\varepsilon_s = -\frac{1.565}{58.7} \times 193 \times 10^{-3} + \frac{483 \times 10^{-3}}{9.7}$$

$$= -1.662 \times 10^{-4}$$

Example 4B.5

Determine the equivalent stress system along the material axis 1-2, for a lamina with ply angle $\theta = 45°$ if stresses along the reference axis are

$\sigma_x = 200$ GPa, $\sigma_y = 50$ GPa, $\sigma_s = 70$ GPa

$m = \cos 45° = 0.707$, $n = \sin 45° = 0.707$

Applying eqn. (4.48), we get

$$\begin{Bmatrix} \sigma_1 \\ \sigma_2 \\ \sigma_6 \end{Bmatrix} = \begin{bmatrix} 0.5 & 0.5 & 1 \\ 0.5 & 0.5 & -1 \\ -0.5 & 0.5 & 0 \end{bmatrix} \begin{Bmatrix} 200 \\ 50 \\ 70 \end{Bmatrix}$$

or,

$$\begin{Bmatrix} \sigma_1 \\ \sigma_2 \\ \sigma_6 \end{Bmatrix} = \begin{Bmatrix} 195 \\ 55 \\ -75 \end{Bmatrix} \text{GPa}$$

Example 4B.6

Determine the strain along the material axis 1-2 for a lamina with ply orientation $\theta = 30°$, if strains along the reference axis xy are

$\varepsilon_x = 0.002$, $\varepsilon_y = 0.0015$, $\varepsilon_s = 0.00008$

$m = \cos 30° = 0.866$, $n = \sin 30° = 0.5$

Applying (4.60), yields

$$\begin{Bmatrix} \varepsilon_1 \\ \varepsilon_2 \\ \varepsilon_6 \end{Bmatrix} = \begin{bmatrix} 0.75 & 0.25 & 0.866 \\ 0.25 & 0.75 & -0.866 \\ -0.433 & 0.433 & 0.5 \end{bmatrix} \begin{Bmatrix} 0.002 \\ 0.0015 \\ 0.00008 \end{Bmatrix}$$

$$= \begin{Bmatrix} 1.944 \times 10^{-3} \\ 1.556 \times 10^{-3} \\ -1.765 \times 10^{-4} \end{Bmatrix}$$

Example 4B.7

A high strength composite has the following elastic constants

$E_1 = 145$ GPa, $E_2 = 12$ GPa, $E_6 = 6$ GPa, $\nu_{12} = 0.25$

Determine the transformed reduced stiffness matrix for the lamina with ply angle $\theta = 45°$.

$\nu_{21} = 0.25 \times \dfrac{12}{145} = 0.0207$

$Q_{11} = \dfrac{145}{1 - 0.25 \times 0.0207} = 145.75$

$Q_{22} = \dfrac{12}{1 - 0.25 \times 0.0207} = 12.06$

$Q_{12} = \dfrac{0.25 \times 2}{1 - 0.25 \times 0.0207} = 3.016$

$Q_{66} = E_6 = 6$

Therefore, the stiffness matrix in the material axis system is

$$[Q]_{1,2} = \begin{bmatrix} 145.75 & 3.016 & 0 \\ 3.016 & 12.06 & 0 \\ 0 & 0 & 6 \end{bmatrix}$$

$m = \cos 45° = 0.707, \qquad n = \sin 45° = 0.707$

From (4.68), we get

$\begin{aligned} Q_{xx} &= (0.707)^4 \times 145.75 + (0.707)^4 \times 12.06 \\ &\quad + 2\,(0.707)^2\,(0.707)^2 \times 3.016 + 4 \times (0.707)^2\,(0.707)^2 \times 6 \\ &= 46.96 \text{ GPa} \end{aligned}$

$\begin{aligned} Q_{yy} &= (0.707)^4 \times 145.75 + (0.707)^4 \times 12.06 \\ &\quad + 2\,(0.707)^2\,(0.707)^2 \times 3.016 + 4 \times (0.707)^2\,(0.707)^2 \times 6 \\ &= 46.96 \text{ GPa} \end{aligned}$

$\begin{aligned} Q_{xy} &= (0.707)^2\,(0.707)^2 \times 145.75 + (0.707)^2\,(0.707)^2 \times 12.06 \\ &\quad + [(0.707)^4 + (0.707)^4] \times 3.016 - 4\,(0.707)^2\,(0.707)^2 \times 6 \\ &= 34.94 \quad \text{GPa} \end{aligned}$

$$Q_{xs} = (0.707)^3 (0.707) \, 145.75 - (0.707)(0.707)^3 \times 12.06$$
$$+ [(0.707) + (0.707)^3 - (0.707)^3 (0.707)] \, 3.016$$
$$+ 2 \, [(0.707)(0.707)^3 - (0.707)^3 (0.707)] \, 6$$
$$= 33.40 \quad \text{GPa}$$

$$Q_{ys} = (0.707)(0.707)^3 \, 145.75 - (0.707)^3 (0.707) \, 12.06 + 0 + 0$$
$$= 33.40 \text{ GPa}$$

$$Q_{ss} = (0.707)^2 (0.707)^2 \, 145.75 + (0.707)^2 (0.707)^2 \, 12.06$$
$$- 2 \, (0.707)^2 (0.707)^2 \, 3.016 + \left[(0.707)^2 - (0.707)^2 \right] 6$$
$$= 37.93 \text{ GPa}$$

Therefore, the transformed reduced stiffness matrix is

$$[Q]_{x,y} = \begin{bmatrix} 46.96 & 34.94 & 33.40 \\ 34.94 & 46.96 & 33.40 \\ 33.40 & 33.40 & 37.93 \end{bmatrix} \text{ GPa}$$

Example 4B.8

The elastic properties along material axes 1-2 of a unidirectional FRP lamina are

$$E_1 = 145 \text{ GPa}, \quad E_2 = 12 \text{ GPa}, \quad E_6 = 6 \text{ GPa}, \quad \nu_{12} = 0.25$$

Determine the variation of elastic properties along the reference axis xy.

Let us calculate the variation of elastic properties at an increment of 10°. A sample calculation is presented for $\theta = 30°$.

$$\frac{1}{E_x} = \frac{m^4}{E_1} + \frac{n^4}{E_2} + m^2 n^2 \left(\frac{1}{E_6} - \frac{2\nu_{12}}{E_1} \right)$$

$$m = \cos 30° = 0.866, \quad n = \sin 30° = 0.5$$

$$\frac{1}{E_x} = \frac{(0.866)^4}{145} + \frac{(0.5)^4}{12} + (0.866)^2 (0.5)^2 \left(\frac{1}{6} - \frac{2 \times 0.25}{145} \right)$$

or, $E_x = 25.20$ GPa

Similarly,

$$\frac{1}{E_y} = \frac{(0.5)^4}{145} + \frac{(0.866)^4}{12} + (0.866)^2 (0.5)^2 \left[\frac{1}{6} - \frac{2 \times 0.24}{145} \right]$$

$$E_y = 12.8 \text{ GPa}$$

$$\frac{1}{E_s} = (0.866)^2 (0.5)^2 + \left[\frac{4}{145} + \frac{4}{12} + \frac{8 \times 0.3}{145} \right] + \left[(0.866)^2 - (0.5)^2 \right] \frac{1}{6}$$

or, $E_s = 1.23$ GPa

$$\frac{v_{xy}}{E_x} = \left[\{(0.866)^4 + (0.5)^4\} \frac{0.3}{145} - (0.866)^2 (0.5)^2 \left(\frac{1}{145} + \frac{1}{12} - \frac{1}{6} \right) \right]$$

or, $v_{xy} = 0.388$

$$v_{yx} = 0.388 \times \frac{12.84}{25.20} = 0.197$$

Similarly,

$$\eta_{xs} = 25.20 \left[(0.866)^3 (0.50) \left(-\frac{1}{6} + \frac{2 \times 0.25}{145} + \frac{2}{12} \right) \right.$$

$$\left. + (0.866)(0.5)^3 \left(\frac{1}{6} - \frac{2 \times 0.25}{145} - \frac{2}{12} \right) \right]$$

$$= -1.23$$

$\eta_{sx} = -0.222$

The elastic properties at various angles have been put in a tabular form in Table 4.2. The variation of different elastic constants with the orientation angle is shown in Fig. 4.13 to Fig. 4.17.

Table 4.2 Variation of unidirectional ply elastic constants with the ply angle in GPa.

θ	E_x	E_y	E_s	v_{xy}	η_{sx}	η_{sy}
0	145.0	12.0	6.00	0.250	0.000	0.000
10	88.2	12.0	6.32	0.340	2.190	0.016
20	42.8	12.2	7.33	0.396	1.820	0.081
30	25.2	12.8	8.93	0.388	1.230	0.222
40	17.8	14.4	10.40	0.346	0.781	0.452
45	15.8	15.8	10.70	0.315	0.603	0.603
50	14.4	17.8	10.40	0.280	0.452	0.783
60	12.8	25.2	8.93	0.198	0.222	1.230
70	12.2	47.8	7.33	0.113	0.081	1.820
80	12.0	88.2	6.32	0.046	0.016	2.190
90	12.0	145.0	6.00	0.0207	0.000	0.000

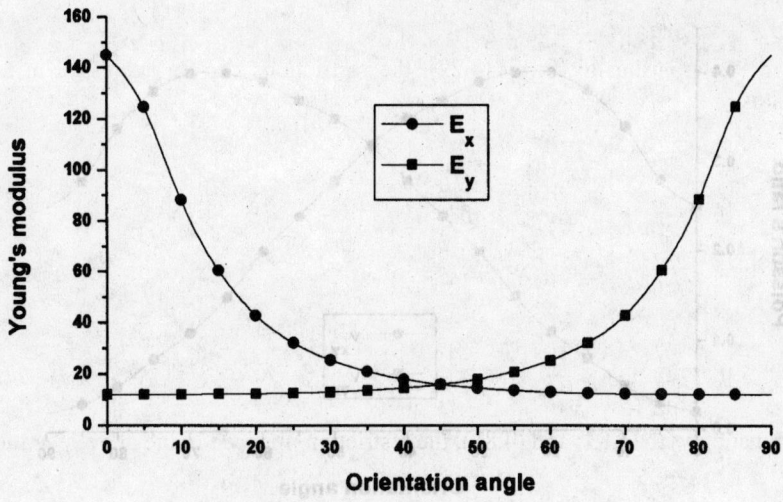

Fig. 4.13 Variation of Young's modulus with θ

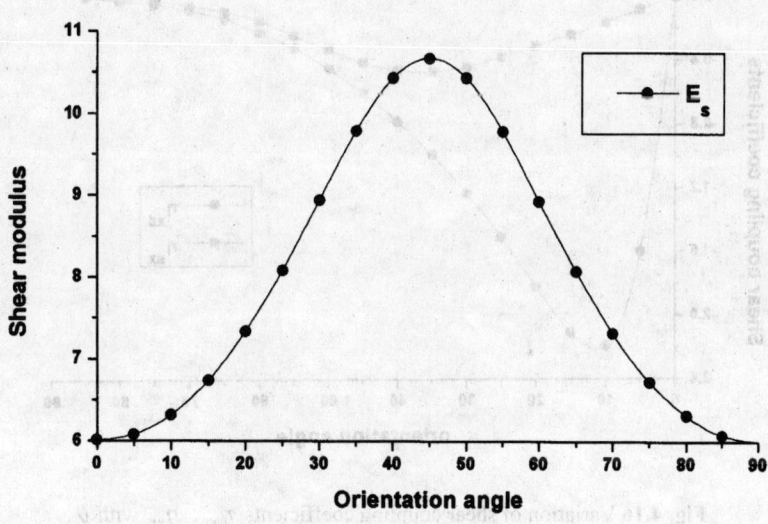

Fig. 4.14 Variation of shear modulus with θ

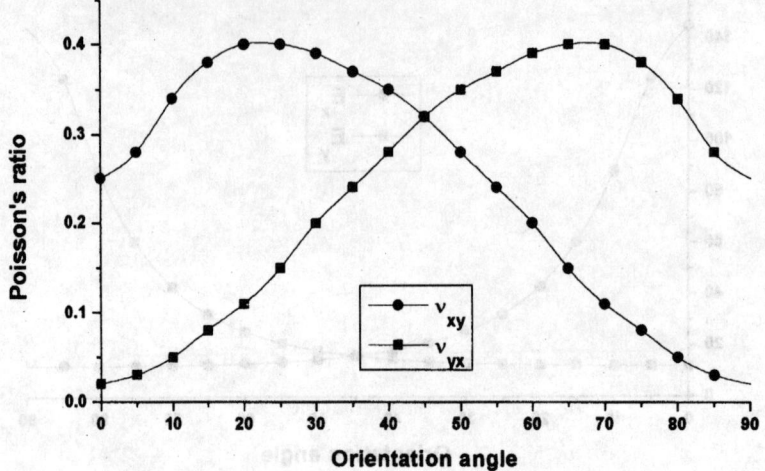

Fig. 4.15 Variation of Poisson's ratio with θ

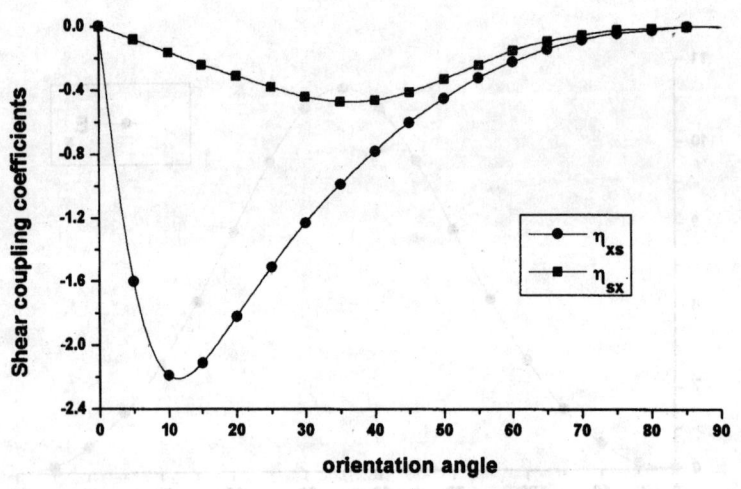

Fig. 4.16 Variation of shear coupling coefficients η_{xs}, η_{sx} with θ

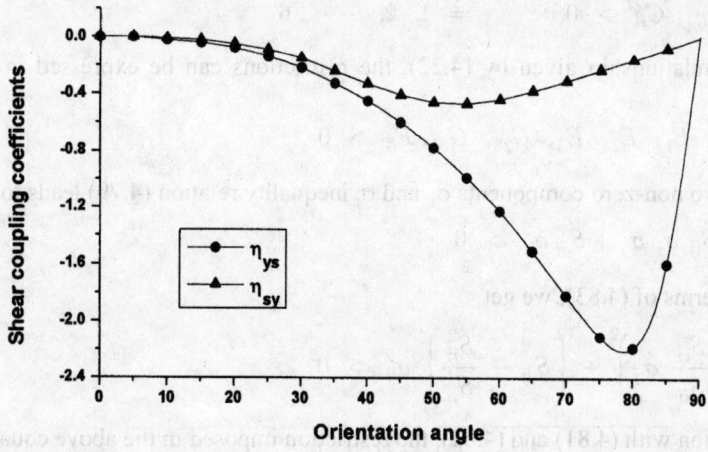

Fig. 4.17 Variation of shear coupling coefficients η_{ys}, η_{sy} with θ

Some observations are made for the nature of variation.

1. E_s is maximum when $\theta = 45°$. The variation is also symmetric about its maximum value. This suggests that the greatest resistance to shear is offered by 45° ply.
2. Young's modulus is maximum in the fibre direction. The reduction of the value is somewhat sharp with small change in the orientation of the fibre. Therefore, the greatest resistance to the membrane action is offered in the direction of the fibre.
3. Shear coupling effects are absent in specially orthotropic lamina, but a deviation from orthotropy results in the occurrence of shear coupling effects.

4.11 RESTRICTIONS ON ELASTIC CONSTANTS

According to the law of physics, a positive energy input is required to bring a material from the unstressed to the stressed condition. This addition of energy is independent of the order in which the load is applied. Specific strain energy per unit volume of the material corresponding to the particular state of stress is U.

For linear elasticity, the specific strain energy U is given by

$$U = \frac{1}{2} \sum_{i=1}^{6} \sigma_i \, \varepsilon_i = \frac{1}{2} \{\sigma\}^T \{\varepsilon\} = \frac{1}{2} \{\varepsilon\}^T \{\sigma\} > 0 \tag{4.78}$$

Substituting Hooke's law using the same notation as given earlier in the chapter

$$U = \frac{1}{2} \{\sigma\}^T [S] \{\sigma\} = \frac{1}{2} \{\varepsilon\}^T [C] \{\varepsilon\} > 0 \tag{4.79}$$

For a single stress component σ_i, the inequality equation (4.79) reduces to

$$S_{ii} \, \sigma_i^2 > 0 \tag{4.80}$$

This indicates that all diagonal elements of $[S]$ must be positive. So are the diagonal elements of $[C]$.

Therefore,
$$S_{ii} > 0, \quad C_{ii} > 0, \quad i = 1, 2, \cdots\cdots, 6 \tag{4.81}$$

Noting the relationship given by (4.22), the restrictions can be expressed in terms of elastic constants, that is,
$$E_i > 0, \quad E_1, E_2, E_3, G_{23}, G_{31}, E_6 > 0 \tag{4.82}$$

In case of two non-zero components σ_i and σ_j inequality relation (4.79) leads to
$$S_{ii} \sigma_i^2 + 2 S_{ij} \sigma_i \sigma_j + S_{jj} \sigma_j^2 > 0 \tag{4.83}$$

Rearranging the terms of (4.83), we get
$$S_{ii}\left(\sigma_i + \frac{S_{ij}}{S_{ii}}\sigma_j\right)^2 + \left(S_{jj} - \frac{S_{ij}^2}{S_{ii}}\right)\sigma_j^2 > 0 \tag{4.84}$$

In combination with (4.81) and (4.82), the restriction imposed in the above equation is
$$S_{ii} S_{jj} - S_{ij}^2 > 0 \tag{4.85}$$

This implies that
$$\begin{aligned} |S_{23}| &< (S_{22} S_{33})^{1/2} \\ |S_{31}| &< (S_{33} S_{11})^{1/2} \\ |S_{12}| &< (S_{11} S_{22})^{1/2} \end{aligned} \tag{4.86}$$

As the compliance matrix is symmetric, we get
$$\frac{v_{ij}}{E_i} = \frac{v_{ji}}{E_j}, \quad i, j = 1, 2, 6 \tag{4.87}$$

Based on the positive definiteness of compliance and stiffnesses, (4.81) and noting the relationship given by (4.27),
$$(1 - v_{23} v_{32}), (1 - v_{13} v_{31}), (1 - v_{12} v_{21}) > 0 \tag{4.88}$$

Combining (4.87) and (4.88) yields
$$\begin{aligned} |v_{21}| &< \left(\frac{E_2}{E_1}\right)^{1/2}, & |v_{12}| &< \left(\frac{E_1}{E_2}\right)^{1/2} \\ |v_{32}| &< \left(\frac{E_3}{E_2}\right)^{1/2}, & |v_{23}| &< \left(\frac{E_2}{E_3}\right)^{1/2} \\ |v_{13}| &< \left(\frac{E_1}{E_3}\right)^{1/2}, & |v_{31}| &< \left(\frac{E_3}{E_1}\right)^{1/2} \end{aligned} \tag{4.89}$$

Similarly, Δ in (4.28) is
$$\Delta = 1 - v_{12} v_{21} - v_{23} v_{32} - v_{31} v_{13} - 2 v_{21} v_{32} v_{13} > 0 \tag{4.90}$$

Rearranging different terms of eqn. (4.90), yields

$$\nu_{21}\,\nu_{32}\,\nu_{13} < \frac{1 - \nu_{21}^2\left(\dfrac{E_1}{E_2}\right) - \nu_{32}^2\left(\dfrac{E_2}{E_3}\right) - \nu_{13}^2\left(\dfrac{E_3}{E_1}\right)}{2} < \frac{1}{2} \qquad (4.91)$$

It can further be rearranged to read

$$\left[1 - \nu_{32}^2\left(\dfrac{E_2}{E_3}\right)\right]\left[1 - \nu_{13}^2\left(\dfrac{E_3}{E_1}\right)\right] - \left[\nu_{21}\left(\dfrac{E_1}{E_2}\right)^{1/2} + \nu_{32}\,\nu_{13}\left(\dfrac{E_3}{E_1}\right)^{1/2}\right]^2 > 0 \qquad (4.92)$$

4.12 TYPICAL ELASTIC PROPERTIES OF A UNIDIRECTIONAL LAMINA

Typical elastic properties of a unidirectional lamina and woven rovings are given in Table 4.3 and 4.4 respectively. These properties refer to room temperature and dry conditions. For unidirectional laminas in Table 4.3, fibre volume is generally 60% whereas it is 50% for woven rovings in Table 4.4.

Table 4.3 Elastic properties of common unidirectional plies

Type of composite	E_1 GPa	E_2 GPa	G_{12} GPa	ν_{12}
High strength carbon/epoxy	140	10	5	0.3
High modulus carbon/epoxy	220	86.9	4.8	0.3
E-glass/epoxy	40	8	4	0.25
Kevlar/epoxy	75	6	2	0.34
E-glass/polyester	33	7.2	2.9	0.29

Table 4.4 Typical elastic properties of plane woven roving

Type of composite	E_1 GPa	E_2 GPa	G_{12} GPa	ν_{12}
High strength carbon/epoxy	70	70	5	0.10
High modulus carbon/epoxy	85	85	5	0.10
E-glass/epoxy	25	25	4	0.20
Kevlar/epoxy	30	30	5	0.20

REFERENCES AND SUGGESTED READINGS

4.1 A. E. Green and W. Zerna, *Theoretical Elasticity*, 2nd Edition, Oxford University Press, New York, 1968

4.2 R. M. Christiansen, *Mechanics of Composite Materials*, John Wiley & Sons, New York, 1979.

4.3 S. W. Tsai and H.T. Hahn, *Introduction to Composite Materials*, Technomic Publishing, Lancaster, USA, 1980.

4.4 I. M. Daniel and O. Ishai, *Engineering Mechanics of Composite Materials*, Oxford University Press, New York, 1994.

EXERCISE 4

4.1 A unidirectional lamina is subjected to stresses as follows

$\sigma_1 = 500$ MPa, $\sigma_2 = 80$ MPa and $\sigma_6 = 25$ MPa

The compliance components of the material are

$S_{11} = 5.525$ (TPa)$^{-1}$; $S_{22} = 97.05$ (TPa)$^{-1}$

$S_{12} = -1.547$ (TPa)$^{-1}$; $S_{66} = -139.5$ (TPa)$^{-1}$

Determine the strain components.

4.2 The reduced stiffness matrix is given by

$$\begin{bmatrix} 181.8 & 2.897 & 0 \\ 2.897 & 10.34 & 0 \\ 0 & 0 & 7.17 \end{bmatrix} \text{GPa}$$

Determine E_1, E_2, G_{12} and ν_{12} of the orthotropic lamina.

4.3 Determine the stiffness and compliance components of aluminium with the following material properties

$E = 69$ GPa and $\nu = 0.3$

What will be the changes in the components if ν is changed to 0.22?

4.4 Write the expressions of compliance components in terms of stiffness components for a unidirectional lamina.

4.5 Derive the transformation relation given by (4.68).

4.6 The values of E_1, E_2 and ν_{12} of an orthotropic lamina are known. Determine G_{12} if the modulus at an angle 30° to the fibre direction is known.

4.7 Stresses in a (1-2) system are given by {10, 5, 4}. Find the stress components for the *xy* system for $\theta = 45°$.

4.8 Strains in a (1-2) system are given by {10, 5, 4}. Find the strain components for the *xy* system for $\theta = 45°$.

4.9 Derive the transformation of a unidirectional lamina for 1-2 axis system to *xy* axis system.

4.10 Is there any bound for Poisson's ratio of an off-axis unidirectional lamina? What will be the effect if the fibre stiffness approaches infinity or matrix stiffness zero?

4.11 The following are the material properties of a unidirectional lamina

$E_1 = 140$ GPa, $E_2 = 10$ GPa, $G_{12} = 6$ GPa, $\nu_{12} = 0.25$

Determine the material properties in a direction 45° inclined to the fibre.

4.12 Determine Poisson's ratio ν_{xy} at an angle $\theta = 45°$ with the fibre direction for a material with the following properties

$\dfrac{E_1}{E_2} = 2.8$, $\dfrac{G_{12}}{E_2} = 0.45$ and $\nu_{12} = 0.3$

4.13 A lamina is loaded at angles $\theta = 30°$ and $45°$ with the fibre direction, and the corresponding modulii obtained are $E_{x_{30}}$ and $E_{x_{60}}$ respectively. Determine a relationship between $E_1, E_2, E_{x_{30}}$ and $E_{x_{45}}$. Determine an expression for E_2 in terms of $E_{x_{30}}$ and $E_{x_{45}}$ for $E_1 \gg E_2$.

4.14 The principal strains for a uniaxially loaded unidirectional lamina ($\sigma_1 = \sigma_0$) are ε_1 and ε_2. The same lamina is now subjected to a biaxial loading $\sigma_1 = \sigma_2 = 0$. Determine the strain in the 2-direction of the same lamina in terms of $\varepsilon_1, \varepsilon_2$ and the ratio E_1/E_2.

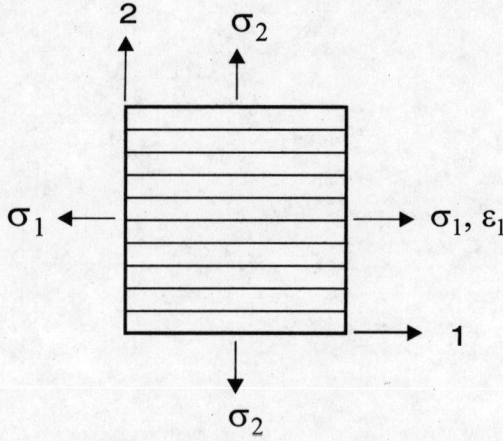

Fig. Prob. 4.14

4.15 An off-axis unidirectional lamina is loaded as shown. Determine normal strains ε_x and ε_y in terms of engineering constants.

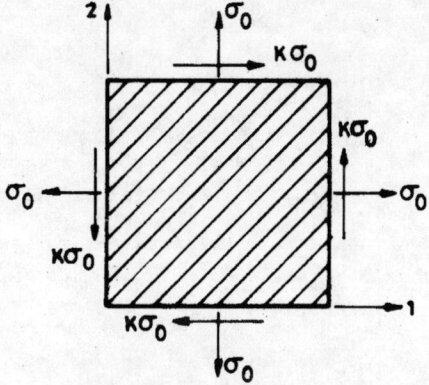

Fig. Prob. 4.15

4.16 Show that Δ in (4.90) for orthotropic materials reduces to $v \leq \frac{1}{2}$ for isotropic materials.

4.17 Show that $Q_{11} + Q_{22} + 2Q_{12}$ is invariant under rotation about the z- axis.

4.18 Write a computer programme to determine $E_x, E_y, G_{xy}, v_{xy}, \eta_{xs}$ and η_{ys} as a functions of θ. Plot their variation with θ from $\theta = 0°$ to $\theta = 90°$ at an interval of $30°$ for an orthotropic material with the following properties

$E_1 = 205$ GPa, $E_2 = 5$ GPa, $G_{12} = 2.5$ GPa, $v = 0.25$.

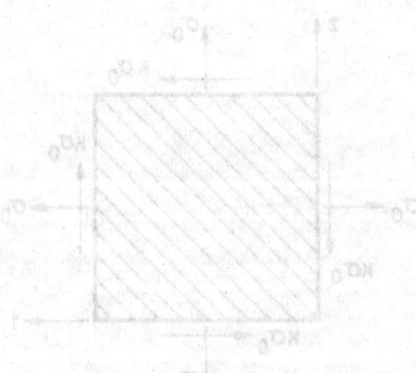

CHAPTER 5

ANALYSIS OF LAMINATED COMPOSITES

5.1 LAMINATES

In the last chapter we have dealt with topics related to a lamina or a single ply. We now pass on to the treatment of a laminate. A laminate is a layered construction shaped as a plate or a shell. A FRP laminate consists of a series of laminae or plies that are bonded together to act as an integral structural element. Individual plies are oriented at different ply angles to produce different strength and stiffness in the required direction of the laminate.

A laminate code is used to identify a laminate. If all individual laminae are unidirectional, then the laminae are denoted by their orientation angle in the sequence from bottom to top. As an example

$$[90|90|45|45|-45|-45|-45|-45|45|45|90|90]$$

Successive laminae with the same orientation can be taken together, the number being put as subscripts. The above laminate can thus be written as

$$[90_2|45_2|-45_4|45_2|90_2]$$

The laminate above is symmetric. In the case of a symmetric construction, only the bottom half of the laminate about the symmetry line need be specified as shown below

$$[90_2|45_2|-45_2]_s$$

It should appear natural that the overall behavior of a multidirectional laminate is a function of the properties and stacking sequence of individual plies.

When a laminate is made of more than one composite material, it is called a hybrid laminate. A similar notation as stated above is used except that a subscript is put on the ply angle to indicate the type of material used. Thus a laminate

$$[90_K, 60_G, 0_K, 0_C, 0_C, 0_K, 60_G, 90_K]$$

gives a sequence of ply angles indicated with the type of composite material. Thus, if we indicate K as the kevlar/epoxy composite, G as the E-glass/epoxy composite, C as the carbon /epoxy composite, then 90_K denotes a lamina whose ply angle is $90°$ and the material of the ply is kevlar/epoxy.

5.2 BASIC ASSUMPTIONS

The following basic assumptions are made

1. Each lamina or ply of the laminate is quasi-homogeneous and orthotropic, but the orientation of the fibre may change from lamina to lamina.
2. All displacements are continuous throughout the laminate.
3. All deformations in the laminate are considered to be small.
4. The laminate is thin and loaded in its plane only. The laminate and its layers are assumed to be

in a plane stress condition except the edges ($\sigma_z = \tau_{xz} = \tau_{yz} = 0$).

5. Transverse shear strains γ_{xz} and γ_{yz} are negligible. This implies that a line originally straight and perpendicular to the laminate mid-plane remains straight and perpendicular to the deformed state.
6. The bond between plies in a laminae are perfect, that is, plies will not slip over each other, and displacements and strains are continuous across interfaces of plies.
7. Strain–displacement and stress–strain relations are linear.

5.3 STRAIN–DISPLACEMENT RELATIONSHIP

Figure 5.1 shows a section of the laminate normal to the y-axis before and after deformation. The middle plane of the laminate is the xy-plane which is equidistant from the top and bottom surfaces and it is also the reference plane.

The inplane displacements of the middle plane are denoted as u_o and v_o in the direction parallel to the x and y-axis system.

Referring to Fig. 5.1, inplane displacements at any arbitrary point B are given by

$$u = u_0 - z \frac{\partial w}{\partial x} \tag{5.1}$$

$$v = v_0 - z \frac{\partial w}{\partial y}$$

where z is the coordinate of the arbitrary point in the z-direction and w is the transverse displacement.

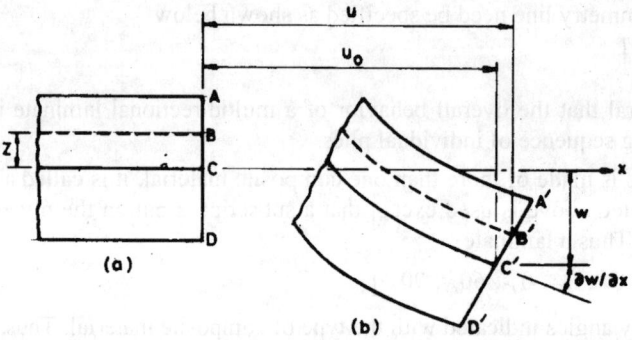

Fig. 5.1 A laminate before and after deformation

We apply the strain–displacement relationship

$$\varepsilon_x = \frac{\partial u}{\partial x} = \frac{\partial u_0}{\partial x} - z \frac{\partial^2 w}{\partial x^2}$$

$$\varepsilon_y = \frac{\partial v}{\partial y} = \frac{\partial v_0}{\partial y} - z \frac{\partial^2 w}{\partial y^2}$$

$$\gamma_{xy} = \varepsilon_s = \frac{\partial u}{\partial y} + \frac{\partial v}{\partial x} = \frac{\partial u_0}{\partial y} + \frac{\partial v_0}{\partial x} - 2z\frac{\partial^2 w}{\partial x \partial y} \tag{5.2}$$

The strain components on the reference plane are indicated as follows

$$\varepsilon_x^0 = \frac{\partial u_o}{\partial x}$$
$$\varepsilon_y^0 = \frac{\partial v_0}{\partial y} \tag{5.3}$$
$$\varepsilon_s^0 = \frac{\partial u_0}{\partial y} + \frac{\partial v_0}{\partial x}$$

Equation (5.2) can now be written as

$$\begin{Bmatrix} \varepsilon_x \\ \varepsilon_y \\ \varepsilon_s \end{Bmatrix} = \begin{Bmatrix} \varepsilon_x^0 \\ \varepsilon_y^0 \\ \varepsilon_s^0 \end{Bmatrix} + z \begin{Bmatrix} \kappa_x \\ \kappa_y \\ \kappa_s \end{Bmatrix} \tag{5.4}$$

where the curvatures of the laminate are

$$\kappa_x = -\frac{\partial^2 w}{\partial x^2}, \quad \kappa_y = -\frac{\partial^2 w}{\partial y^2}, \quad \kappa_s = -2\frac{\partial^2 w}{\partial x \partial y} \tag{5.5}$$

Equation (5.4) indicates that strains vary linearly through the thickness.

5.4 STRESS–STRAIN RELATIONS

Consider a lamina at a distance $\bar{z}_k$ from the laminate reference plane (Fig. 5.2). Stress–strain relations for the kth lamina with reference to the principal material axes are [(4.34)]

$$\begin{Bmatrix} \sigma_1 \\ \sigma_2 \\ \sigma_6 \end{Bmatrix}_k = \begin{bmatrix} Q_{11} & Q_{12} & 0 \\ Q_{12} & Q_{22} & 0 \\ 0 & 0 & Q_{66} \end{bmatrix}_k \begin{Bmatrix} \varepsilon_1 \\ \varepsilon_2 \\ \varepsilon_6 \end{Bmatrix}_k \tag{5.6}$$

The transformed stresses in the reference axis for the kth lamina is given by [(4.67)]

$$\begin{Bmatrix} \sigma_x \\ \sigma_y \\ \sigma_s \end{Bmatrix}_k = \begin{bmatrix} Q_{xx} & Q_{xy} & Q_{xs} \\ Q_{yx} & Q_{yy} & Q_{ys} \\ Q_{sx} & Q_{sy} & Q_{ss} \end{bmatrix}_k \begin{Bmatrix} \varepsilon_x \\ \varepsilon_y \\ \varepsilon_s \end{Bmatrix}_k \tag{5.7}$$

Combining (5.4) and (5.7), yields

$$\begin{Bmatrix} \sigma_x \\ \sigma_y \\ \sigma_s \end{Bmatrix}_k = \begin{bmatrix} Q_{xx} & Q_{xy} & Q_{xs} \\ Q_{yx} & Q_{yy} & Q_{ys} \\ Q_{sx} & Q_{sy} & Q_{ss} \end{bmatrix}_k \begin{Bmatrix} \varepsilon_x^0 \\ \varepsilon_y^0 \\ \varepsilon_s^0 \end{Bmatrix} + z \begin{bmatrix} Q_{xx} & Q_{xy} & Q_{xs} \\ Q_{yx} & Q_{yy} & Q_{ys} \\ Q_{sx} & Q_{sy} & Q_{ss} \end{bmatrix}_k \begin{Bmatrix} \kappa_x \\ \kappa_y \\ \kappa_s \end{Bmatrix} \tag{5.8}$$

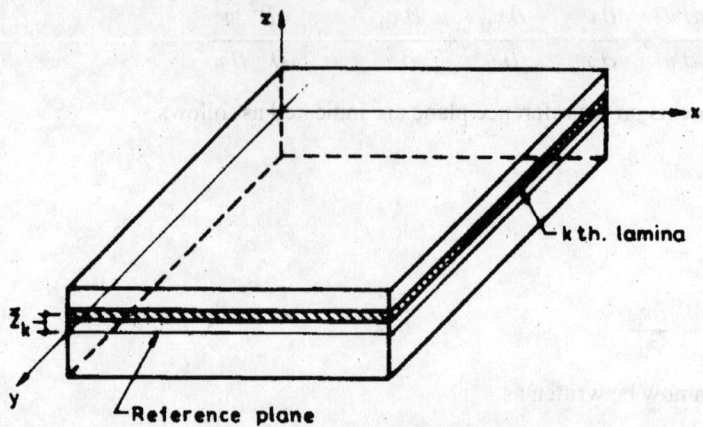

Fig. 5.2 *k*th lamina of the laminate

or, in compact form,

$$\{\sigma\}^k_{x,y} = [Q]^k_{x,y}\{\varepsilon^0\}_{x,y} + z[Q]^k_{x,y}\{\kappa\}_{x,y} \tag{5.9}$$

Strains in a laminate vary linearly across the thickness. Stresses, however, do not. This is mainly because of the discontinuous variation of the transformed stiffness matrix $[Q]^k_{x,y}$ from lamina to lamina [Fig. 5.3].

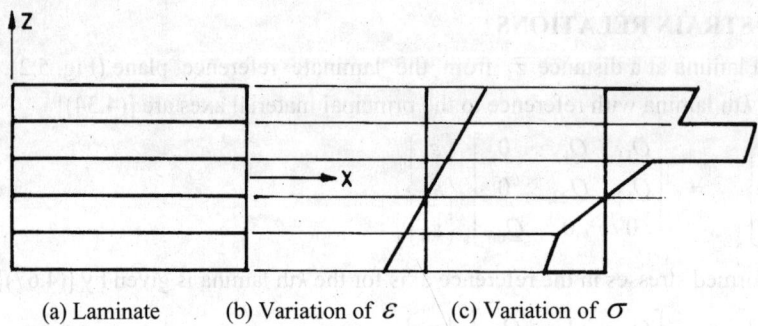

(a) Laminate (b) Variation of ε (c) Variation of σ

Fig. 5.3 Stress and strain variation in a laminate

5.5 EQUILIBRIUM EQUATIONS

There are abrupt changes of slope for the variation of stresses over the depth. Stress varies from lamina to lamina in a discontinuous manner. Stress of the *k*th lamina of Fig. 5.2 is given by (5.8) and is shown in Fig. 5.3. Stress resultants of the *k*th lamina are shown in Fig. 5.4. They can be obtained from the stresses acting on the *k*th lamina by integrating over the depth.

$$N^k_x = \int_{d_k} \sigma_x dz \qquad M^k_x = \int_{d_k} \sigma_x . z\, dz \tag{5.10}$$

$$N^k_y = \int_{d_k} \sigma_y dz \qquad M^k_y = \int_{d_k} \sigma_y . z\, dz$$

$$N_s^k = \int_{d_k} \sigma_s \, dz \qquad M_s^k = \int_{d_k} \sigma_s \cdot z \cdot dz$$

where
- z = coordinate along the depth
- d_k = thickness of the kth lamina
- N_x^k, N_y^k = normal forces per unit length
- N_s^k = inplane shear force per unit length
- M_x^k, M_y^k = bending moment per unit length
- M_s^k = twisting moment per unit length

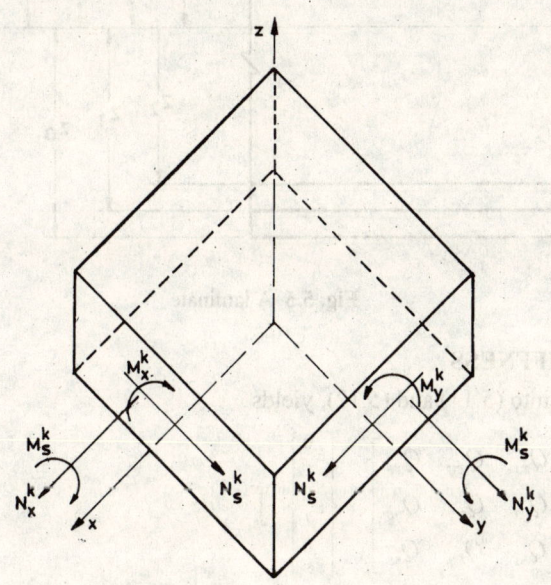

Fig. 5.4 Stress resultants in a lamina

The z-coordinate corresponding to the top and bottom surfaces for each ply is indicated in Fig. 5.5. For the n-ply laminate of Fig. 5.5, the stress resultants are given below.

$$\begin{Bmatrix} N_x \\ N_y \\ N_s \end{Bmatrix} = \sum_{k=1}^{n} \int_{z_{k-1}}^{z_k} \begin{Bmatrix} \sigma_x \\ \sigma_y \\ \sigma_s \end{Bmatrix}_k dz \tag{5.11}$$

and

$$\begin{Bmatrix} M_x \\ M_y \\ M_s \end{Bmatrix} = \sum_{k=1}^{n} \int_{z_{k-1}}^{z_k} \begin{Bmatrix} \sigma_x \\ \sigma_y \\ \sigma_s \end{Bmatrix}_k z \, dz \tag{5.12}$$

where z_k and z_{k-1} are the z-coordinates of the upper and lower surfaces of the kth lamina.

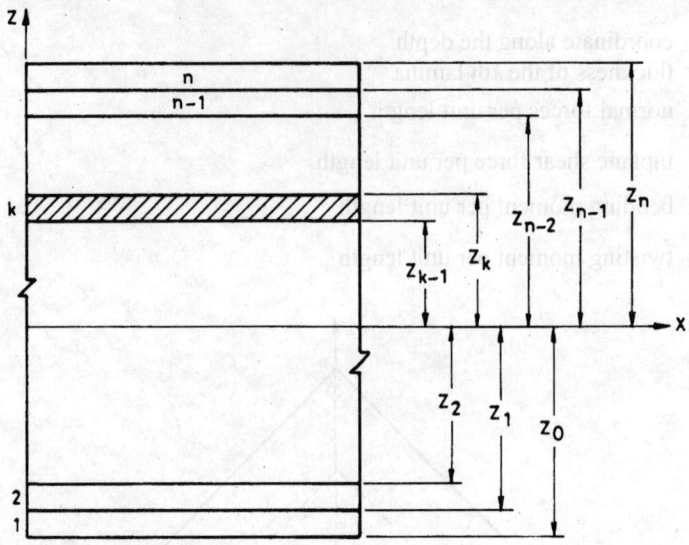

Fig. 5.5 A laminate

5.6 LAMINATE STIFFNESS

Substituting (5.8) into (5.11) and (5.12), yields

$$\begin{Bmatrix} N_x \\ N_y \\ N_s \end{Bmatrix} = \sum_{k=1}^{n} \left\{ \begin{bmatrix} Q_{xx} & Q_{xy} & Q_{xs} \\ Q_{yx} & Q_{yy} & Q_{ys} \\ Q_{sx} & Q_{sy} & Q_{ss} \end{bmatrix}_k \begin{Bmatrix} \varepsilon_x^0 \\ \varepsilon_y^0 \\ \varepsilon_s^0 \end{Bmatrix} \int_{z_{k-1}}^{z_k} dz \right.$$

$$\left. + \begin{bmatrix} Q_{xx} & Q_{xy} & Q_{xs} \\ Q_{yx} & Q_{yy} & Q_{ys} \\ Q_{sx} & Q_{sy} & Q_{ss} \end{bmatrix}_k \begin{Bmatrix} \kappa_x \\ \kappa_y \\ \kappa_s \end{Bmatrix} \int_{z_{k-1}}^{z_k} z\, dz \right\} \quad (5.13)$$

Similarly,

$$\begin{Bmatrix} M_x \\ M_y \\ M_s \end{Bmatrix} = \sum_{k=1}^{n} \left\{ \begin{bmatrix} Q_{xx} & Q_{xy} & Q_{xs} \\ Q_{yx} & Q_{yy} & Q_{ys} \\ Q_{sx} & Q_{sy} & Q_{ss} \end{bmatrix}_k \begin{Bmatrix} \varepsilon_x^0 \\ \varepsilon_y^0 \\ \varepsilon_s^0 \end{Bmatrix} \int_{z_{k-1}}^{z_k} z\, dz \right.$$

$$\left. + \begin{bmatrix} Q_{xx} & Q_{xy} & Q_{xs} \\ Q_{yx} & Q_{yy} & Q_{ys} \\ Q_{sx} & Q_{sy} & Q_{ss} \end{bmatrix}_k \begin{Bmatrix} \kappa_x \\ \kappa_y \\ \kappa_s \end{Bmatrix} \int_{z-1}^{z_k} z^2\, dz \right\} \quad (5.14)$$

Equations (5.13) and (5.14) can be written in compact form as

$$\{N\}_{x,y} = \sum_{k=1}^{n} \left\{ [Q]_{x,y}^{k} \int_{z_{k-1}}^{z_k} dz \, \{\varepsilon^0\}_{x,y} + [Q]_{x,y}^{k} \int_{z_{k-1}}^{z_k} z \, dz \, \{\kappa\}_{x,y} \right\} \qquad (5.15)$$

$$\{M\}_{x,y} = \sum_{k=1}^{n} \left\{ [Q]_{x,y}^{k} \int_{z_{k-1}}^{z_k} dz \, \{\varepsilon^0\}_{x,y} + [Q]_{x,y}^{k} \int_{z_{k-1}}^{z_k} z^2 \, dz \, \{\kappa\}_{x,y} \right\} \qquad (5.16)$$

In (5.13) to (5.16), $\{\varepsilon^0\}_{x,y}$ and $\{\kappa\}_{x,y}$, the reference plane's strains and curvatures are not functions of z and they are valid for the entire laminate.

Performing the integration of eqn. (5.15), yields

$$\{N\}_{x,y} = \left[\sum_{k=1}^{n} [Q]_{x,y}^{k} \left(z_k - z_{k-1} \right) \right] \{\varepsilon^0\}_{x,y}$$

$$+ \left[\frac{1}{2} \sum_{k=1}^{n} [Q]_{x,y}^{k} \left(z_k^2 - z_{k-1}^2 \right) \right] \{\kappa\}_{x,y} \qquad (5.17)$$

Similarly,

$$\{M\}_{x,y} = \left[\frac{1}{2} \sum_{k=1}^{n} [Q]_{x,y}^{k} \left(z_k^2 - z_{k-1}^2 \right) \right] \{\varepsilon^0\}_{x,y}$$

$$+ \left[\frac{1}{3} \sum_{k=1}^{n} [Q]_{x,y}^{k} \left(z_k^3 - z_{k-1}^3 \right) \right] \{\kappa\}_{x,y} \qquad (5.18)$$

Equations (5.17) and (5.18) can be written as

$$\{N\}_{x,y} = [A]_{x,y} \{\varepsilon^0\}_{x,y} + [B]_{x,y} \{\kappa\}_{x,y} \qquad (5.19)$$

$$\{M\}_{x,y} = [B]_{x,y} \{\varepsilon^0\}_{x,y} + [D]_{x,y} \{\kappa\}_{x,y}$$

Thus,

$$A_{ij} = \sum_{k=1}^{n} Q_{ij}^{k} \left(z_k - z_{k-1} \right)$$

$$B_{ij} = \frac{1}{2} \sum_{k=1}^{n} Q_{ij}^{k} \left(z_k^2 - z_{k-1}^2 \right) \qquad i,j = x, y, s \qquad (5.20)$$

$$D_{ij} = \frac{1}{3} \sum_{k=1}^{n} Q_{ij}^{k} \left(z_k^3 - z_{k-1}^3 \right)$$

Thus, from (5.19), the relations of all stress-resultants to the deformation is given by

$$\begin{Bmatrix} N_x \\ N_y \\ N_s \\ \hline M_x \\ M_y \\ M_s \end{Bmatrix} = \begin{bmatrix} A_{xx} & A_{xy} & A_{xs} & B_{xx} & B_{xy} & B_{xs} \\ A_{yx} & A_{yy} & A_{ys} & B_{yx} & B_{yy} & B_{ys} \\ A_{sx} & A_{sy} & A_{ss} & B_{sx} & B_{sy} & B_{ss} \\ \hline B_{xx} & B_{xy} & B_{xs} & D_{xx} & D_{xy} & D_{xs} \\ B_{yx} & B_{yy} & B_{ys} & D_{yx} & D_{yy} & D_{ys} \\ B_{sx} & B_{sy} & B_{ss} & D_{sx} & D_{sy} & D_{ss} \end{bmatrix} \begin{Bmatrix} \varepsilon_x^0 \\ \varepsilon_y^0 \\ \varepsilon_s^0 \\ \hline \kappa_x \\ \kappa_y \\ \kappa_s \end{Bmatrix} \qquad (5.21)$$

Equation (5.19) when expanded for inplane forces and moments separately are written as

$$\begin{Bmatrix} N_x \\ N_y \\ N_s \end{Bmatrix} = \begin{bmatrix} A_{xx} & A_{xy} & A_{xs} \\ A_{yx} & A_{yy} & A_{ys} \\ A_{sx} & A_{sy} & A_{ss} \end{bmatrix} \begin{Bmatrix} \varepsilon_x^0 \\ \varepsilon_y^0 \\ \varepsilon_s^0 \end{Bmatrix} + \begin{bmatrix} B_{xx} & B_{xy} & B_{xs} \\ B_{yx} & B_{yy} & B_{ys} \\ B_{sx} & B_{sy} & B_{ss} \end{bmatrix} \begin{Bmatrix} \kappa_x \\ \kappa_y \\ \kappa_s \end{Bmatrix} \qquad (5.22)$$

and

$$\begin{Bmatrix} M_x \\ M_y \\ M_s \end{Bmatrix} = \begin{bmatrix} B_{xx} & B_{xy} & B_{xs} \\ B_{yx} & B_{yy} & B_{ys} \\ B_{sx} & B_{sy} & B_{ss} \end{bmatrix} \begin{Bmatrix} \varepsilon_x^0 \\ \varepsilon_y^0 \\ \varepsilon_s^0 \end{Bmatrix} + \begin{bmatrix} D_{xx} & D_{xy} & D_{xs} \\ D_{yx} & D_{yy} & D_{ys} \\ D_{sx} & D_{sy} & D_{ss} \end{bmatrix} \begin{Bmatrix} \kappa_x \\ \kappa_y \\ \kappa_s \end{Bmatrix} \qquad (5.23)$$

Equation (5.21) can be written in compact form with respect to reference axes as given below

$$\begin{Bmatrix} \{N\} \\ \{M\} \end{Bmatrix} = \begin{bmatrix} [A] & [B] \\ [B] & [D] \end{bmatrix} \begin{Bmatrix} \{\varepsilon^0\} \\ \{\kappa\} \end{Bmatrix} \qquad (5.24)$$

$[A]$, $[B]$ and $[D]$ matrices are symmetric and the total stiffness matrix of (5.24) is also symmetric.

$[A]$ is referred to as the inplane stiffness matrix. The A_{ij} terms are the inplane stiffness terms which relate the inplane forces to inplane strains. A_{ij} is termed as the inplane stiffness coefficient of the laminate. They are independent of the stacking sequence of the laminate. The weighting factor is the thickness of the lamina, $z_k - z_{k-1}$.

$[B]$ is the coupling stiffness matrix. The B_{ij} terms are the coupling terms which relate the inplane forces to curvatures and moments to inplane strains. The elements B_{ij} are called the coupling stiffness coefficients of the laminate. Physically, this means that if $B_{ij} \neq 0$, inplane forces produce flexural and twisting deformations in addition to inplane deformations. Similarly for $B_{ij} \neq 0$, moments produce inplane deformations in addition to flexural and twisting deformations. The elements B_{ij} are dependent on the stacking sequence.

$[D]$ is the bending stiffness matrix. The D_{ij} terms are bending stiffness terms which relate moments to curvatures. The elements D_{ij} are the bending stiffness coefficients of the laminate. They are strongly dependent on the stacking sequence of the laminate.

5.7 DETERMINATION OF LAMINA STRESSES AND STRAINS

Suppose loads $\{N\}$ and $\{M\}$ are imposed along the reference axes of the laminate, then the deformation along the mid-plane of the laminate is to be calculated. This can be obtained from the inversion of (5.24).

Equation (5.24) when written in terms of submatrices yields

$$\{N\} = [A]\{\varepsilon^0\} + [B]\{\kappa\} \tag{5.25}$$

$$\{M\} = [B]\{\varepsilon^0\} + [D]\{\kappa\} \tag{5.26}$$

Pre-multiplying both sides of (5.25) by $[A]^{-1}$ and adjusting the terms, we get

$$\{\varepsilon^0\} = [A]^{-1}\{N\} - [A]^{-1}[B]\{\kappa\} \tag{5.27}$$

Combining (5.26) and (5.27) yields

$$\{M\} = [B][A]^{-1}\{N\} - [B][A]^{-1}[B]\{\kappa\} + [D]\{\kappa\} \tag{5.28}$$

Equations (5.27) and (5.28) can now be written in terms of the following matrix notations

$$\left\{\frac{\{\varepsilon^0\}}{\{M\}}\right\} = \left[\begin{array}{c|c}[A^*] & [B^*] \\ \hline [C^*] & [D^*]\end{array}\right]\left\{\frac{\{N\}}{\{\kappa\}}\right\} \tag{5.29}$$

where $[A^*] = [A]^{-1}$, $[B^*] = -[A]^{-1}[B]$, $[C^*] = [B][A]^{-1}$

$$[D^*] = [D] - [B][A]^{-1}[B] \tag{5.30}$$

From the second set of equations of the partitioned matrix of (5.29), we get

$$\{\kappa\} = [D^*]^{-1}\{M\} - [D^*]^{-1}[C^*]\{N\} \tag{5.31}$$

Substituting $\{\kappa\}$ from (5.31) into first set of equations of the partitioned matrix of (5.29), we get

$$\{\varepsilon^0\} = ([A^*] - [B^*][D^*]^{-1}[C^*])\{N\} + [B^*][D^*]^{-1}\{M\} \tag{5.32}$$

Equations (5.31) and (5.32) can now be combined as follows

$$\left\{\frac{\{\varepsilon^0\}}{\{\kappa\}}\right\} = \left[\begin{array}{c|c}[a] & [b] \\ \hline [c] & [d]\end{array}\right]\left\{\frac{\{N\}}{\{M\}}\right\} \tag{5.33}$$

where

$$[a] = [A^*] - [B^*][D^*]^{-1}[C^*]$$

$$[b] = [B^*][D^*]^{-1}$$

$$[c] = -[D^*]^{-1}[C^*] = [b]^T \tag{5.34}$$

$$[d] = [D^*]^{-1}$$

Thus, the compliance matrix of the laminate is

$$\left[\begin{array}{c|c}[a] & [b] \\ \hline [b] & [d]\end{array}\right]$$

Since, the stiffness matrix is symmetric, the compliance matrix is also symmetric.

Combining (5.30) and (5.34), we can write

$$[a] = [A]^{-1} - [b][B][A]^{-1}$$

$$[b] = -[A]^{-1}[B][d] \tag{5.35}$$

$$[d] = ([D] - [B][A]^{-1}[B])^{-1}$$

$[a]$ is the inplane compliance matrix

$[b]$ is the coupling compliance matrix

$[d]$ is the flexural compliance matrix

It may be noted that the individual compliance matrices are symmetric. The 6×6 matrix of (5.33) is symmetric. Further, the compliances that relate reference plane strains to applied moments are identical to those that relates curvatures to inplane loads.

With (5.33), inplane strains and curvatures at the middle plane of the laminate in the direction of the reference axis are calculated. For the kth lamina at a distance z_k from the middle plane, inplane strains are given by (5.4),

$$\begin{Bmatrix} \varepsilon_x \\ \varepsilon_y \\ \varepsilon_s \end{Bmatrix}_k = \begin{Bmatrix} \varepsilon_x^0 \\ \varepsilon_y^0 \\ \varepsilon_s^0 \end{Bmatrix} + z_k \begin{Bmatrix} \kappa_x \\ \kappa_y \\ \kappa_s \end{Bmatrix} \qquad (5.36)$$

Strains in the principal material direction can be obtained by transformation using (4.60)

$$\begin{Bmatrix} \varepsilon_1 \\ \varepsilon_2 \\ \varepsilon_6 \end{Bmatrix}_k = \begin{bmatrix} m^2 & n^2 & mn \\ n^2 & m^2 & -mn \\ -2mn & 2mn & (m^2 - n^2) \end{bmatrix} \begin{Bmatrix} \varepsilon_x \\ \varepsilon_y \\ \varepsilon_s \end{Bmatrix}_k \qquad (5.37)$$

Note that the first matrix on the right of (5.37) is not $[T]$, the transformation matrix of the previous chapter.

Once strains in the principal material axes are obtained for the kth lamina, stresses for kth lamina in the principal material axes can be obtained as

$$\begin{Bmatrix} \sigma_1 \\ \sigma_2 \\ \sigma_6 \end{Bmatrix}_k = \begin{bmatrix} Q_{11} & Q_{12} & 0 \\ Q_{12} & Q_{22} & 0 \\ 0 & 0 & Q_{66} \end{bmatrix}_k \begin{Bmatrix} \varepsilon_1 \\ \varepsilon_2 \\ \varepsilon_6 \end{Bmatrix}_k \qquad (5.38)$$

5.8 COUPLING EFFECTS

Coupling effects are absent in isotropic materials. But in an anisotropic laminate construction, various types of coupling effects between stresses and strains exist. Coupling effects are not very desirable features unless otherwise required for special purposes.

Let us first deal with the coupling compliance matrix $[b]$ of (5.33). The equation clearly indicates that axial forces give rise to curvatures and bending and twisting moments will cause axial strains. These effects can be eliminated if $[B] = [0]$ in (5.24)

From (5.20), we get

$$B_{ij} = \frac{1}{2} \sum_{k=1}^{n} Q_{ij}^k \left(z_k^2 - z_{k-1}^2 \right) \qquad i, j = x, y, s \qquad (5.39)$$

From Fig. 5.6, it can be seen that

$$z_k^2 - z_{k-1}^2 = -(z_{k'}^2 - z_{k'-1}^2) \qquad (5.40)$$

for two identical laminae, both situated at identical distances from the middle plane–one below and the other above the plane. Substituting eqn. (5.40) into eqn. (5.39) for these two laminae, will yield a null value of B_{ij} with Q_{ij} same for both. Thus, if a symmetrical laminate is constructed, it will always make $[B] = [0]$. Therefore, the easiest solution to get rid of the coupling effect is to construct the laminate symmetrical about the mid-plane. Also see (5.47).

Another type of coupling effect occurs in the stiffness matrix related to the inplane stiffness as shown in (5.22). Coupling occurs due to the presence of non-zero values of A_{xs} and A_{ys}. This suggests that the inplane axial force gives rise to inplane shear strain and vice versa. Let us investigate the possibility of eliminating this coupling effect. The first equation of (5.20) is reproduced below.

$$A_{ij} = \sum_{k=1}^{n} Q_{ij}^{k} (z_k - z_{k-1}) \qquad i, j = x, y, s \tag{5.41}$$

It may be noted that (from Fig. 5.6)

$$Q_{is}^{k}(-\theta) = -Q_{is}^{k}(\theta) \qquad i = x, y \tag{5.42}$$

and

$$z_k - z_{k-1} = z_{k'} - z_{k'-1} = d_k \tag{5.43}$$

for a pair of identical laminae equidistant from the mid-plane of the plate-one above having fibre orientation θ and the other below having fibre orientation $-\theta$ with d_k as the thickness of the kth lamina. This has been proved later in Section 5.10.2. Therefore, if the laminate is anti-symmetrical, its effects will be additive and $A_{xs} = A_{ys} \neq 0$.

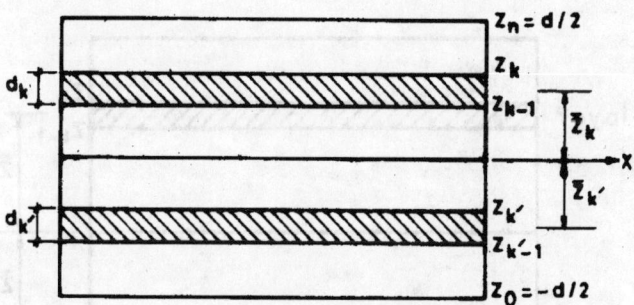

Fig. 5.6 Symmetrical position of the laminate

One approach of taking care of this coupling effect or making the laminate orthotropic is to construct a laminate with pairs of identical orthotropic laminae in a way that if one lamina of the pair has a ply angle θ, the other will have a ply angle $-\theta$. In fact, the laminate will be anti-symmetric for such a case.

For the construction of a laminate, if the ply angles are restricted to only two angles, that is, θ and $-\theta$, then the laminate is referred to as *angle-ply laminate*. If identical laminae are placed, one at $+\theta$, and the next at $-\theta$, then the laminate is called a *regular angle-ply laminate*. If this regular angle-ply laminate is made symmetric by placing the two middle laminae in the same orientation, then it is referred to as a *symmetric regular angle-ply laminate*.

Similarly, coupling occurs due to non-zero D_{xs} and D_{ys} terms which physically mean that a bending moment will create an angle of twisting and a twisting moment will create curvatures. The last of (5.20) gives

$$D_{ij} = \frac{1}{3} \sum_{k=1}^{n} Q_{ij}^k \left(z_k^3 - z_{k-1}^3 \right) \tag{5.44}$$

The term

$$\frac{1}{3}\left(z_k^3 - z_{k-1}^3\right)$$

$$= (z_k - z_{k-1}) \frac{z_k^2 + z_k \cdot z_{k-1} + z_{k-1}^2}{3}$$

$$= d_k \cdot \left\{ \left(\frac{z_k + z_{k-1}}{2}\right)^2 + \frac{(z_k - z_{k-1})^2}{12} \right\}$$

$$= d_k \cdot \bar{z}_k^2 + \frac{d_k^3}{12}$$

Therefore, D_{ij} can be written as [Fig. 5.7]

$$D_{ij} = \sum_{k=1}^{n} Q_{ij}^k \left(d_k \cdot \bar{z}_k^2 + \frac{d_k^3}{12} \right) \tag{5.45}$$

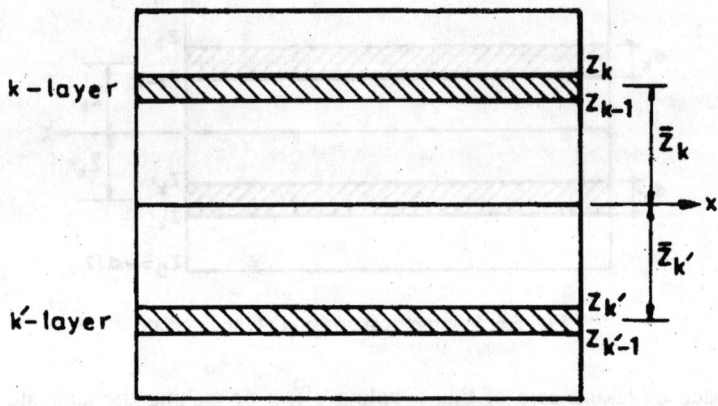

Fig. 5.7 Symmetric laminate

In order to bring in the orthotropy of the $[D]$ matrix, that is, to make $D_{xs} = D_{ys} = 0$, the laminate should be constructed anti-symmetrically with respect to the mid-plane, similar to that discussed above for dealing with orthotropy of inplane stiffness.

5.9 TYPES OF LAMINATE CONFIGURATION

All possible types of laminates may be constructed by varying the combination of ply thickness, ply angle, stacking sequence, number of plies, ply elastic constants and other similar parameters. Instead of constructing laminates in a haphazard manner, it is usually done on a more systematic basis. Generally, laminate stiffnesses assume simplified values as against those given by (5.21). However, the laminate configuration can be broadly classified into three types : symmetric, anti-symmetric and unsymmetric. Types of laminate configuration are given in Fig. 5.8.

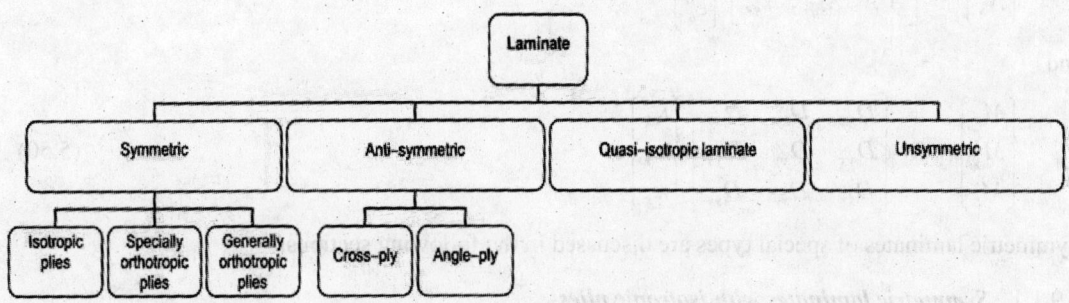

Fig. 5.8 Classification of the laminate

5.9.1 Symmetric Laminates

The middle surface of the laminate is considered as the reference plane of the laminate. A laminate is termed as symmetric when plies on one side of the reference plane are identical to those on the other side in terms of thickness, orientation, properties and position. The laminate is symmetric in both geometry and material properties about the reference plane. A symmetric laminate can have an odd or even number of plies. A symmetric laminate can be made from individual plies whose fibres are either singly-oriented or multi-oriented. The plies can all be of one composite material or it may be made a hybrid when more than one composite material is used.

An n-layer symmetric laminate is shown in Fig. 5.7. Two equidistant identical layers k and k' one below and the other above the reference plane are shown in the figure. The thickness, location of centroid and stiffness properties are numerically same for both plies [Fig. 5.6].

$$Q_{ij}^k = Q_{ij}^{k'} \tag{5.46}$$

$$d_k = d_{k'}$$

From the second equation of (5.20), we get

$$B_{ij} = \frac{1}{2} \sum_{k=1}^{n} Q_{ij}^k \left(z_k^2 - z_{k-1}^2\right)$$

$$= \frac{1}{2} \sum_{k=1}^{n} Q_{ij}^k \left(z_k + z_{k-1}\right)\left(z_k - z_{k-1}\right)$$

or $\quad B_{ij} = \sum_{k=1}^{n} Q_{ij}^k \, d_k \, \bar{z}_k \tag{5.47}$

For the two identical plies which we have considered, two terms of (5.47) will cancel each other, as they are of opposite sign. Thus, for a symmetric laminate

$$B_{ij} = 0 \qquad i, j = x, y, s \tag{5.48}$$

Stress resultants are related to 'strains' for a symmetric laminate

$$\begin{Bmatrix} N_x \\ N_y \\ N_s \end{Bmatrix} = \begin{bmatrix} A_{xx} & A_{xy} & A_{xs} \\ A_{yx} & A_{yy} & A_{ys} \\ A_{sx} & A_{sy} & A_{ss} \end{bmatrix} \begin{Bmatrix} \varepsilon_x^0 \\ \varepsilon_y^0 \\ \varepsilon_s^0 \end{Bmatrix} \quad (5.49)$$

and

$$\begin{Bmatrix} M_x \\ M_y \\ M_s \end{Bmatrix} = \begin{bmatrix} D_{xx} & D_{xy} & D_{xs} \\ D_{yx} & D_{yy} & D_{ys} \\ D_{sx} & D_{sy} & D_{ss} \end{bmatrix} \begin{Bmatrix} \kappa_x \\ \kappa_y \\ \kappa_s \end{Bmatrix} \quad (5.50)$$

Symmetric laminates of special types are discussed in the following sections.

5.9.1.1 *Symmetric laminates with isotropic plies*

Isotropic plies of various thicknesses are arranged symmetrically about the reference axes from the geometric and material point of view. All layers may not be of the same material.

Let us consider a pair of symmetric plies k and k'. The ply stiffness for each ply is given by

$$Q_{xx}^k = Q_{yy}^k = Q_{xx}^{k'} = Q_{yy}^{k'} = \frac{E_k}{1 - v_k^2}$$

$$Q_{xs}^k = Q_{ys}^k = Q_{xs}^{k'} = Q_{ys}^{k'} = 0 \quad (5.51)$$

$$Q_{xy}^k = Q_{xy}^{k'} = \frac{v_k \cdot E_k}{2(1 - v_k^2)}$$

$$Q_{ss}^k = Q_{ss}^k = \frac{E_k}{2(1 + v_k)}$$

From (5.20), (5.21) and (5.51),

$$\begin{aligned} A_{xx} &= A_{yy} \\ A_{xs} &= A_{ys} = 0 \\ D_{xx} &= D_{yy} \\ D_{xs} &= D_{ys} = 0 \end{aligned} \quad (5.52)$$

The stress resultants are related to the deformations as follows

$$\begin{Bmatrix} N_x \\ N_y \\ N_s \end{Bmatrix} = \begin{bmatrix} A_{xx} & A_{xy} & 0 \\ A_{xy} & A_{xx} & 0 \\ 0 & 0 & A_{ss} \end{bmatrix} \begin{Bmatrix} \varepsilon_x^0 \\ \varepsilon_y^0 \\ \varepsilon_s^0 \end{Bmatrix} \quad (5.53)$$

and

$$\begin{Bmatrix} M_x \\ M_y \\ M_s \end{Bmatrix} = \begin{bmatrix} D_{xx} & D_{xy} & 0 \\ D_{xy} & D_{xx} & 0 \\ 0 & 0 & D_{ss} \end{bmatrix} \begin{Bmatrix} \kappa_x \\ \kappa_y \\ \kappa_s \end{Bmatrix} \quad (5.54)$$

5.9.1.2 Symmetric laminates with specially orthotropic plies (symmetric cross-ply laminates)

In specially orthotropic laminates, plies are placed at 0° and 90°. Let us consider symmetric laminates having alternate layers of 0° and 90° – it may be of the type

$[0\,|\,90\,|\,0\,|\,90\,|\,0]$ or $[0\,|\,90]_{ns}$

If the kth layer is oriented in the principal 1-direction along the x-axis, then we get

$$Q_{xx}^k = Q_{11}^k = \frac{E_1^k}{1 - \nu_{12}^k \nu_{21}^k}$$

$$Q_{xy}^k = Q_{12}^k = \frac{\nu_{21}^k E_1^k}{1 - \nu_{12}^k \nu_{21}^k}$$

$$Q_{yy}^k = Q_{22}^k = \frac{E_2^k}{1 - \nu_{12}^k \nu_{21}^k} \qquad (5.55)$$

$$Q_{xs}^k = Q_{16}^k = 0$$

$$Q_{ys}^k = Q_{26}^k = 0$$

$$Q_{ss}^k = Q_{66}^k = G_{12}^k = E_6^k$$

It is evident from (5.55) that

$$A_{xs} = A_{ys} = 0$$
$$D_{xs} = D_{ys} = 0 \qquad (5.56)$$

Stress resultants are related to 'strains' as follows

$$\begin{Bmatrix} N_x \\ N_y \\ N_s \end{Bmatrix} = \begin{bmatrix} A_{xx} & A_{xy} & 0 \\ A_{xy} & A_{yy} & 0 \\ 0 & 0 & A_{ss} \end{bmatrix} \begin{Bmatrix} \varepsilon_x^0 \\ \varepsilon_y^0 \\ \varepsilon_s^0 \end{Bmatrix} \qquad (5.57)$$

and

$$\begin{Bmatrix} M_x \\ M_y \\ M_s \end{Bmatrix} = \begin{bmatrix} D_{xx} & D_{xy} & 0 \\ D_{xy} & D_{yy} & 0 \\ 0 & 0 & D_{ss} \end{bmatrix} \begin{Bmatrix} \kappa_x \\ \kappa_y \\ \kappa_s \end{Bmatrix} \qquad (5.58)$$

5.9.1.3 Symmetric laminates with generally orthotropic layers (symmetric angle-ply laminate)

It has already been stated that if the plies of laminate are alternately oriented at θ and $-\theta$, then the laminate is called an angle-ply laminate. If such a laminate consists of an odd number of alternating $+\theta$ and $-\theta$ plies of equal thickness, then it is symmetric, e.g., $[\theta\,|-\theta\,|\,\theta\,|-\theta\,|\,\theta]_s$. For these cases A_{xs}, A_{ys}, D_{xs} and D_{ys} are not zero and they can be shown to be largest when $n = 3$, where n is the number of plies and decrease in proportion to $1/n$, as n increases.

5.9.1.4 Symmetric laminates with anistropic layers

Being symmetric $B_{ij} = 0$ for the laminate. Many of the stiffness simplifications possible for other laminates cannot be achieved for this case.

5.9.2 Balanced Laminates

If a laminate consists of pairs of layers with identical thickness and elastic properties, but the fibres are placed with orientation of $+\theta$ and $-\theta$ with respect to the laminate reference axes, then the laminate is known as a balanced laminate. They need not be placed at the same distance from the reference plane. For such a lamina, the inplane shear coupling terms are

$$A_{is} = \sum_{k=1}^{n} Q_{is}^k (z_k - z_{k-1}) = \sum_{k=1}^{n} Q_{is}^k \cdot d_k \qquad i=x, y \qquad (5.59)$$

From (4.68), a coupling stiffness associated with the shear is given by

$$Q_{xs} = m^3 n Q_{11} - mn^3 Q_{22} + (mn^3 - m^3 n) Q_{12} + 2(mn^3 - m^3 n) Q_{66}$$

or, $\quad Q_{xs} = m^3 n (Q_{11} - Q_{12} - 2Q_{66}) + mn^3 (Q_{12} - Q_{22} + 2Q_{66})$ \hfill (5.60)

where $m = \cos\theta$ and $n = \sin\theta$.

When we consider a ply having ply angle $-\theta$, $n (= \sin\theta)$ will be of opposite sign, then

$$Q_{is}(\theta) = -Q_{is}(-\theta) \qquad (5.61)$$

For each balanced pair of plies k and k' we get

$$d_k = d_{k'}$$
$$\theta_k = -\theta_{k'} \qquad (5.62)$$

Then from (5.59) and (5.61), we get

$$A_{is} = 0 \qquad (5.63)$$

Thus, for a balanced stiffness, the inplane shear stiffnesses A_{is}s are zero.

A balanced laminate may be any of the three types – symmetric, anti-symmetric or asymmetric. For example, a lay-up of the above three types can be shown as follows

Symmetric : $[\pm\theta_1 | \pm\theta_2 | \pm\theta_3]_s$

Anti-symmetric : $[-\theta_1 | -\theta_2 | -\theta_2 | \theta_2 | \theta_2 | \theta_1]$

Asymmetric : $[-\theta_1 | -\theta_2 | -\theta_3 | \theta_1 | \theta_2 | \theta_3]$

Only for an anti-symmetric laminate D_{is}, the bending twisting coupling stiffness will be zero.

5.9.3 Anti-Symmetric Laminates

A symmetric laminate is the obvious choice for avoiding coupling between bending and extension. Symmetric laminates are used in practice to overcome the problem of warping of laminates during the curing process. However, non-symmetric laminates are required in certain applications. To make jet turbine fan blades with pretwists, coupling is necessary. The same reasoning is valid for helicopter blades. If the shear stiffness of a laminate consisting of laminae with unidirectional fibres is to be increased, it can be so achieved by placing the plies at different orientations with respect to the laminate axis. If an even number of such plies are placed in alternate layers with $-\theta$ and θ orientation, e.g., $-\theta | \theta | -\theta | \theta$, then the symmetry of the laminate is lost and it will as such behave in a different fashion from a symmetric laminate.

An anti-symmetric laminate is made up of a number of plies in such a way that for each ply below the middle surface, there is a ply at the same distance above it having identical thickness and material properties, but opposite ply angle. As such, the number of plies in an anti-symmetric laminate has to be an even number. A hybrid anti-symmetric laminate such as $[90_G \,|\, \alpha_C \,|-\alpha_C \,|-90_G]$ is possible. Anti-symmetric laminates are thus a special case of balanced laminates.

It has already been shown that for a balanced laminate, $A_{is} = 0$ (5.63).

Further, the bending /twisting coupling matrix is

$$D_{is} = \frac{1}{3} \sum_{k=1}^{n} Q_{is}^k (z_k^3 - z_{k-1}^3) = 0$$

as $\quad z_k^3 - z_{k-1}^3 = (z_{k'}^3 - z_{k'-1}^3)$

and $\quad Q_{is}^k = - Q_{is}^{k'}$

for a balanced pair of plies k and k' situated at $+\theta$ and $-\theta$.

The coupling stiffnesses B_{ij} vary for different classes of anti-symmetric laminates and they are in general, not zero. Stress resultants are related to 'strains' in the following form in general for anti-symmetric laminates

$$\begin{Bmatrix} N_x \\ N_y \\ N_s \\ \hline M_x \\ M_y \\ M_\theta \end{Bmatrix} = \begin{bmatrix} A_{xx} & A_{xy} & 0 & B_{xx} & B_{xy} & B_{xs} \\ A_{yx} & A_{yy} & 0 & B_{yx} & B_{yy} & B_{ys} \\ 0 & 0 & A_{ss} & B_{sx} & B_{sy} & B_{ss} \\ \hline B_{xx} & B_{xy} & B_{xs} & D_{xx} & D_{xy} & 0 \\ B_{yx} & B_{yy} & B_{ys} & D_{yx} & D_{yy} & 0 \\ B_{sx} & B_{sy} & B_{ss} & 0 & 0 & D_{\theta\theta} \end{bmatrix} \begin{Bmatrix} \varepsilon_x^0 \\ \varepsilon_y^0 \\ \varepsilon_s^0 \\ \hline \kappa_x \\ \kappa_y \\ \kappa_s \end{Bmatrix} \quad (5.65)$$

5.9.3.1 Anti-symmetric cross-ply laminates

Cross-ply laminates consist of plies 0° and 90°. Consider that for a 0° ply situated at a distance z above the middle surface, there is one 90° ply of identical characteristics, made of the same material, having the same thickness at the same distance below the middle surface and having the same elastic properties. Also consider a pair of plies k and k' at 0° and 90°. The following relationships are valid

$$z_k = - z_{k'}$$
$$d_k = d_{k'}$$
$$Q_{xx}^k = Q_{yy}^{k'}$$
$$Q_{yy}^k = Q_{xx}^{k'} \quad (5.66)$$
$$Q_{xy}^k = Q_{xy}^{k'}$$
$$Q_{xs}^k = Q_{ys}^k = Q_{xs}^{k'} = Q_{ys}^{k'} = 0$$
$$Q_{ss}^k = Q_{ss}^{k'}$$

Based on the relationship of (5.66), the laminate stiffness coefficients assume the following relations

$$A_{xx} = A_{yy}$$
$$A_{xs} = A_{ys} = 0$$
$$B_{xx} = -B_{yy}$$
$$B_{xy} = B_{xs} = B_{ys} = B_{ss} = 0 \qquad (5.67)$$
$$D_{xx} = D_{yy}$$
$$D_{xs} = D_{ys} = 0$$

Equation (5.65) for this case becomes

$$\begin{Bmatrix} N_x \\ N_y \\ N_s \\ M_x \\ M_y \\ M_s \end{Bmatrix} = \begin{bmatrix} A_{xx} & A_{xy} & 0 & B_{xx} & 0 & 0 \\ A_{yx} & A_{yy} & 0 & 0 & -B_{xx} & 0 \\ 0 & 0 & A_{ss} & 0 & 0 & 0 \\ B_{xx} & 0 & 0 & D_{xx} & D_{xy} & 0 \\ 0 & -B_{xx} & 0 & D_{yx} & D_{yy} & 0 \\ 0 & 0 & 0 & 0 & 0 & D_{ss} \end{bmatrix} \begin{Bmatrix} \varepsilon_x^0 \\ \varepsilon_y^0 \\ \varepsilon_s^0 \\ \kappa_x \\ \kappa_y \\ \kappa_s \end{Bmatrix} \qquad (5.68)$$

With the increase of a number of plies within a constant laminate thickness, coupling stiffness B_{xx} tends to zero.

5.9.3.2 *Anti-symmetric angle-ply laminate*

An anti-symmetric angle-ply laminate consists of plies situated in pairs in such a way that for a ply at a distance z oriented at $+\theta$ on one side of the middle surface, there is a corresponding ply of equal thickness and material properties, but oriented at $-\theta$ at the same distance below the middle surface.

As the lamina is anti-symmetric

$$A_{is} = D_{is} = 0 \qquad i = x, y \qquad (5.69)$$

Consider a pair of plies k and k' at ply angles θ and $-\theta$ respectively. The following relation results

$$\bar{z}_k = -\bar{z}_{k'}$$
$$d_k = d_{k'}$$
$$Q_{xx}^k = Q_{xx}^{k'}$$
$$Q_{yy}^k = Q_{yy}^{k'} \qquad (5.70)$$
$$Q_{xy}^k = Q_{xy}^{k'}$$
$$Q_{xs}^k = -Q_{xs}^{k'}$$
$$Q_{ys}^k = -Q_{ys}^k$$
$$Q_{ss}^k = Q_{ss}^{k'}$$

Substituting the above relations in (5.47) yields

$$B_{xx} = B_{yy} = B_{xy} = B_{ss} = 0 \qquad (5.71)$$

Stress resultants for the anti–symmetric angle-ply laminate are given by

$$\begin{Bmatrix} N_x \\ N_y \\ N_s \\ \hline M_x \\ M_y \\ M_s \end{Bmatrix} = \begin{bmatrix} A_{xx} & A_{xy} & 0 & 0 & 0 & B_{xs} \\ A_{yx} & A_{yy} & 0 & 0 & 0 & B_{ys} \\ 0 & 0 & A_{ss} & B_{sx} & B_{sy} & 0 \\ \hline 0 & 0 & B_{xs} & D_{xx} & D_{xy} & 0 \\ 0 & 0 & B_{ys} & D_{yx} & D_{yy} & 0 \\ B_{sx} & B_{sy} & 0 & 0 & 0 & D_{ss} \end{bmatrix} \begin{Bmatrix} \varepsilon_x^o \\ \varepsilon_y^o \\ \varepsilon_s^o \\ \hline \kappa_x \\ \kappa_y \\ \kappa_s \end{Bmatrix} \quad (5.72)$$

The coupling stiffnesses B_{xs} and B_{ys} tend to become zero with the increase of number of plies for a fixed laminate thickness.

5.9.4 Quasi-Isotropic Laminate

We have so far dealt with cross-ply and angle-ply laminates, made either of specially orthotropic or generally orthotropic plies. A third possibility exists. This is a quasi-isotropic laminate which is a combination of specially and generally orthotropic plies. The net result is that the inplane stiffnesses and compliances and engineering elastic constants are identical in all directions.

With the help of orthotropic laminae it is possible to manufacture a laminate which exhibits some elements of isotropic behaviour. As an example a laminate having three or more identical orthotropic laminae are considered. These laminae have the same material and geometric properties and are oriented at the same angle relative to the adjacent laminae. For this case, the extensional stiffness $[A]$ will be isotropic, but not so for $[B]$ and $[D]$. The adjacent laminae are oriented at π/n, where n is the total number of laminae.

Some examples of quasi-isotropic laminates are as follows

$$(-36 \mid -72 \mid 72 \mid 36 \mid 0)$$
$$(0 \mid \mp 45 \mid 90)_s$$
$$(60 \mid -60 \mid 0)_s$$
$$(-45 \mid 0 \mid 45 \mid 90)_s$$

The inplane force-displacement relationship is given by

$$\begin{Bmatrix} N_x \\ N_y \\ N_s \end{Bmatrix} = \begin{bmatrix} A_{xx} & A_{xy} & 0 \\ A_{xy} & A_{xx} & 0 \\ 0 & 0 & (A_{xx} - A_{xy})/2 \end{bmatrix} \begin{Bmatrix} \varepsilon_x^0 \\ \varepsilon_y^0 \\ \varepsilon_s^0 \end{Bmatrix} \quad (5.73)$$

5.9.5 Unsymmetric Laminates With Isotropic Layers

Consider an unsymmetric laminate consisting of multiple isotropic layers. For the kth layer, the stiffnesses are given by

$$Q_{xx}^k = Q_{yy}^k = \frac{E_k}{1 - \nu_k^2}$$

$$Q_{xs}^k = Q_{ys}^k = 0$$

$$Q_{ss}^k = \frac{E_k}{2(1+\nu_k)} \tag{5.74}$$

When d_k is arbitrary, special reduction of stiffnesses does not occur. Due to different material properties and thicknesses of plies about the middle surface of the laminate, coupling occurs between inplane actions and bending displacements and vice versa. It is the laminate heterogeneity that gives rise to this coupling and not due to material orthotropy. Stress resultants are given by

$$\begin{Bmatrix} N_x \\ N_y \\ N_s \\ \hline M_x \\ M_y \\ M_s \end{Bmatrix} = \begin{bmatrix} A_{xx} & A_{xy} & 0 & B_{xx} & B_{xy} & 0 \\ A_{xy} & A_{xx} & 0 & B_{xy} & B_{xx} & 0 \\ 0 & 0 & A_{ss} & 0 & 0 & B_{ss} \\ \hline B_{xx} & B_{xy} & 0 & D_{xx} & D_{xy} & 0 \\ B_{xy} & B_{xx} & 0 & D_{xy} & D_{xx} & 0 \\ 0 & 0 & B_{ss} & 0 & 0 & D_{ss} \end{bmatrix} \begin{Bmatrix} \varepsilon_x^0 \\ \varepsilon_y^0 \\ \varepsilon_s^0 \\ \hline \kappa_x \\ \kappa_y \\ \kappa_s \end{Bmatrix} \tag{5.75}$$

If specially orthotropic layers are used for unsymmetric laminates, $A_{xx} \neq A_{yy}$, $B_{xx} \neq B_{yy}$ and $D_{xx} \neq D_{yy}$, then eqn. (5.75) will have to be changed accordingly.

5.10 EXAMPLES: SERIES A

Example 5A.1

What are the types of laminates given below? Mention which elements of $[A]$, $[B]$, $[D]$ are zero for each one of them

a. $[\pm 45 | \pm 45]$

b. $[30 | -45 | -30 | 45]$

c. $[\pm \theta]$

d. $[0 | 90 | 0 | 90]$

e. $[0 | 45 | 90 | -45]$

f. $[0 | 90]_s$

a. $[\pm 45 | \pm 45]$ is a symmetric angle-ply laminate.

$B_{ij} = 0$

b. $[30 | -45 | -30 | 45]$ is an asymmetric balanced laminate.

$A_{xs} = A_{ys} = 0$

c. $[\pm \theta]$ is an anti-symmetric angle-ply laminate.

$A_{xs} = A_{ys} = D_{xs} = D_{ys} = 0$

and $B_{xx} = B_{xy} = B_{yy} = B_{ss} = 0$

d. $[0 | 90 | 0 | 90]$ is an anti-symmetric cross-ply laminate

$A_{xs} = A_{ys} = D_{xs} = D_{ys} = 0$

$B_{xy} = B_{xs} = B_{ys} = B_{ss} = 0$

e. $[0\,|\,45\,|\,90\,|-45]$ is a quasi-isotropic laminate

$A_{xs} = A_{ys} = 0$

f. $[0\,|\,90]_s$ is a symmetric cross-ply laminate

$B_{ij} = 0, \quad A_{xs} = A_{ys} = D_{xs} = D_{ys} = 0$

Example 5A.2

Which terms of $[A]$, $[B]$ and $[D]$ matrices are zero for the following laminates

a. $[\alpha\,|-\alpha\,|\,\alpha\,|-\alpha\,|\,\alpha]$
b. $[\alpha\,|-\alpha\,|-\alpha\,|\,\alpha]$
c. $[\alpha\,|-\alpha\,|\,\alpha\,|-\alpha]$

a. $[\alpha\,|-\alpha\,|\,\alpha\,|-\alpha\,|\,\alpha]$ is a symmetric angle-ply laminate.

$B_{ij} = 0$ for this case

b. $[\alpha\,|-\alpha\,|-\alpha\,|\,\alpha]$ is a symmetric angle-ply laminate.

$B_{ij} = 0$ for this case

c. $[\alpha\,|-\alpha\,|\,\alpha\,|-\alpha]$ is an anti-symmetric angle-ply laminate

$A_{is} = D_{is} = 0 \quad \text{with} \quad i = x, y$
$B_{xx} = B_{yy} = B_{xy} = B_{ss} = 0$

Example 5A.3

Compute all terms of $[A]$, $[B]$ and $[D]$ matrices for a $[0\,|\,90]$ laminate with the following lamina properties

$E_1 = 140$ GPa, $\quad E_2 = 10$ GPa, $\quad E_6 = 5$ GPa, $\quad \nu_{12} = 0.3$

Ply thickness $d = 0.125$ mm

For 0° ply,

$$\nu_{21} = \frac{\nu_{12} \cdot E_2}{E_1} = \frac{0.3 \times 10}{140} = 0.0214$$

The reduced stiffness coefficients for this ply are

$$Q_{11} = \frac{140}{1 - 0.3 \times 0.0214} = 140.9$$

$$Q_{22} = \frac{10}{1 - 0.3 \times 0.0214} = 10.06$$

$$Q_{12} = \frac{0.3 \times 10}{1 - 0.3 \times 0.0214} = 3.01 = Q_{21}$$

For 0° ply, the reference axis and the principal material axes are in the same direction, the reduced stiffness matrix for this ply is

$$[Q]_{x,y}^1 = \begin{bmatrix} 140.9 & 3.01 & 0 \\ 3.01 & 10.06 & 0 \\ 0 & 0 & 5 \end{bmatrix} \text{GPa}$$

For 90° ply (the second one), $[Q]_{1,2}$ is similar to that given for 0° ply. But the reduced stiffnesses in the reference axes are to be obtained from transformation.

$m = \cos 90° = 0 \quad n = \sin 90° = 1$
$Q_{xx} = Q_{22} = 10.06; \quad Q_{yy} = Q_{11} = 140.9; \quad Q_{xy} = 3.01$
$Q_{xs} = Q_{ys} = 0, \quad Q_{ss} = 5$

Therefore,

$$[Q]_{x,y}^2 = \begin{bmatrix} 10.06 & 3.01 & 0 \\ 3.01 & 140.9 & 0 \\ 0 & 0 & 5 \end{bmatrix} \text{GPa}$$

According to the first of (5.20), we get

$$A_{ij} = \sum Q_{ij}^k (z_k - z_{k-1})$$

$$[A] = \begin{bmatrix} 140.9 & 3.01 & 0 \\ 3.01 & 10.06 & 0 \\ 0 & 0 & 5 \end{bmatrix} \times 0.125 + \begin{bmatrix} 10.06 & 3.01 & 0 \\ 3.01 & 140.9 & 0 \\ 0 & 0 & 5 \end{bmatrix} \times 0.125$$

$$= \begin{bmatrix} 18.77 & 0.75 & 0 \\ 0.75 & 18.77 & 0 \\ 0 & 0 & 1.25 \end{bmatrix}$$

Similarly, from (5.47), we get

$$[B] = \begin{bmatrix} 140.9 & 3.01 & 0 \\ 3.01 & 10.06 & 0 \\ 0 & 0 & 5 \end{bmatrix} (0.125) \left(-\frac{0.125}{2} \right)$$

$$+ \begin{bmatrix} 10.96 & 3,01 & 0 \\ 3.01 & 140.9 & 0 \\ 0 & 0 & 5 \end{bmatrix} (0.125) \left(\frac{0.125}{2} \right)$$

$$= \begin{bmatrix} -1.02 & 0 & 0 \\ 0 & 1.02 & 0 \\ 0 & 0 & 0 \end{bmatrix}$$

From (5.45), [D] matrix is given by

$$[D] = \begin{bmatrix} 140.9 & 3.01 & 0 \\ 3.01 & 10.06 & 0 \\ 0 & 0 & 5 \end{bmatrix} \left[0.125 \times \left(-\frac{0.125}{2} \right)^2 + \frac{(0.125)^3}{12} \right]$$

$$+ \begin{bmatrix} 10.06 & 3.01 & 0 \\ 3.01 & 140.9 & 0 \\ 0 & 0 & 5 \end{bmatrix} \left[0.125 \times \left(\frac{0.125}{2} \right)^2 + \frac{(0.125)^3}{12} \right]$$

$$= \begin{bmatrix} 0.0983 & 0.0039 & 0 \\ 0.00390 & 0.0983 & 0 \\ 0 & 0 & 0.0065 \end{bmatrix}$$

Example 5A.4

Calculate [A], [B], and [D] for $[+45|-45]$ laminate with the following lamina properties

$E_1 = 140$ GPa, $E_2 = 10$ GPa, $E_6 = G_{12} = 5$ GPa, $\nu_{12} = 0.3$, $d = 0.125$ mm

For the first lamina, $\theta = 45°$

$m = \cos 45 = 0.707$

$n = \sin 45 = 0.707$

$\nu_{21} = 0.3 \times \dfrac{10}{140} = 0.0214$

From the previous example, we know that

$Q_{11} = 140.9$ GPa, $Q_{22} = 10.1$ GPa, $Q_{12} = Q_{21} = 3$ GPa, $Q_{66} = 5$ GPa

$Q_{xx} = (0.707)^4 \times 140.9 + (0.707)^4 \times 10.1 + 2(0.707)^2 (0.707)^2 \times 3$
$\qquad + 4(0.707)^2 (0.707)^2 5 = 44.25$

$Q_{yy} = 44.25$

$Q_{xy} = (10.1 + 140.9 + 2 \times 3 - 4 \times 5)(0.707)^4 = 34.25$

$Q_{xs} = (140.9 - 10.1 + 0 + 0)(0.707)^4 = 32.7$

$Q_{ys} = 32.7$

$Q_{ss} = (140.910.1 - 2 \times 3 + 0)(0.707)^4 = 36.25$

Therefore, for a 45° ply transformed reduced stiffness in the reference axis system is

$$[Q]_{x,y}^{45} = \begin{bmatrix} 44.25 & 34.25 & 32.70 \\ 34.25 & 44.25 & 32.70 \\ 32.70 & 32.70 & 36.23 \end{bmatrix}$$

Similarly, for a $-45°$ ply, stiffnesses in the reference axis direction are

$$[Q]_{x,y}^{-45} = \begin{bmatrix} 44.25 & 34.25 & -32.70 \\ 34.25 & 44.25 & -32.70 \\ -32.70 & -32.70 & 36.23 \end{bmatrix}$$

$[A]$ matrix is given by

$$[A] = \begin{bmatrix} 44.25 & 34.25 & 32.70 \\ 34.25 & 44.25 & 32.70 \\ 32.70 & 32.70 & 36.23 \end{bmatrix} (0.125) + \begin{bmatrix} 44.25 & 34.25 & -32.70 \\ 34.25 & 44.25 & -32.70 \\ -32.70 & -32.70 & 36.23 \end{bmatrix} (0.125)$$

$$= \begin{bmatrix} 11.06 & 8.56 & 0 \\ 8.56 & 11.06 & 0 \\ 0 & 0 & 9.08 \end{bmatrix}$$

$[B]$ matrix is given by

$$[B] = \begin{bmatrix} 44.25 & 34.25 & 32.70 \\ 34.25 & 44.25 & 32.70 \\ 32.70 & 32.70 & 36.23 \end{bmatrix} (0.125) \frac{(-0.125)}{2}$$

$$+ \begin{bmatrix} 44.25 & 34.25 & -32.70 \\ 34.25 & 44.25 & -32.70 \\ -32.70 & -32.70 & 36.23 \end{bmatrix} (0.125) \frac{(0.125)}{2}$$

$$= \begin{bmatrix} 0 & 0 & -0.511 \\ 0 & 0 & -0.511 \\ -0.511 & -0.511 & 0 \end{bmatrix}$$

Similarly, $[D]$ matrix can be similarly obtained

$$[D] = \begin{bmatrix} 44.25 & 34.25 & 32.70 \\ 34.25 & 44.25 & 32.70 \\ 32.70 & 32.70 & 36.23 \end{bmatrix} \left[(0.125) \cdot \left(-\frac{0.125}{2} \right)^2 + \frac{(0.125)^3}{12} \right]$$

$$+ \begin{bmatrix} 44.25 & 34.25 & -32.70 \\ 34.25 & 44.25 & -32.70 \\ -32.70 & -32.70 & 36.23 \end{bmatrix} \left[(0.125) \left(\frac{0.125}{2} \right)^2 + \frac{(0.125)^3}{12} \right]$$

$$= \begin{bmatrix} 0.0576 & 0.0446 & 0 \\ 0.0446 & 0.0576 & 0 \\ 0 & 0 & 0.0472 \end{bmatrix}$$

Example 5A.5

Compute [A] matrix for a [0|±45] laminate with the following laminate properties

$E_1 = 145$ GPa, $E_2 = 10.5$ GPa, $E_6 = 7.5$ GPa, $\nu_{12} = 0.28$

Thickness of each lamina is 0.25 mm

$$\nu_{21} = 0.28 \times \frac{10.5}{145} = 0.0203$$

For the 0° lamina (1st lamina), elements of reduced stiffness matrix is as follows

$$Q_{xx} = \frac{145}{1 - 0.0203 \times 0.28} = 145.83$$

$$Q_{yy} = \frac{10.5}{1 - 0.0203 \times 0.28} = 10.56$$

$$Q_{xy} = \frac{0.28 \times 10.5}{1 - 0.0203 \times 0.28} = 2.94$$

$$Q_{ss} = E_s = 7.5$$

$$[Q]^1_{x,y} = \begin{bmatrix} 145.83 & 2.94 & 0 \\ 2.94 & 10.56 & 0 \\ 0 & 0 & 7.5 \end{bmatrix}$$

For a 45° lamina

$$[Q]^2_{x,y} = \begin{bmatrix} 48.07 & 33.07 & 33.82 \\ 33.07 & 48.07 & 33.82 \\ 33.82 & 33.82 & 37.63 \end{bmatrix}$$

For a −45° lamina

$$[Q]^3_{x,y} = \begin{bmatrix} 48.07 & 33.07 & -33.82 \\ 33.07 & 48.07 & -33.82 \\ -33.82 & -33.82 & 37.63 \end{bmatrix}$$

Therefore, [A] matrix is given by

$$[A] = \begin{bmatrix} 145.83 & 33.07 & 0 \\ 2.94 & 10.56 & 0 \\ 0 & 0 & 7.5 \end{bmatrix}(0.25) + \begin{bmatrix} 48.07 & 33.07 & 33.82 \\ 33.07 & 48.07 & 33.82 \\ 33.82 & 33.82 & 37.63 \end{bmatrix}(0.25)$$

$$+ \begin{bmatrix} 48.07 & 33.07 & -33.82 \\ 33.07 & 48.07 & -33.82 \\ -33.82 & -33.82 & 37.63 \end{bmatrix}(0.25)$$

$$= \begin{bmatrix} 60.49 & 17.27 & 0 \\ 17.27 & 60.49 & 0 \\ 0 & 0 & 20.69 \end{bmatrix} \text{ MN/m}$$

Example 5A.6

Determine the stiffness matrices for a quasi-isotropic $[-60\,|\,0\,|+60]$ laminate with the following material properties.

$E_1 = 140$ GPa, $\quad E_2 = 10$ GPa, $\quad E_6 = 6$ GPa, $\quad \nu_{12} = 0.3$

The thickness of the lamina is 0.2 mm.

$$\nu_{21} = \frac{0.3 \times 10}{140} = 0.0214$$

For 0° lamina, different components of stiffnesses are

$$Q_{xx} = \frac{140}{1 - 0.3 \times 0.0214} = 140.9 \quad \text{GPa}$$

$$Q_{yy} = \frac{10}{1 - 0.3 \times 0.0214} = 10.06 \quad \text{GPa}$$

$$Q_{ss} = 6 \text{ GPa}$$

$$Q_{xx} = \frac{10 \times 0.3}{1 - 0.3 \times 0.0214} = 3.02 \quad \text{GPa}$$

Therefore, the stiffness matrix for 0° lamina is

$$[Q]_{x,y}^0 = \begin{bmatrix} 140.9 & 3.02 & 0 \\ 3.02 & 10.06 & 0 \\ 0 & 0 & 6 \end{bmatrix} \text{ GPa}$$

For the 60° lamina, different components of stiffnesses

$m = \cos 60 = 0.5$
$n = \sin 60 = 0.866$

$$Q_{xx} = (0.5)^4\, 14.09 + (0.866)^4\, 10.06 + 2\,(0.5)^2\,(0.866)^2\, 3.02$$
$$\quad + 4\,(0.5)^2\,(0.866)^2\, 6$$
$$= 20.1$$

$$Q_{yy} = (0.8666)^4\, 140.9 + (0.5)^4\, 10.06 + 2\,(0.5)^2\,(0.866)^2\, 3.02$$
$$\quad + 4\,(0.5)^2\,(0.866)^2\, 6$$
$$= 85.51$$

$$Q_{xy} = (0.5)^2\,(0.866)^2\, 14.09 + (0.5)^2\,(0.866)^2\, 10.06$$
$$\quad + \left[(0.5)^4 + (0.866)^4\right] 3.02 - 4\,(0.5)^2\,(0.866)^2\, 6$$
$$= 25.69$$

$$Q_{xs} = (0.5)^3 (0.866)\ 140.9 - (0.5)(0.866)^3\ 10.06$$
$$+ \left[(0.5)(0.866)^3 - (0.5)^3(0.866)\right] 3.02$$
$$+ 2\left[(0.5)(0.866)^3 - (0.5)^3(0.866)\right] 6$$
$$= 15.24$$

$$Q_{ys} = (0.5)(0.866)^3\ 14.09 - (0.5)^3(0.866)\ 10.06$$
$$+ \left[(0.5)^3 (0.866) - (0.5)(0.866)^3\right] 3.02$$
$$+ 2\left[(0.5)^3 (0.866) - (0.5)(0.866)^3\right] 6$$
$$= 41.41$$

$$Q_{ss} = (0.5)^2 (0.866)^2\ 140.9 + (0.5)^2 (0.866)^2\ 10.06$$
$$- 2(0.5)^2 (0.866)\ 2.02 + \left[(0.5)^2 - (0.866)^2\right] 6$$
$$= 28.67$$

For a $-60°$ lamina, different components of stiffnesses are calculated

$m = \cos(-60°) = 0.5$
$n = \sin(-60) = -0.866$
$Q_{xx} = 20.10$
$Q_{yy} = 85.51$
$Q_{xy} = 25.69$
$Q_{xs} = -15.24$
$Q_{ys} = -41.41$
$Q_{ss} = 28.67$

Different sub-matrices for the laminate stiffness are given by

$$[A] = \begin{bmatrix} 140.9 & 3.02 & 0 \\ 3.02 & 10.06 & 0 \\ 0 & 0 & 6 \end{bmatrix}(0.2) + \begin{bmatrix} 20.1 & 25.68 & 15.24 \\ 25.68 & 85.51 & 41.41 \\ 15.24 & 41.41 & 28.67 \end{bmatrix}(0.2)$$

$$+ \begin{bmatrix} 20.1 & 25.68 & -15.24 \\ 25.68 & 85.51 & -41.41 \\ -15.24 & -41.41 & 28.67 \end{bmatrix}(0.2)$$

$$= \begin{bmatrix} 36.22 & 10.88 & 0 \\ 10.88 & 36.22 & 0 \\ 0 & 0 & 12.67 \end{bmatrix} \text{GPa-mm}$$

Similarly,

$$[B] = \begin{bmatrix} 140.9 & 3.02 & 0 \\ 3.02 & 10.06 & 0 \\ 0 & 0 & 6 \end{bmatrix}(0.2)(0) + \begin{bmatrix} 20.1 & 25.68 & 15.24 \\ 25.68 & 85.51 & 41.41 \\ 15.24 & 41.41 & 28.67 \end{bmatrix}(0.2)(0.2)$$

$$+ \begin{bmatrix} 20.1 & 25.68 & -15.24 \\ 25.68 & 85.51 & -41.41 \\ -15.24 & -41.41 & 28.67 \end{bmatrix}(0.2)(0.2)$$

$$= \begin{bmatrix} 0 & 0 & -1.22 \\ 0 & 0 & -3.32 \\ -1.22 & -3.22 & 0 \end{bmatrix} \text{ GPa - mm}^2$$

Similarly,

$$[D] = \begin{bmatrix} 140.9 & 3.02 & 0 \\ 3.02 & 10.06 & 0 \\ 0 & 0 & 6 \end{bmatrix}\left(0.2 \times 0 + \frac{0.2^3}{12}\right)$$

$$+ \begin{bmatrix} 20.1 & 25.68 & 15.24 \\ 25.68 & 85.51 & 41.41 \\ 15.24 & 41.41 & 28.67 \end{bmatrix}\left(0.2 \times (-0.2)^2 + \frac{0.2^3}{12}\right)$$

$$+ \begin{bmatrix} 20.1 & 25.68 & -15.24 \\ 25.68 & 85.51 & -41.41 \\ -15.24 & -41.41 & 28.67 \end{bmatrix}\left[0.2 \times (0.2)^2 + \frac{0.2^3}{12}\right]$$

$$= \begin{bmatrix} 0.442 & 0.447 & 0 \\ 0.447 & 1.482 & 0 \\ 0 & 0 & 0.497 \end{bmatrix} \text{ GPa- mm}^3$$

Example 5A.7

Determine $[A]$, $[B]$ and $[D]$ matrices for an anti-symmetric $[-45|45|-45|45]$ angle-ply laminate. Each ply has the same thickness of 0.25 mm. The material properties are

$E_1 = 138$ GPa, $\quad E_2 = 9$ GPa, $\quad E_6 = 6.9$ GPa, $\quad \nu_{12} = 0.3$

$\nu_{21} = 0.3 \times \dfrac{9}{138} = 0.0196$

$Q_{11} = \dfrac{138}{1 - 0.3 \times 0.0196} = 138.82$

$Q_{22} = \dfrac{9}{1 - 0.3 \times 0.0196} = 9.05$

$Q_{66} = 6.9$

$Q_{12} = Q_{21} = \dfrac{0.3 \times 9}{1 - 0.3 \times 0.0196} = 2.72$

For a 45° lamina

$m = \cos 45° = 0.707, \quad n = \sin 45° = 0.707$

$Q_{xx} = (0.707)^4\, 138.82 + (0.707)^4\, 9.05 + 2\,(0.707)^2 (0.707)^2\, 2.72$
$\quad + 4\,(0.707)^2 (0.707)^2\, 6.9$
$\quad = 45.23$

$Q_{yy} = 45.23$

$Q_{xy} = [9.05 + 138.82 + 2 \times 2.72 - 4 \times 6.9]\,(0.707)^4$
$\quad = 31.43$

$Q_{xs} = [138.82 - 9.05 + 0 + 0]\,(0.707)^4$
$\quad = 32.44$

$Q_{ys} = 32.44$

$Q_{ss} = [138.82 + 9.05 - 2 \times 2.70 - 0]\,(0.707)^4$
$\quad = 35.62$

Therefore,

$[Q]^{45}_{x,y} = \begin{bmatrix} 45.23 & 31.43 & 32.44 \\ 31.43 & 45.23 & 32.44 \\ 32.44 & 32.44 & 35.62 \end{bmatrix}$

Thus, for a $-45°$ lamina, we get

$[Q]^{-45}_{x,y} = \begin{bmatrix} 45.23 & 31.43 & -32.44 \\ 31.43 & 45.23 & -32.44 \\ -32.44 & -32.44 & 35.62 \end{bmatrix}$ GPa

$[A] = 2\,[Q]^{45}_{x,y} \times (0.25) + [Q]^{-45}_{x,y}\,(0.25) = \begin{bmatrix} 45.23 & 31.43 & 0 \\ 31.43 & 45.23 & 0 \\ 0 & 0 & 35.62 \end{bmatrix}$ GPa-mm

$[B] = [Q]^{45}_{x,y} \times (0.25)\,[-0.375 + 0.125] + [Q]^{-45}_{x,y}\,[0.375 - 0.125]\,(0.25)$

$= \begin{bmatrix} 0 & 0 & 4.06 \\ 0 & 0 & 4.06 \\ 4.06 & 4.06 & 0 \end{bmatrix}$ GPa-mm^2

$$[D] = [Q]_{x,y}^{-45} \left[(0.25) \times (-0.375)^2 + \frac{(0.25)^3}{12} \right] + [Q]_{x,y}^{45} \left[(0.25) \times (-0.125)^2 + \frac{(0.25)^3}{12} \right]$$

$$+ [Q]_{x,y}^{-45} \left[(0.25)(0.125)^2 + \frac{(0.25)^3}{12} \right] + [Q]_{x,y}^{45} \left[(0.25)(0.375)^2 + \frac{(0.25)^3}{12} \right]$$

$$= \begin{bmatrix} 3.77 & 2.62 & 0 \\ 2.62 & 3.77 & 0 \\ 0 & 0 & 2.97 \end{bmatrix} \text{GPa-mm}^3$$

Example 5A.8

Determine $[A]$, $[B]$ and $[D]$ matrices for a $[0\,|\,45\,|\,-45\,|\,90]_s$ laminate configuration. The thickness of each ply is 0.25 mm. The material properties for each ply are identical and they are
$E_1 = 140$ GPa, $E_2 = 10$ GPa, $E_6 = 5$ GPa, $\nu_{12} = 0.3$
This is a quasi-isotropic laminate (Fig. 5.9).
$\nu_{21} = 0.3 \times \dfrac{10}{140} = 0.0214$

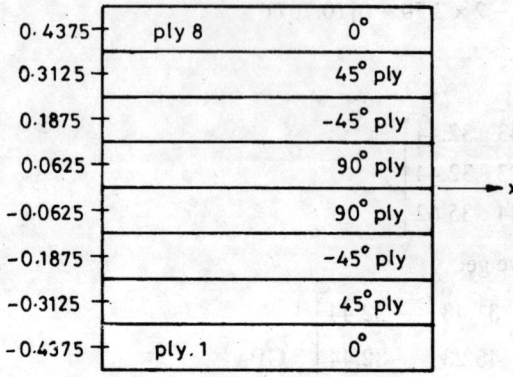

Fig. 5.9 Example 5.A8

From Example 5A.3, the stiffness matrix of any ply in the principal material axis direction is given by

$$[Q]_{1,2} = \begin{bmatrix} 140.9 & 3 & 0 \\ 3 & 10.1 & 0 \\ 0 & 0 & 5 \end{bmatrix}$$

For 0° plies,
$[Q]_{x,y}^0 = [Q]_{1,2}$
For 90° (Example 5A.3) plies,

$$[Q]_{x,y}^{90} = \begin{bmatrix} 10.1 & 3 & 0 \\ 3 & 140.9 & 0 \\ 0 & 0 & 5 \end{bmatrix}$$

For 45° plies (Example 5A.4),

$$[Q]_{x,y}^{45} = \begin{bmatrix} 44.25 & 34.25 & 32.70 \\ 34.25 & 44.25 & 32.70 \\ 32.70 & 32.70 & 36.23 \end{bmatrix}$$

For −45° plies,

$$[Q]_{x,y}^{-45} = \begin{bmatrix} 44.25 & 34.25 & -32.70 \\ 34.25 & 44.25 & -32.70 \\ -32.70 & -32.70 & 36.23 \end{bmatrix}$$

Therefore,

$$[A] = \left[\left([Q]_{x,y}^{0} + [Q]_{x,y}^{45} + [Q]_{x,y}^{-45} + [Q]_{x,y}^{90}\right) 2 \times 90.25 \right]$$

$$= \begin{bmatrix} 59.9 & 18.7 & 0 \\ 18.7 & 59.9 & 0 \\ 0 & 0 & 20.7 \end{bmatrix}$$

Because of symmetry,

$$[B] = [0]$$

$$[D] = \left([Q]_{x,y}^{0} \left\{ (0.25)(-0.4375)^2 + \frac{(0.25)^3}{12} \right\} + [Q]_{x,y}^{0} \left\{ (0.25)(0.4375)^2 + \frac{(0.25)^3}{12} \right\} \right.$$

$$+ [Q]_{x,y}^{90} \left\{ (0.25)(-0.3125)^2 + \frac{(0.25)^3}{12} \right\} + [Q]_{x,y}^{90} \left\{ (0.25)(0.3125)^2 + \frac{(0.25)^3}{12} \right\}$$

$$+ [Q]_{x,y}^{45} \left\{ (0.25)(-0.1875)^2 + \frac{(0.25)^3}{12} \right\} + [Q]_{x,y}^{45} \left\{ (0.25)(0.1875)^2 + \frac{(0.25)^3}{12} \right\}$$

$$\left. + [Q]_{x,y}^{-45} \left\{ (0.25)(-0.0625)^2 + \frac{(0.25)^3}{12} \right\} + [Q]_{x,y}^{-45} \left\{ (0.25)(0.0625)^2 + \frac{(0.25)^3}{12} \right\} \right)$$

$$= \begin{bmatrix} 8.31 & 1.32 & 0.51 \\ 1.32 & 2.19 & 0.51 \\ 0.51 & 0.51 & 1.48 \end{bmatrix}$$

Example 5A.9

A $[+45|-45|-45|45]$ symmetric laminate is subjected to $N_x = 100$ MPa–mm

The material properties are

$E_1 = 140$ GPa, $E_2 = 10$ GPa, $E_6 = 5$ GPa, $\nu_{12} = 0.3$ $d_k = 0.25$ mm (for all)

Determine the resulting stresses along the reference axis for each lamina.
For the laminate $[45|-45|-45|45]$, the following are the stiffness matrices

$$[A] = \begin{bmatrix} 44.25 & 34.25 & 0 \\ 34.25 & 44.25 & 0 \\ 0 & 0 & 36.25 \end{bmatrix}$$

$$[B] = [0]$$

$$[D] = \begin{bmatrix} 3.69 & 2.85 & 0 \\ 2.85 & 3.69 & 0 \\ 0 & 0 & 3.02 \end{bmatrix}$$

$$[a] = [A^*] = [A]^{-1} = \begin{bmatrix} 0.0226 & -0.0175 & 0 \\ -0.0175 & 0.0226 & 0 \\ 0 & 0 & 0.276 \end{bmatrix}$$

The inplane strains in the middle plane are

$$\begin{Bmatrix} \varepsilon_x^0 \\ \varepsilon_y^0 \\ \varepsilon_s^0 \end{Bmatrix} \begin{bmatrix} 0.0226 & -0.0175 & 0 \\ -0.0175 & 0.0226 & 0 \\ 0 & 0 & 0.02809 \end{bmatrix} \begin{Bmatrix} 100 \\ 0 \\ 0 \end{Bmatrix} \times 10^{-3} = \begin{Bmatrix} 0.00226 \\ -0.00175 \\ 0 \end{Bmatrix}$$

For a 45° bottom lamina, stresses along the reference axis are

$$\begin{Bmatrix} \sigma_x \\ \sigma_y \\ \sigma_s \end{Bmatrix} = \begin{bmatrix} 44.25 & 34.25 & 32.70 \\ 34.25 & 44.25 & 32.70 \\ 32.70 & 32.70 & 36.23 \end{bmatrix} \begin{Bmatrix} 0.00226 \\ -0.00175 \\ 0 \end{Bmatrix} = \begin{Bmatrix} 40.07 \\ 0 \\ 16.68 \end{Bmatrix}$$

For a $-45°$ bottom lamina, stresses along the reference axes are

$$\begin{Bmatrix} \sigma_x \\ \sigma_y \\ \sigma_s \end{Bmatrix}_{-45°} = \begin{bmatrix} 44.25 & 34.25 & -32.70 \\ 34.25 & 44.25 & -32.70 \\ -32.70 & -32.70 & 36.23 \end{bmatrix} \begin{Bmatrix} 0.00226 \\ -0.00175 \\ 0 \end{Bmatrix} = \begin{Bmatrix} 40.07 \\ 0 \\ -16.68 \end{Bmatrix}$$

Example 5A.10

A $[-45|45|-45|45]$ anti-symmetric angle-ply laminate is subjected to a uniaxial force per unit length $N_x = 50$ MPa–mm. Determine the resulting stresses associated with a x- and y-axis of each lamina. Material properties of the ply are same as Example 5A.9.
The following relationship can be obtained

$$\begin{Bmatrix} \varepsilon_x^0 \\ \varepsilon_y^0 \\ \varepsilon_s^0 \\ \kappa_x^0 \\ \kappa_y^0 \\ \kappa_s^0 \end{Bmatrix} = \begin{bmatrix} 0.04386 & -0.02861 & 0 & 0 & 0 & 0.02083 \\ -0.02861 & 0.04386 & 0 & 0 & 0 & 0.02083 \\ 0 & 0 & 0.03284 & 0.02083 & 0.02083 & 0 \\ 0 & 0 & 0.02083 & 0.52625 & -0.34331 & 0 \\ 0 & 0 & 0.02083 & -0.34331 & 0.52625 & 0 \\ 0.02083 & 0.02083 & 0 & 0 & 0 & 0.39356 \end{bmatrix}$$

$$\begin{Bmatrix} 50 \\ 0 \\ 0 \\ 0 \\ 0 \\ 0 \end{Bmatrix} \times 10^{-3} = \begin{Bmatrix} 0.002193 \\ -0.001430 \\ 0 \\ 0 \\ 0 \\ 0.001042 \end{Bmatrix} \text{ mm/mm}$$

There is at least one non-zero curvature term. Total strains and stresses are dependent on the distance z. For the first ply having ply angle $-45°$, $z = -0.5$ mm (considering the bottom of the ply), resulting strains are

$$\begin{Bmatrix} \varepsilon_x \\ \varepsilon_y \\ \varepsilon_s \end{Bmatrix} = \begin{Bmatrix} \varepsilon_x^0 \\ \varepsilon_y^0 \\ \varepsilon_s^0 \end{Bmatrix} + z \begin{Bmatrix} \kappa_x \\ \kappa_y \\ \kappa_s \end{Bmatrix} = \begin{Bmatrix} 0.002193 \\ -0.001430 \\ 0 \end{Bmatrix} + (-0.5) \begin{Bmatrix} 0 \\ 0 \\ 0.001042 \end{Bmatrix}$$

$$= \begin{Bmatrix} 0.002193 \\ -0.001430 \\ -0.000521 \end{Bmatrix} \text{ mm/mm}$$

The stresses at the bottom surface in the reference axis direction are

$$\begin{Bmatrix} \sigma_x \\ \sigma_y \\ \sigma_s \end{Bmatrix} = \begin{bmatrix} 45.23 & 31.42 & -32.44 \\ 31.42 & 45.23 & -32.44 \\ -32.44 & -32.44 & 35.62 \end{bmatrix} \begin{Bmatrix} 0.002193 \\ -0.001430 \\ -0.000521 \end{Bmatrix} \times 10^{-3} = \begin{Bmatrix} 71.16 \\ 21.43 \\ -6.19 \end{Bmatrix} \text{ MPa}$$

Likewise, stresses can be obtained at any level.

Location	σ_x MPa	σ_y MPa	σ_s MPa
Ply 1, top	37.3	−12.7	−6.2
Ply 1, bottom	71.66	21.43	−6.19
Ply 2, top	62.7	12.7	34.0
Ply 2, bottom	54.2	4.2	24.7
Ply 3, top	54.2	4.2	−24.7
Ply 3, bottom	62.7	12.7	−34.0
Ply 4, top	71.66	21.43	6.19
Ply 4, bottom	37.3	−12.7	6.2

Example 5A.11

A cross-ply laminate $[0 \mid 90]_s$ made from high strength carbon/epoxy unidirectional plies and subjected to a tensile membrane longitudinal force of $N_x = 100$ N/mm. Each ply is 0.125 mm thick and have identical properties as given below [Fig. 5.10].

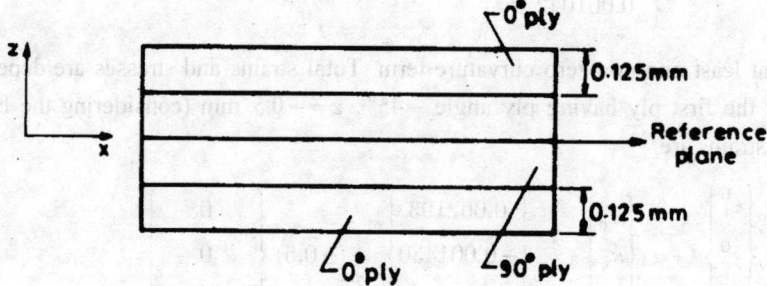

Fig. 5.10 A cross-ply laminate

$E_1 = 140$ GPa, $E_2 = 10$ GPa, $E_s = 5$ GPa, and $\nu_{12} = 0.3$

Calculate the stresses in the lamina in the principal material direction.

The following are the results of stiffness calculation. For the ply, the stiffness in the principal directions are

For 0° ply,

$$[Q]^0_{1,2} = \begin{bmatrix} 140.9 & 3.0 & 0 \\ 3.0 & 10.1 & 0 \\ 0 & 0 & 5.0 \end{bmatrix} \text{ GPa}$$

For the given laminate,

$$[A] = \begin{bmatrix} 37.8 & 1.5 & 0 \\ 1.5 & 37.8 & 0 \\ 0 & 0 & 2.5 \end{bmatrix} \text{ GPa/mm}$$

$$[\alpha] = [A]^{-1} = \begin{bmatrix} 0.0265 & -0.0011 & 0 \\ -0.0011 & 0.0265 & 0 \\ 0 & 0 & 0.4 \end{bmatrix} \text{ mm/GPa}$$

Therefore,

$$\begin{Bmatrix} \varepsilon_x^0 \\ \varepsilon_y^0 \\ \varepsilon_s^0 \end{Bmatrix} = \begin{bmatrix} 0.0265 & -0.0011 & 0 \\ -0.0011 & 0.0265 & 0 \\ 0 & 0 & 0.4 \end{bmatrix} \begin{Bmatrix} 100 \\ 0 \\ 0 \end{Bmatrix} = \begin{Bmatrix} 2650 \\ -110 \\ 0 \end{Bmatrix} 10^{-6}$$

As there are no moments acting on the laminate which is symmetric, there are no curvatures. The strain distribution is constant.

For a 0° ply, the principal material axes coincides with the reference axes

$$\begin{Bmatrix} \sigma_x \\ \sigma_x \\ \sigma_x \end{Bmatrix} = \begin{Bmatrix} \sigma_1 \\ \sigma_2 \\ \sigma_6 \end{Bmatrix} = \begin{bmatrix} 140.9 & 3.0 & 0 \\ 3.0 & 10.1 & 0 \\ 0 & 0 & 5.0 \end{bmatrix} \begin{Bmatrix} 2650 \\ -110 \\ 0 \end{Bmatrix} \times 10^{-6} \times 10^{-3} \text{ N/mm}^2$$

$$= \begin{Bmatrix} 373 \\ 7 \\ 0 \end{Bmatrix} \text{ N/mm}^2$$

Ply angle for ply 2 is 90°

$$m = \cos \theta = \cos 90° = 0$$
$$n = \sin 90° = 1$$

Therefore, for the second ply, strains in the principal material axes direction are

$$\begin{Bmatrix} \varepsilon_1 \\ \varepsilon_2 \\ \varepsilon_6 \end{Bmatrix}_2 = \begin{bmatrix} 0 & 1 & 0 \\ 1 & 0 & 0 \\ 0 & 0 & 2 \end{bmatrix} \begin{Bmatrix} 2650 \\ -110 \\ 0 \end{Bmatrix} \times 10^{-6}$$

$$= \begin{Bmatrix} -110 \\ 2650 \\ 0 \end{Bmatrix} 10^{-6}$$

Therefore, the stresses in the principal material directions for the second ply are

$$\begin{Bmatrix} \sigma_1 \\ \sigma_2 \\ \sigma_3 \end{Bmatrix} = \begin{bmatrix} 140.9 & 3.0 & 0 \\ 3.0 & 10.1 & 0 \\ 0 & 0 & 5.0 \end{bmatrix} \begin{Bmatrix} -110 \\ 2650 \\ 0 \end{Bmatrix} 10^{-6} \times 10^{-3} \text{ N/mm}^2$$

$$= \begin{Bmatrix} 7 \\ 26 \\ 0 \end{Bmatrix} \text{ N/mm}^2$$

5.11 LAMINATE ENGINEERING CONSTANTS

It is sometimes useful to deal with effective laminate engineering constants rather than laminate stiffnesses. This is mainly used for initial estimates as detailed calculation procedures are not needed at that stage.

For a balanced symmetric laminate subjected to uniaxial loading N_x, Young's modulus $\overline{E}_x$ and Poisson's ratio $\overline{v}_{xy}$ are given by

$$\overline{E}_x = \frac{N_x}{\varepsilon_x^0 \cdot d}, \quad \overline{v}_{xy} = -\frac{\varepsilon_y^0}{\varepsilon_x^0} \tag{5.76}$$

where ε_x^0 and ε_y^0 are the inplane strains in the x- and y-directions respectively and d is the laminate thickness.

x and y are the principal laminate axes and the bar symbol denotes the effective laminate properties.

The 'stress–strain' relations are given by

$$\begin{Bmatrix} N_x \\ 0 \\ 0 \end{Bmatrix} = \begin{bmatrix} A_{xx} & A_{xy} & 0 \\ A_{xy} & A_{yy} & 0 \\ 0 & 0 & A_{ss} \end{bmatrix} \begin{Bmatrix} \varepsilon_x^0 \\ \varepsilon_y^0 \\ \varepsilon_s^0 \end{Bmatrix} \tag{5.77}$$

The laminate is considered to be balanced; as such normal forces will not create any shearing strain. Equation (5.77) results in two equations.

$$N_x = A_{xx} \varepsilon_x^0 + A_{xy} \varepsilon_y^0 \tag{5.78}$$

$$0 = A_{xy} \varepsilon_x^o + A_{yy} \varepsilon_y^0$$

From (5.78), after eliminating ε_y^0 from the first equation with the help of the second equation, we can write

$$N_x = \left(A_{xx} - \frac{A_{xy}^2}{A_{yy}} \right) \varepsilon_x^0 \tag{5.79}$$

Combining (5.76) and (5.79), yields

$$\overline{E}_x = \frac{1}{d} \left[A_{xx} - \frac{A_{xy}^2}{A_{yy}} \right] \tag{5.80}$$

$$\overline{v}_{xy} = -\frac{A_{xy}}{A_{yy}}$$

Similarly, for uniaxial loading N_y only, we get

$$\overline{E}_y = \frac{1}{d}\left[A_{yy} - \frac{A_{xy}^2}{A_{xx}}\right] \tag{5.81}$$

$$\overline{v}_{yx} = -\frac{A_{xy}}{A_{xx}}$$

Likewise, for shear loading N_s, we get

$$\overline{E}_s = \frac{A_{ss}}{d} \tag{5.82}$$

It may be noted that shear coupling terms are zero for a balanced laminate.

If the laminate is not balanced, the determination of effective engineering constants by the above approach becomes difficult. It turns out simpler to work with laminate compliances. Let us restrict ourselves to symmetric laminates only; the main simplification is $[B] = [0]$.

Force–deformation relationship for a symmetric laminate for inplane loads is

$$\begin{Bmatrix}\varepsilon_x^0 \\ \varepsilon_y^0 \\ \varepsilon_s^0\end{Bmatrix} = \begin{bmatrix}a_{xx} & a_{xy} & a_{xs} \\ a_{yx} & a_{yy} & a_{ys} \\ a_{sx} & a_{sy} & a_{ss}\end{bmatrix}\begin{Bmatrix}N_x \\ N_y \\ N_s\end{Bmatrix} \tag{5.83}$$

When the force–deformation relations are written in terms of engineering constants, we get

$$\begin{Bmatrix}\varepsilon_x^0 \\ \varepsilon_y^0 \\ \varepsilon_s^0\end{Bmatrix} = \begin{bmatrix}\dfrac{1}{\overline{E}_x} & -\dfrac{\overline{v}_{yx}}{\overline{E}_y} & \dfrac{\overline{\eta}_{sx}}{\overline{E}_s} \\ -\dfrac{\overline{v}_{xy}}{\overline{E}_x} & \dfrac{1}{\overline{E}_y} & \dfrac{\overline{\eta}_{sy}}{\overline{E}_s} \\ \dfrac{\overline{\eta}_{xs}}{\overline{E}_x} & \dfrac{\overline{\eta}_{ys}}{\overline{E}_y} & \dfrac{1}{\overline{E}_s}\end{bmatrix}\begin{Bmatrix}N_x \\ N_y \\ N_s\end{Bmatrix}\frac{1}{d} \tag{5.84}$$

Comparing (5.83) with (5.84), we get

$$\overline{E}_x = \frac{1}{da_{xx}}; \quad \overline{E}_y = \frac{1}{da_{yy}}, \quad \overline{E}_s = \frac{1}{da_{ss}} \tag{5.85}$$

$$\overline{v}_{xy} = -\frac{a_{yx}}{a_{xx}}; \quad \overline{v}_{yx} = -\frac{a_{xy}}{a_{yy}}, \quad \overline{\eta}_{sx} = -\frac{a_{xs}}{a_{ss}}$$

$$\overline{\eta}_{xs} = \frac{a_{sx}}{a_{xx}}, \quad \overline{\eta}_{ys} = \frac{a_{sy}}{a_{yy}}, \quad \overline{\eta}_{sy} = \frac{a_{ys}}{a_{ss}}$$

The laminate is symmetric. Taking advantage of the symmetry, the following relationship holds good

$$\frac{\overline{v}_{xy}}{\overline{E}_x} = \frac{\overline{v}_{yx}}{\overline{E}_y}$$

$$\frac{\overline{\eta}_{xs}}{\overline{E}_x} = \frac{\overline{\eta}_{sx}}{\overline{E}_s} \tag{5.86}$$

$$\frac{\overline{\eta}_{ys}}{\overline{E}_y} = \frac{\overline{\eta}_{sy}}{\overline{E}_s}$$

Following similar procedures, effective laminate flexural moduli can be determined. For a symmetric laminate subjected to flexure only, laminate moment curvature relationship is given by

$$\begin{Bmatrix} \kappa_x \\ \kappa_y \\ \kappa_s \end{Bmatrix} = \begin{bmatrix} d_{xx} & d_{xy} & d_{xs} \\ d_{yx} & d_{yy} & d_{ys} \\ d_{sx} & d_{sy} & d_{ss} \end{bmatrix} \begin{Bmatrix} M_x \\ M_y \\ M_s \end{Bmatrix} \tag{5.87}$$

If in a laminate of width b, M_x only acts with $M_y = M_s = 0$, the resulting curvature is [Fig. 5.11]

$$\kappa_x = d_{xx} \cdot M_x = d_{xx} \cdot \frac{M}{b} = \frac{1}{\rho_x} \tag{5.88}$$

where $M = M_x \cdot b$ = total bending moment

b = laminate width

ρ_x = radius of curvature = $\dfrac{1}{\kappa_x}$

Moment–curvature relationship for an equivalent homogeneous beam is

$$\frac{M}{\overline{E}_{fx} I_{yy}} = \frac{1}{\rho_x} \tag{5.89}$$

where

$\overline{E}_{fx}$ is the effective flexural modulus

$$I_{yy} = \frac{bd^3}{12} \tag{5.90}$$

Combining (5.88) and (5.90), yields

$$\overline{E}_{fx} = \frac{12}{d^3 d_{xx}} \tag{5.91}$$

Similarly, effective flexural modulus in the y-direction is

$$\overline{E}_{fy} = \frac{12}{d^3 d_{yy}} \tag{5.92}$$

General asymmetric laminate reduces to eqn. (5.83) for inplane loading and effective moduli can be obtained in terms of engineering constants. However, the force-displacement relationship for antisymmetric laminate is more complicated than symmetric laminates and the above approach may not be appropriate for this case.

5.12 EXAMPLES: SERIES B

Example 5B.1

For a symmetric angle-ply laminate $[\pm 45]_s$, find the relation of E_6 with effective laminate modulus $\overline{E}_x$ and effective laminate Poisson's ratio $\overline{v}_{xy}$.

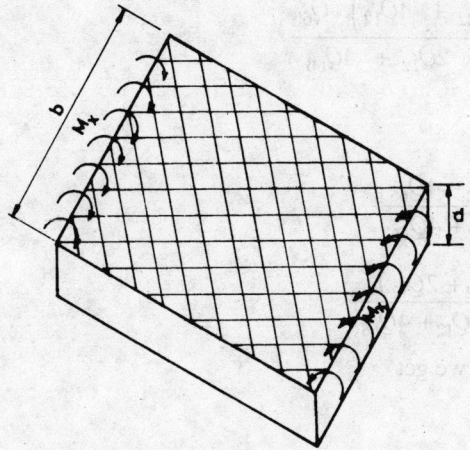

Fig. 5.11 Pure bending of a symmetric laminate

We know from (5.80) that

$$\overline{E}_x = \frac{1}{d}\left[A_{xx} - \frac{A_{xy}^2}{A_{yy}}\right]$$

For the symmetric angle-ply laminate,

$$A_{xx} = \sum_{k=1}^{4} Q_{xx}^k \; d_k = d \cdot Q_{xx}$$

Similarly, $A_{xy} = d \cdot Q_{xy}$, $\quad A_{yy} = d \cdot Q_{yy}$

For both $+45°$ and $-45°$ lamina, we have

$$Q_{xx} = \frac{1}{4}(Q_{11} + Q_{22} + 2Q_{12} + 4Q_{66}) = Q_{yy}$$

$$Q_{xy} = \frac{1}{4}(Q_{11} + Q_{22} + 2Q_{12} - 4Q_{66})$$

Therefore,

$$A_{xx} = A_{yy} = \frac{d}{4}(Q_{11} + Q_{22} + 2Q_{12} + 4Q_{66})$$

$$A_{xy} = \frac{d}{4}(Q_{11} + Q_{22} + 2Q_{12} - 4Q_{66})$$

$$\overline{E}_x = \frac{1}{d}\left(\frac{A_{xx}^2 - A_{xy}^2}{A_{xx}}\right)$$

$$= \frac{1}{d}(A_{xx} + A_{xy})(A_{xx} - A_{xy})/A_{xx}$$

$$= \frac{4 \cdot d \cdot 4}{4 \cdot 4 \cdot 4 \cdot d} \frac{(2Q_{11} + 2Q_{22} + 4Q_{12})8Q_{66}}{Q_{11} + Q_{22} + 2Q_{12} + 4Q_{66}}$$

$$= \frac{2\left(2Q_{11} + 2Q_{22} + 4Q_{12}\right)Q_{66}}{Q_{11} + Q_{22} + 2Q_{12} + 4Q_{66}} \qquad (a)$$

$$\bar{\nu}_{xy} = \frac{A_{xy}}{A_{xx}}$$

$$= \frac{Q_{11} + Q_{22} + 2Q_{12} - 4Q_{66}}{Q_{11} + Q_{22} + 2Q_{12} + 4Q_{66}}$$

$$1 + \bar{\nu}_{xy} = \frac{2\left(Q_{11} + Q_{22} + 2Q_{12}\right)}{Q_{11} + Q_{22} + 2Q_{12} + 4Q_{66}} \qquad (b)$$

Comparing (a) with eqn. (b), we get

$$2\left(1 + \bar{\nu}_{xy}\right)Q_{66} = \bar{E}_x$$

or,
$$\frac{\bar{E}_x}{2\left(1 + \bar{\nu}_{xy}\right)} = Q_{66} = E_6$$

Example 5B.2

Determine $\bar{E}_x$, $\bar{E}_s$ and $\bar{\nu}_{xy}$ for a $[\pm 45]_s$ laminate. The thickness of each ply is 0.25 mm and material properties are $E_1 = 145$ GPa, $E_2 = 10.5$ GPa, $E_6 = 7.5$ GPa, $\nu_{12} = 0.28$

$$\nu_{21} = 0.28 \times \frac{10.5}{145} = 0.0203$$

From example 5A.5, we get

$$[Q]_{x,y}^{45} = \begin{bmatrix} 48.07 & 33.07 & 33.82 \\ 33.07 & 48.07 & 33.82 \\ 33.82 & 33.82 & 37.63 \end{bmatrix}$$

$$[Q]_{x,y}^{-45} = \begin{bmatrix} 48.07 & 33.07 & -33.82 \\ 33.07 & 48.07 & -33.82 \\ -33.82 & -33.82 & 37.63 \end{bmatrix}$$

$$[A] = \begin{bmatrix} 48.07 & 33.07 & 33.82 \\ 33.07 & 48.07 & 33.82 \\ 33.82 & 33.82 & 37.63 \end{bmatrix} \times 2 \times 0.25$$

$$+ \begin{bmatrix} 48.07 & 33.07 & -33.82 \\ 33.07 & 48.07 & -33.82 \\ -33.82 & -33.82 & 37.63 \end{bmatrix} \times 2 \times 0.25$$

$$= \begin{bmatrix} 48.07 & 33.07 & 0 \\ 33.07 & 48.07 & 0 \\ 0 & 0 & 37.63 \end{bmatrix}$$

Total thickness of the laminate, $d = 0.25 \times 4 = 1$

$$\bar{E}_x = \frac{1}{d}\left[A_{xx} - \frac{A_{xy}^2}{A_{yy}}\right]$$

$$= \frac{1}{1}\left[48.07 - \frac{(33.07)^2}{48.07}\right] = 25.32 \text{ GPa}$$

$$\bar{E}_s = \frac{A_{ss}}{d} = \frac{37.63}{1} = 37.63 \quad \text{GPa}$$

$$\bar{\nu}_{xy} = \frac{A_{xy}}{A_{yy}} = \frac{33.07}{48.07} = 0.688$$

Example 5B.3

A laminate of $[0|\pm 45°|90°]_s$ lay-up is loaded under inplane biaxial loading [Fig. 5.9]. $N_x = 2N_0$, $N_y = 4N_0$, $N_s = 0$ and the strains occurred have the following value

$$\varepsilon_x^0 = 2.6 \times 10^{-3}$$

$$\varepsilon_y^0 = 10.2 \times 10^{-3}$$

$$\varepsilon_s^0 = 0$$

Determine Poisson's ratio $\bar{\nu}_{xy}$.

'Stress–strain' relationship for this case is

$$\begin{Bmatrix} N_x \\ N_y \\ 0 \end{Bmatrix} = \begin{bmatrix} A_{xx} & A_{xy} & 0 \\ A_{yx} & A_{yy} & 0 \\ 0 & 0 & A_{ss} \end{bmatrix} \begin{Bmatrix} \varepsilon_x^0 \\ \varepsilon_y^0 \\ 0 \end{Bmatrix}$$

Equation (a) when expanded gives

$$N_x = A_{xx}\varepsilon_x^0 + A_{xy}\varepsilon_y^0$$

$$N_y = A_{yx}\varepsilon_x^0 + A_{yy}\varepsilon_y^0 \tag{b}$$

or, $\quad 2N_0 = A_{xx}\varepsilon_x^0 + A_{xy}\varepsilon_y^0$

$$4N_0 = A_{yx}\varepsilon_x^0 + A_{yy}\varepsilon_y^0 \tag{c}$$

For this laminate $A_{xx} = A_{yy}$

Solving (c), yields

$$A_{xx} = \frac{4 N_0 \, \varepsilon_y^0 - 2 N_0 \, \varepsilon_x^0}{\varepsilon_y^{0\,2} - \varepsilon_x^{0\,2}} = \frac{(4 \times 10.2 - 2 \times 2.6) \, 10^{-3} \, N_0}{(10.2^2 - 2.6^2) \, 10^{-3}} = 0.366 \, N_0$$

$$A_{xy} = \frac{4 N_0 \, \varepsilon_x^0 - 2 N_0 \, \varepsilon_y^0}{\varepsilon_x^{0\,2} - \varepsilon_y^{0\,2}} = \frac{(4 \times 2.6 - 2 \times 10.2) \, 10^{-3} \, N_0}{(2.6^2 - 10.2^2) \, 10^{-3}} = 0.103 \, N_0$$

$$\bar{\nu}_{xy} = \frac{A_{xy}}{A_{xx}} = \frac{1.316 \, N_0}{4.684 \, N_0} = 0.2809$$

REFERENCES AND SUGGESTED READINGS

5.1 I. M. Daniel and O. Ishai, *Engineering Mechanics of Composite Materials*, Oxford University Press, New York, 1994.

5.2 M. H. Datoo, *Mechanics of Fibrous Composites*, Elsevier Applied Science, London, 1991.

5.3 R. F. Gibson, *Principles of Composite Materials Mechanics*, McGraw Hill Book Co., New York, 1993.

5.4 R. M. Jones, *Mechanics of Composite Materials*, 2nd Edition, McGraw Hill, 1999.

5.5 B. D. Agarwal and L. J. Broutman, *Analysis and Performance of Fibre Composites*, 2nd Edition, John Wiley & Sons, 1990.

5.6 M. W. Hyer, *Stress Analysis of Fiber-Reinforced Composite Materials*, McGraw Hill International Edition, Massachussets, USA, 1998.

EXERCISE 5

5.1 What are the types of laminate given below. Mention which elements $[A]$, $[B]$ and $[D]$ are zero for each one of them
  (a) $[\pm \alpha]$
  (b) $[90/0/0]_s$
  (c) $[0/\pm 45/\mp 45/0]$
  (d) $[60/-60/-60/60]$
  (e) $[60/-60/60/-60/60]$
  (f) $[45/-45/45/-45]$
  (g) $[0/\pm 45]$
  (h) $[0/60/-30]$

5.2 Compute all terms of $[A]$, $[B]$ and $[D]$ matrices for a $[-45/45]$ laminate with the following lamina properties

   $E_1 = 145$ GPa, $E_2 = 8$ GPa, $E_6 = 6$ GPa, $\nu_{12} = 0.3$

   $d = 0.1$ mm (thickness of each lamina)

5.3 A cross-ply laminate $[0/90]$ consists of two layers, thickness of each is d. The elastic properties of the lamina are as follows

$E_1 = 10 E_0$, $E_2 = E_0$, $E_6 = 0.8 E_0$, $\nu_{12} = 0.25$

Compute the laminate stiffness in terms of E_0 and d. What simplification in the values will result by assuming $E_1 \gg E_2 \gg E_6$.

5.4 Compute $[D]$ matrix for a $[0/\pm 45]$ laminate with the following laminate properties

$E_1 = 140$ GPa, $E_2 = 10$ GPa, $E_6 = 6$ GPa, $\nu_{12} = 0.3$, $d = 0.1$ mm

5.5 Determine the stiffness matrices for a quasi-isotropic laminate $[0/\pm 45/90]_s$ with the following material properties

$E_1 = 140$ GPa, $E_2 = 10$ GPa, $E_6 = 6$ GPa, $\nu_{12} = 0.3$, $d = 0.2$ mm

5.6 Prove that bending–extension coupling stiffnesses are zero for a symmetric laminate

5.7 (a) A cross-ply $[0/90]$ laminate having equal thickness of plies is not a quasi-isotropic laminate–why?

(b) Do quasi-isotropic laminates behave as an isotropic homogenous material? Explain.

5.8 Show that for a regular anti-symmetric laminate

$A_{xs} = A_{ys} = D_{xs} = D_{ys} = 0$

5.9 A $[\mp 45/-45]$ laminate is subjected to $N_x = 50$ MPa.

The material properties are

$E_1 = 140$ GPa, $E_2 = 10$ GPa, $E_6 = 5$ GPa, $\nu_{12} = 0.3$, $d = 0.2$ mm

Determine the resulting stresses along the reference axis of each lamina.

5.10 A uniaxial force $N_x = 100$ MPa-mm is applied to a symmetric cross-ply laminate $[0/90/90/0]$.

Determine the resulting stress in each lamina in the direction of reference axes. The material properties of each ply are

$E_1 = 140$ GPa, $E_2 = 10$ GPa, $E_6 = 5$ GPa, $\nu_{12} = 0.3$, $d = 0.2$ mm

5.11 A laminate of $[0/60/-60]_s$ is loaded under inplane biaxial loading

$N_x = N_0$
$N_y = 2 N_0$
$N_s = 0$

and the resulting strains are

$\varepsilon_x = 1.5 \times 10^{-3}$
$\varepsilon_y = 5 \times 10^{-3}$

Determine Poisson's ratio $\overline{\nu}_{xy}$.

5.12 Derive the following relationship for a quasi-isotropic laminate

$$\overline{G}_{xy} = \frac{\overline{E}_x}{2(1+\overline{\nu}_{xy})}$$

CHAPTER 6

ANALYTICAL METHODS OF LAMINATED PLATE

6.1 INTRODUCTION

There has been a spate of increase in the use of composite materials in structural engineering applications. This has resulted in a widespread use of composites, which has made necessary the analysing of laminated plates. The study of the response of the laminated plates follows one of the following two approaches

1. The composite laminate can be considered as a homogeneous material where the effect of constituent materials and laminae is treated as average properties of composites.
2. A layered analysis in which layerwise displacements are allowed.

In the first approach, two-dimensional theories are used to analyze composite laminates for stresses. The two-dimensional theories are obtained from three-dimensional elasticity theory by making assumptions concerning the variation of displacements and stresses through the thickness of the laminate. The classical laminate plate theory (CLPT) is an extension of the classical plate theory to isotropic plates. In the layerwise theory the displacements are expanded within each layer using the Lagrangian interpolation [6.1].

6.2 CLASSICAL LAMINATE PLATE THEORY (CLPT)

In the earlier chapters, stress resultants – axial force, shear force, twisting moment and bending moment have been expressed in terms of material and geometric properties of the laminate.

6.2.1 Basic Assumptions

The following assumptions are made in the classical laminate plate theory

1. The plate is thin, that is, the thickness is much smaller in comparison to other physical dimensions.
2. The plate is made of an arbitrary number of orthotropic layers. However, the principal material direction of each layer need not coincide with the plate axes.
3. Behaviour of each layer is linear and elastic.
4. Each ply and the plate have constant thickness.
5. Transverse normal stress σ_z is neglected.
6. Transverse shear strains ε_{xz} and ε_{yz} are negligible.
7. Transverse shear stresses σ_{xz} and σ_{yz} are assumed to be zero.

Assumptions 6 and 7 are general and have nothing specific about fibre reinforced composite material. Kirchhoff's assumptions of negligible transverse shear strains ε_{xz} and ε_{yz} and negligible transverse normal strain constitute a case of non-deformable normal to the middle surface.

The derivation that follows is essentially similar to the classical plate theory.

6.2.2 Equilibrium Equations of Laminated Plates

The plate of Fig.6.1 is subjected to arbitrary transverse load q.
The equilibrium equations of elasticity for a ply are given by [6.2].

$$\frac{\partial \sigma_x}{\partial x} + \frac{\partial \sigma_{xy}}{\partial y} + \frac{\partial \sigma_{xz}}{\partial z} + F_x = 0$$

$$\frac{\partial \sigma_{yx}}{\partial x} + \frac{\partial \sigma_y}{\partial y} + \frac{\partial \sigma_{yz}}{\partial z} + F_y = 0 \qquad (6.1)$$

$$\frac{\partial \sigma_{zx}}{\partial x} + \frac{\partial \sigma_{zy}}{\partial y} + \frac{\partial \sigma_z}{\partial z} + F_z = 0$$

where σ_x, σ_y and σ_z are normal stress components, σ_{xy}, σ_{yz} and σ_{zx} are shear stress components. We revert back to the conventional notation and F_x, F_y and F_z are the body force components.
We know from Chapter 5 that stress resultants are given as follows:

$$\begin{Bmatrix} N_x \\ N_y \\ N_{xy} \\ Q_x \\ Q_y \end{Bmatrix} = \int_{-d/2}^{d/2} \begin{Bmatrix} \sigma_x \\ \sigma_y \\ \sigma_{xy} \\ \sigma_{xz} \\ \sigma_{yz} \end{Bmatrix} dz = \sum_{k=1}^{n} \int_{z_{k-1}}^{z_k} \begin{Bmatrix} \sigma_x \\ \sigma_y \\ \sigma_{xy} \\ \sigma_{xz} \\ \sigma_{yz} \end{Bmatrix}_k dz \qquad (6.2)$$

$$\begin{Bmatrix} M_x \\ M_y \\ M_{xy} \end{Bmatrix} = \int_{-d/2}^{d/2} \begin{Bmatrix} \sigma_x \\ \sigma_y \\ \sigma_{xy} \end{Bmatrix} z\, dz = \sum_{k=1}^{n} \int_{z_{k-1}}^{z_k} \begin{Bmatrix} \sigma_x \\ \sigma_y \\ \sigma_z \end{Bmatrix}_k z\, dz \qquad (6.3)$$

where d is the thickness of the laminate and n is the number of plies.

Turning now to the first of (6.1), neglecting the body force term F_x, integrating term by term across each ply and summing up across the plate provides

$$\sum_{k=1}^{n} \int_{z_{k-1}}^{z_k} \frac{\partial \sigma_{x_k}}{\partial x} dz + \sum_{k=1}^{n} \int_{z_{k-1}}^{z_k} \frac{\partial \sigma_{xy_k}}{\partial y} dz + \sum_{k=1}^{n} \int_{z_{k-1}}^{z_k} \frac{\partial \sigma_{xz_k}}{\partial z} dz = 0 \qquad (6.4a)$$

In the first two terms, differentiation and integration can be interchanged, hence

$$\frac{\partial}{\partial x}\left[\sum_{k=1}^{n} \int_{z_{k-1}}^{z_k} \sigma_{x_k} dz\right] + \frac{\partial}{\partial y}\left[\sum_{k=1}^{n} \int_{z_{k-1}}^{z_k} \sigma_{xy_k} dz\right] + \left[\sum_{k=1}^{n} \sigma_{xz_k}\right]_{z_{k-1}}^{z_k} = 0 \qquad (6.4b)$$

From (6.2), the first two terms of (6.4a) are N_x and N_{xy}. The third term indicates the summation of the inter-laminar shear stresses. At the interface between plies, the direction of the shear stresses is opposite to each other in two distinct plies, though the magnitude of the shear stress is same for both. Hence, the inter-laminar shear stresses will get cancelled making the summation of the third term of eqn. (6.4b) zero.

Equation (6.3) can be written as

$$\frac{\partial N_x}{\partial x} + \frac{\partial N_{xy}}{\partial y} = 0 \qquad (6.5)$$

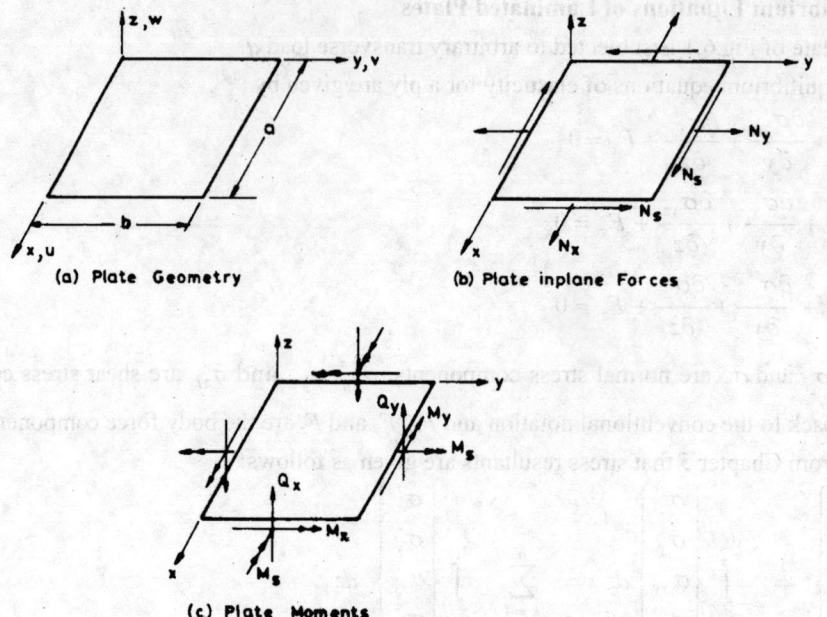

Fig. 6.1 Nomenclature for Stress Resultants

Similarly, integrating equilibrium equations in *y*-direction provides

$$\frac{\partial N_{xy}}{\partial x} + \frac{\partial N_y}{\partial y} = 0 \tag{6.6}$$

Now,

$$(Q_x, Q_y) = \int_{-d/2}^{d/2} \left(\sigma_{xz_k}, \sigma_{yz_k} \right) dz \tag{6.7}$$

Likewise, equilibrium in the *z*-direction upon integration and the resulting summation provides

$$\frac{\partial Q_x}{\partial x} + \frac{\partial Q_y}{\partial y} + q = 0 \tag{6.8}$$

where q is the intensity of the transverse load on the plate.

In addition to the integrated force equilibrium equations above, two moment equilibrium equations are also needed.

Multiplying the first of (6.1) by '$z\,dz$', integrating across each ply and summing across all the laminae results in the following

$$\sum_{k=1}^{n} \int_{z_{k-1}}^{z_k} \frac{\partial \sigma_{x_k}}{\partial x} z\,dz + \sum_{k=1}^{n} \int_{z_{k-1}}^{z_k} \frac{\partial \sigma_{xy_k}}{\partial y} z\,dz + \sum_{k=1}^{n} \int_{z_{k-1}}^{z_k} \frac{\partial \sigma_{xz_k}}{\partial z} z\,dz = 0 \tag{6.9}$$

Again, integration and summation can be interchanged in the first two terms of differentiation with the result that the first two terms become $(\partial M_x/\partial x) + (\partial M_{xy}/\partial y)$.

However, the third term is integrated by parts as follows

$$\sum_{k=1}^{n} \int_{z_{k-1}}^{x_k} \frac{\partial \sigma_{xz_k}}{\partial z} z\, dz = \sum_{k=1}^{n} \left\{ \left[z \sigma_{xz_k} \right]_{z_{k-1}}^{z_k} - \int_{z_{k-1}}^{z_k} \sigma_{xz_k}\, dz \right\} \tag{6.10}$$

Here, the last term is clearly $-Q_x$. Again, in the first term on the right, the moments of all interlaminar stresses between plies get cancelled with each other. Therefore, (6.10) reduces to

$$\frac{\partial M_x}{\partial x} + \frac{\partial M_{xy}}{\partial y} - Q_x = 0 \tag{6.11}$$

Similarly, moment equilibrium equation in the y-direction yields

$$\frac{\partial M_{xy}}{\partial x} + \frac{\partial M_y}{\partial y} - Q_y = 0 \tag{6.12}$$

There are thus 5 equilibrium equations in all for a rectangular plate, which are independent of the material properties: (6.5), (6.6), (6.8), (6.11) and (6.12).

Using the relations of stress–resultants in terms of strains or curvatures as given in (5.21) which in turn are related to displacements (5.3) and (5.5) and finally, substituting them into equilibrium eqn. (6.5) results in the following equation

$$\frac{\partial}{\partial x}\left[A_{xx}\frac{\partial u_0}{\partial x} + A_{xy}\frac{\partial v_0}{\partial y} + A_{xs}\left(\frac{\partial u_0}{\partial y} + \frac{\partial v_0}{\partial x}\right) - B_{xx}\frac{\partial^2 w}{\partial x^2} - B_{xy}\frac{\partial^2 w}{\partial y^2} - 2B_{xs}\frac{\partial^2 w}{\partial x \partial y} \right]$$

$$+ \frac{\partial}{\partial y}\left[A_{xs}\frac{\partial u_0}{\partial x} + A_{ys}\frac{\partial v_0}{\partial y} + A_{ss}\left(\frac{\partial u_0}{\partial y} + \frac{\partial u_0}{\partial x}\right) \right.$$

$$\left. - B_{xs}\frac{\partial^2 w}{\partial x^2} - B_{ys}\frac{\partial^2 w}{\partial y^2} - 2B_{ss}\frac{\partial^2 w}{\partial x \partial y} \right] = 0 \tag{6.13}$$

which when arranged becomes

$$A_{xx} u_{0,xx} + 2A_{xs} u_{0,xy} + A_{ss} u_{0,yy} + A_{xs} v_{0,xx} + (A_{xy} + A_{ss}) v_{0,xy}$$
$$+ A_{ys} v_{0,yy} - B_{xx} w_{,xxx} - 3B_{xs} w_{,xxy} - (B_{xy} + 2B_{ss}) w_{,xyy} - B_{ys} w_{,yyy} = 0 \tag{6.14}$$

Similarly, (6.6) will finally become

$$A_{xs} u_{0,xx} + (A_{xy} + A_{ss}) u_{0,xy} + A_{ys} u_{0,yy} + A_{ss} v_{0,xx} + A_{ys} v_{0,xy}$$
$$+ A_{yy} v_{0,yy} - B_{xs} w_{,xxx} - (B_{xy} + 2B_{ss}) w_{,xxy} - 3B_{ys} w_{,xyy} - B_{yy} w_{,yyy} = 0 \tag{6.15}$$

Combining (6.8), (6.11) and (6.12) will yield

$$\frac{\partial^2 M_x}{\partial x^2} + 2\frac{\partial^2 M_{xy}}{\partial x \partial y} + \frac{\partial^2 M_y}{\partial y^2} + q = 0 \tag{6.16}$$

Expressing the stress–resultants in terms of strains and curvatures from (5.23) which when writing in terms of displacements from (5.3) and (5.5) yields

$$D_{xx} w_{,xxxx} + 4D_{xs} w_{,xxxy} + 2(D_{xy} + 2D_{ss}) w_{,xxyy} + 4D_{ys} w_{,xyyy}$$
$$+ D_{yy} w_{,yyyy} - B_{xx} u_{0,xxx} - 3B_{xs} u_{0,xxy} - (B_{xy} + 2B_{ss}) u_{0,xyy} - B_{ys} u_{0,yyy} \quad (6.17)$$
$$- B_{xs} v_{0,xxx} - (B_{xy} + 2B_{ss}) v_{0,xxy} - 3B_{ys} v_{0,xyy} - B_{yy} v_{0,yyy} = q$$

In the above equations, the comma indicates differentiation of the symbol with respect to the subscript.

For the case of symmetric laminates, $B_{ij} = 0$ and if the laminate is specially orthotropic, further simplification occurs for all the terms with xs and ys subscripts which become zero in addition to $B_{ij} = 0$. For homogenous symmetric laminates $B_{ij} = 0$ and $D_{ij} = A_{ij} t^2 / 12$. In all those cases, (6.14) and (6.15) are uncoupled from (6.17), that is, (6.17) is expressed in terms of w whereas (6.14) and (6.15) are in terms of u_0 and v_0, but not w.

6.2.3 Bending of Composite Plates

For a specially orthotropic laminate (all terms with xs and ys vanish in addition to B_{ij}), (6.17) reduces to

$$D_{11} \frac{\partial^4 w}{\partial x^4} + 2(D_{12} + 2D_{66}) \frac{\partial^4 w}{\partial x^2 \partial y^2} + D_{22} \frac{\partial^4 w}{\partial y^4} = q \quad (6.18)$$

The boundary conditions at the edges of the plate are similar to that of the isotropic plate.

6.2.3.1 *A rectangular plate with all edges simply supported: Navier's solution*

Let us consider a rectangular plate having a and b as the dimensions along two sides. All the edges of the plate are simply supported. The boundary conditions for a specially orthotropic laminate are

$$\begin{aligned} x &= 0, a & w &= 0, & M_x &= -D_{11} w_{,xx} - D_{12} w_{,yy} \\ y &= 0, b & w &= 0, & M_y &= -D_{12} w_{,xx} - D_{22} w_{,yy} \end{aligned} \quad (6.19)$$

A solution in the following form of infinite double sine series has been suggested by Navier

$$w(x, y) = \sum_{m=1}^{\infty} \sum_{n=1}^{\infty} A_{mn} \sin \frac{m\pi x}{a} \sin \frac{n\pi y}{b} \quad (6.20)$$

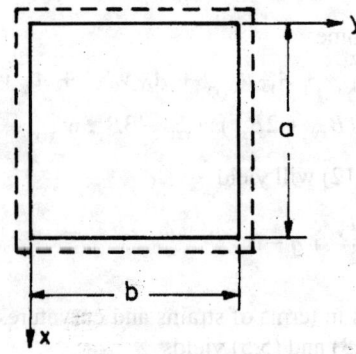

Fig. 6.2 A rectangular plate with all edges simply supported

The expression for w in (6.20) satisfies the boundary conditions given in (6.19). $q(x, y)$ is also expressed in terms of double Fourier sine series

$$q(x, y) = \sum_{m=1}^{\infty} \sum_{n=1}^{\infty} q_{mn} \sin\frac{m\pi x}{a} \sin\frac{n\pi y}{b} \tag{6.21}$$

The Fourier coefficients can be determined from the relationship

$$q_{mn} = \frac{4}{ab} \int_0^b \int_0^a q(x, y) \sin\frac{m\pi x}{a} \sin\frac{n\pi y}{b} \, dx \, dy \tag{6.22}$$

The plate is subjected to uniform load

$$q(x, y) = q_0 \tag{6.23}$$

In the case of a uniform load on an all edges simply supported plate, the solution of (6.18) is

$$w = \frac{16 q_0 a^4}{\pi^6} \sum_{m=1}^{\infty} \sum_{n=1}^{\infty} \frac{1}{mn D_{mn}} \sin\frac{m\pi x}{a} \sin\frac{n\pi y}{b} \tag{6.24}$$

The maximum deflection becomes

$$w_{\max} = \frac{16 q_0 a^4}{\pi^6} \sum_m \sum_n \frac{(-1)^{\frac{m+n+2}{2}}}{mn \, D_{mn}} \tag{6.25}$$

For a [0|90|0|90]s laminate, with $E_1/E_2 = 10$, $\nu_{12} = 0.25$ and $E_6/E_2 = 0.5$, the non-dimensional maximum deflection works out to be

$$\frac{w_{\max} E_2 d^3}{a^4 q_0} \times 10^{-2} = 1.445 \tag{6.26}$$

where

$$D_{mn} = D_{xx}\left(\frac{m}{a}\right)^4 + 2(D_{xy} + 2D_{ss})\left(\frac{mn}{ab}\right)^2 + D_{yy}\left(\frac{n}{b}\right)^4 \tag{6.27}$$

The stresses in the kth lamina is given by

$$\begin{Bmatrix} \sigma_x \\ \sigma_y \\ \sigma_{xy} \end{Bmatrix}_k = \frac{16 q_0 z}{\pi^4} \sum_{m=1,3}^{\infty} \sum_{n=1,3}^{\infty} \frac{1}{mnD} \begin{Bmatrix} \left[-Q_{11}^k \left(\frac{m}{a}\right)^2 - Q_{12}^k \left(\frac{n}{b}\right)^2\right] \sin\frac{m\pi x}{a} \sin\frac{n\pi y}{b} \\ \left[-Q_{12}^k \left(\frac{m}{a}\right)^2 - Q_{22}^k \left(\frac{n}{b}\right)^2\right] \sin\frac{m\pi x}{a} \sin\frac{n\pi y}{b} \\ 2Q_{66}^k \left(\frac{m}{a}\right)\left(\frac{n}{b}\right) \cos\frac{m\pi x}{a} \cos\frac{n\pi y}{b} \end{Bmatrix} \tag{6.28}$$

Similar calculations for the determination of stresses for other laminae can be carried out. A typical normalised stress distribution σ_x/q_0 across the thickness is given in Fig. 6.3.

6.2.4 Bending of Symmetric Angle-Ply Laminates

Symmetric angle-ply laminates are found to be characterized to a full extent of extensional stiffness as well as bending stiffness. However, being symmetric, bending–extension coupling stiffness does not exist. Such laminates are characterized by non-vanishing bending–twisting coupling stiffness terms D_{xs} and D_{ys}. They are mathematically equivalent to homogeneous anisotropic plates. The inclusion of bending–twisting coupling terms, D_{xs} and D_{ys}, in the governing equations significantly increases the complexity of the analysis.

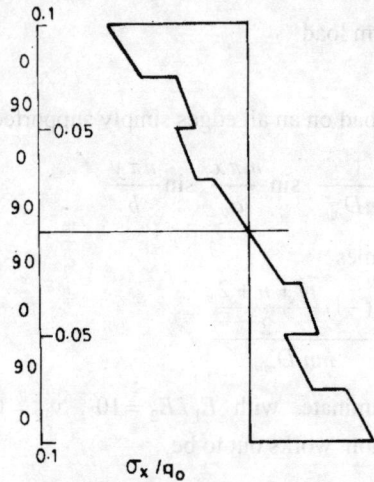

Fig. 6.3 Normalised stress distribution across the thickness

For a rectangular plate with dimensions a and b and subjected to a uniform transverse load q, the governing differential equation for the case of a symmetric angle-ply laminate is

$$D_{xx}\frac{\partial^4 w}{\partial x^4} + 4D_{xs}\frac{\partial^4 w}{\partial x^3 \partial y} + 2(D_{xy} + 2D_{ss})\frac{\partial^4 w}{\partial x^2 \partial y^2} + 4D_{ys}\frac{\partial^4 w}{\partial x \partial y^3} + D_{yy}\frac{\partial^4 w}{\partial y^4} = q \quad (6.29)$$

For simply supported edges, the boundary conditions are as follows

$$\begin{aligned} x = 0, a: \quad w = 0 \quad & M_x = -D_{xx}\frac{\partial^2 w}{\partial x^2} - 2D_{xs}\frac{\partial^2 w}{\partial x \partial y} - D_{xy}\frac{\partial^2 w}{\partial y^2} = 0 \\ y = 0, b: \quad w = 0 \quad & M_y = -D_{xy}\frac{\partial^2 w}{\partial x^2} - 2D_{ys}\frac{\partial^2 w}{\partial x \partial y} - D_{yy}\frac{\partial^2 w}{\partial y^2} = 0 \end{aligned} \quad (6.30)$$

Double Fourier series sine terms used for specially orthotropic laminates will not satisfy the boundary conditions for this case due to the presence of bending–twisting coupling terms. Also the advantage obtained earlier for the separation of variables cannot be achieved for this case by using double Fourier series. The problem however can be solved by other approaches. For example the Ritz method can be applied for the solution of the above plate problem.

Total potential energy of the plate is given by

$$V = \frac{1}{2}\int_0^b\int_0^a\left[D_{xx}\left(\frac{\partial^2 w}{\partial x^2}\right)^2 + 2D_{xy}\frac{\partial^2 w}{\partial x^2}\cdot\frac{\partial^2 w}{\partial y^2} + D_{yy}\left(\frac{\partial^2 w}{\partial y^2}\right)^2\right.$$
$$\left. + 4\left(D_{xs}\frac{\partial^2 w}{\partial x^2} + D_{ys}\frac{\partial^2 w}{\partial y^2}\right)\frac{\partial^2 w}{\partial x\partial y} + 4D_{ss}\left(\frac{\partial^2 w}{\partial x\partial y}\right)^2 - 2qw\right]\partial x\partial y \quad (6.31)$$

The solution is assumed to be of the following form

$$w = \sum_{m=1}^{M}\sum_{n=1}^{N} A_{mn} X_m(x) Y_n(y) \quad (6.32)$$

Substituting this series of eqn. (6.32) into energy eqn. (6.33) and minimizing the potential energy with respect to the coefficients yields the following $M \times N$ conditions

$$\sum_{i=1}^{M}\sum_{j=1}^{N}\left\{D_{xx}\int_0^a\frac{d^2X_i}{dx^2}\cdot\frac{d^2X_m}{dx^2}dx\int_0^b Y_j Y_n\,dy + D_{xy}\left[\int_0^a X_m\frac{d^2X_i}{dx^2}dx\int_0^b Y_j\frac{d^2Y_n}{dy^2}dy\right.\right.$$
$$\left. + \int_0^a X_i\frac{d^2X_m}{dx^2}dx\int_0^b Y_n\frac{dY_j}{dy^2}dy\right] + D_{yy}\int_0^a X_i X_m\,dx\cdot\int_0^b\frac{d^2Y_j}{dy^2}\frac{d^2Y_n}{dy^2}dy$$
$$+ 4D_{ss}\int_0^a\frac{dX_i}{dx}\cdot\frac{dX_m}{dx}dx\int_0^b\frac{dY_j}{dy}\cdot\frac{dY_n}{dy}dy$$
$$+ 2D_{xs}\left[\int_0^a\frac{d^2X_i}{dx^2}\cdot\frac{dX_m}{dx}dx\int_0^b Y_n\frac{dY_j}{dy}dy\right.$$
$$\left. + \int_0^a\frac{dX_i}{dx}\cdot\frac{d^2X_m}{dx^2}dx\cdot\int_0^b Y_n\frac{dY_j}{dy}dy\right]$$
$$+ 2D_{ys}\left[\left\{\int_0^a X_m\cdot\frac{dX_i}{dx}dx\int_0^b\frac{dY_j}{dy}\frac{d^2Y_n}{dy^2}dy\right.\right.$$
$$\left.\left.\left. + \int_0^a X_i\frac{dX_m}{dx}dx\int_0^b\frac{d^2Y_j}{dy^2}\cdot\frac{dY_n}{dy}\cdot dy\right]\right\}A_{ij} = q\int_0^a X_m\,dx\int_0^b Y_n\,dy\right.$$

$$m = 1, 2, \ldots\ldots, M$$
$$n = 1, 2, \ldots\ldots, N \quad (6.33)$$

Equation (6.33) will form a system of $M \times N$ linear simultaneous equations for the determination of $M \times N$ unknowns of A_{mn}.

For symmetric laminates which are simply supported having non-zero D_{xs} and D_{ys} stiffness terms, a solution in the form of separation of variables which satisfies moment boundary conditions does not exist [6.3]. This difficulty can be overcome by the application of either Ritz or Galerkin's method, as only the displacement boundary conditions are required to be satisfied in them. If the geometric boundary conditions are satisfied (i.e., $w = 0$, and $w_x = 0$) even if natural boundary conditions are not satisfied, (that $M_n = \overline{M}_x$, $N_n = \overline{N}_x$, where $\overline{M}_x$ and $\overline{N}_x$ are prescribed moments), the approximate energy will tend to converge to the exact energy with the incorporation of large number of terms of displacement expressions. Thus, for simple supports, it becomes necessary to choose functions in conjunction with (6.32) in which $w = 0$ on the plate boundary and the slope is free to take on any value. It is likely to have cases where moment boundary conditions will not be satisfied. In such cases convergence of (6.32) will be slow. As such, large number of terms are required for satisfactory convergence.

For the symmetric angle-ply laminate, the following solutions are assumed

$$X_m(x) = \sin \frac{m\pi x}{a}, \quad Y_n(y) = \sin \frac{n\pi y}{b} \tag{6.34}$$

Using 49 terms (up through $m = 7$, $n = 7$) in the deflection approximation, for a uniformly loaded square plate with $D_{yy}/D_{xx} = 1$, $(D_{xy} + 2D_{ss})/D_{xx} = 1.5$ and $D_{xs}/D_{xx} = D_{ys}/D_{xx} = -0.5a$, and solving eqn. (6.33), the maximum deflection at the centre is [6.4].

$$w_{\max} = \frac{0.00425 a^4 q_0}{D_{xx}} \tag{6.35}$$

whereas if D_{xs} and D_{ys} are ignored, that is, the symmetric angle-ply laminate is approximated by a specially orthotropic laminate with $D_{yy}/D_{xx} = 1$, $(D_{xy} + 2D_{ss})/D_{xx} = 1.5$ and $D_{xs} = D_{ys} = 0$, then the maximum deflection is

$$w_{\max} = \frac{0.00324 a^4 q_0}{D_{xx}} \tag{6.36}$$

Thus, the error in ignoring twist–coupling terms is 24 percent which is obviously not a small error.

6.2.5 Bending of Anti–Symmetric Cross-Ply Laminate

A rectangular cross-ply laminate is considered for which

$$\begin{aligned} A_{xx} &= A_{yy}, \quad B_{yy} = -B_{xx}, \quad D_{yy} = D_{xx} \\ B_{ij} &= 0 \text{ except } B_{xx} \neq 0, \; B_{yy} \neq 0 \end{aligned} \tag{6.37}$$

The differential equations are all coupled in this case.

$$A_{xx}\frac{\partial^2 u_0}{\partial x^2} + A_{ss}\frac{\partial^2 u_0}{\partial y^2} + (A_{xy} + A_{ss})\frac{\partial^2 v_0}{\partial x \partial y} - B_{xx}\frac{\partial^3 w}{\partial x^3} = 0$$

$$(A_{xy} + A_{ss})\frac{\partial^2 u_0}{\partial x \partial y} + A_{ss}\frac{\partial^2 v_0}{\partial x^2} + A_{xx}\frac{\partial^2 v_0}{\partial y^2} + B_{xx}\frac{\partial^3 w}{\partial y^3} = 0 \tag{6.38}$$

$$D_{xx}\left(\frac{\partial^4 w}{\partial x^4} + \frac{\partial^4 w}{\partial y^4}\right) + 2(D_{xy} + 2D_{ss})\frac{\partial^4 w}{\partial x^2 \partial y^2} - B_{xx}\left(\frac{\partial^3 u_0}{\partial x^3} - \frac{\partial^3 v_0}{\partial y^3}\right) = q$$

Whitney and Leissa [6.5] have solved the problem for rectangular plates with simply supported edge boundary conditions.

$$x = 0, a \;:\; w = 0, \quad M_x = B_{xx} u_{0,x} - D_{xx} w_{,xx} - D_{xy} w_{,yy} = 0$$
$$v_0 = 0, \quad N_x = A_{xx} u_{0,x} + A_{xy} v_{0,y} - B_{xx} w_{,xx} = 0$$
$$y = 0, b \;:\; w = 0, \quad M_y = -B_{xx} v_{0,y} - D_{xy} w_{,xx} - D_{xx} w_{,yy} = 0$$
$$u_0 = 0, \quad N_y = A_{xy} u_{0,x} + A_{yy} v_{0,y} + B_{xx} w_{,xx} = 0 \tag{6.39}$$

and the deflections are assumed as

$$u_0 = \sum_{m=1}^{\infty} \sum_{n=1}^{\infty} A_{mn} \cos\frac{m\pi x}{a} \sin\frac{n\pi y}{b}$$
$$v_0 = \sum_{m=1}^{\infty} \sum_{n=1}^{\infty} B_{mn} \sin\frac{m\pi x}{a} \cos\frac{n\pi y}{b} \tag{6.40}$$
$$w = \sum_{m=1}^{\infty} \sum_{n=1}^{\infty} C_{mn} \sin\frac{m\pi x}{a} \sin\frac{n\pi y}{b}$$

The expressions of u_0, v_0 and w given in (6.40) satisfy the three governing differential equations and the boundary conditions if the transverse loading is represented by Fourier sine series of (6.24).

A rectangular anti-symmetric cross–ply laminated graphite–epoxy plate is subjected to sinusoidal transverse loading $q = q_0 \sin\frac{\pi x}{a} \sin\frac{\pi y}{b}$. The normalised maximum deflection is plotted for a number of layers in Fig. 6.4. A plate with infinite layers corresponds to a specially orthotropic plate in which bending–extension coupling is neglected. Figure 6.4 reveals that the effect of bending–extension coupling considerably reduces with the increase of number of layers [6.6].

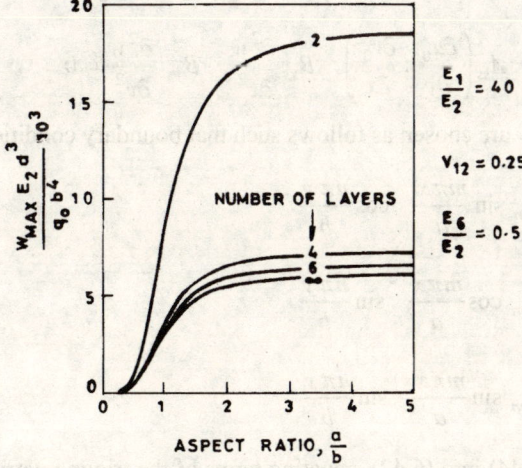

Fig. 6.4 Normalised deflection of an anti-symmetric cross-ply laminate under sinusoidal loading

6.2.6 Bending of Anti–Symmetric Angle-Ply Laminates

Consider a rectanglular anti-symmetric angle-ply laminate.
For this lay-up,
$$A_{xs} = A_{ys} = B_{xx} = B_{xy} = B_{yy} = B_{ss} = D_{xs} = D_{ys} = 0 \tag{6.41}$$

The governing differential equations for equilibrium are [6.7]

$$A_{xx}\frac{\partial^2 u_0}{\partial x^2} + A_{ss}\frac{\partial^2 u_0}{\partial y^2} + (A_{xy} + A_{ss})\frac{\partial^2 v_0}{\partial x \partial y} - 3B_{xs}\frac{\partial^3 w}{\partial x^2 \partial y} - B_{ys}\frac{\partial^3 w}{\partial y^3} = 0$$

$$(A_{xy} + A_{ss})\frac{\partial^2 u_0}{\partial x \partial y} + A_{ss}\frac{\partial^2 v_0}{\partial x^2} + A_{yy}\frac{\partial^2 v_0}{\partial y^2} - B_{xs}\frac{\partial^3 w}{\partial x^3} - 3B_{ys}\frac{\partial^3 w}{\partial x \partial y^2} = 0 \quad (6.42)$$

$$D_{xx}\frac{\partial^4 w}{\partial x^4} + 2(D_{xy} + 2D_{ss})\frac{\partial^4 w}{\partial x^2 \partial y^2} + D_{yy}\frac{\partial^4 w}{\partial y^4} - B_{xs}\left(3\frac{\partial^3 u_0}{\partial x^2 \partial y} + \frac{\partial^3 v_0}{\partial x^3}\right)$$

$$- B_{ys}\left(\frac{\partial^3 u_0}{\partial y^2} + 3\frac{\partial^3 v_0}{\partial x \partial y}\right) = q$$

The boundary conditions for hinged edges, free in the tangential direction are given by
for $x = 0, a$

$$w = 0, \quad M_x = B_{xs}\left(\frac{\partial u}{\partial y} + \frac{\partial v}{\partial x}\right) - D_{xx}\frac{\partial^2 w}{\partial x^2} - D_{xy}\frac{\partial^2 w}{\partial y^2} = 0$$

$$u = 0, \quad N_{xy} = A_{ss}\left(\frac{\partial u}{\partial y} + \frac{\partial v}{\partial x}\right) - B_{xs}\frac{\partial^2 w}{\partial x^2} - B_{ys}\frac{\partial^2 w}{\partial y^2} = 0$$

for $y = 0, b$ \quad (6.43)

$$w = 0, \quad M_y = B_{ys}\left(\frac{\partial u_0}{\partial y} + \frac{\partial v_0}{\partial x}\right) - D_{xy}\frac{\partial^2 w}{\partial x^2} - D_{yy}\frac{\partial^2 w}{\partial y^2} = 0$$

$$v = 0, \quad N_{xy} = A_{ss}\left(\frac{\partial u_0}{\partial y} + \frac{\partial v_0}{\partial x}\right) - B_{xs}\frac{\partial^2 w}{\partial x^2} - B_{ys}\frac{\partial^2 w}{\partial y^2} = 0$$

Expressions for u_0, v_0, w are chosen as follows such that boundary conditions of (6.38) are satisfied

$$u = \sum_{m=1}^{\infty}\sum_{n=1}^{\infty} A_{mn} \sin\frac{m\pi x}{a} \cos\frac{n\pi y}{b}$$

$$v = \sum_{m=1}^{\infty}\sum_{n=1}^{\infty} B_{mn} \cos\frac{m\pi x}{a} \sin\frac{n\pi y}{b} \quad (6.44)$$

$$w = \sum_{m=1}^{\infty}\sum_{n=1}^{\infty} C_{mn} \sin\frac{m\pi x}{a} \sin\frac{n\pi y}{b}$$

Substitution of (6.44) into (6.42), equating terms of the trigonometric functions and then solving the resulting simultaneous algebraic equations for Fourier coefficients, we find that

$$A_{mn} = q_{mn}\frac{R^4 b^3 n}{\pi^3 D_{mn}}\left[\left(A_{ss}m^2 + A_{yy}n^2 R^2\right)\left(3B_{xs}m^2 + B_{ys}n^2 R^2\right)\right.$$

$$\left. - m^2\left(A_{xy} + A_{ss}\right)\left(B_{xs}m^2 + 3B_{ys}n^2 R^2\right)\right]$$

$$B_{mn} = q_{mn} \frac{R^3 b^3 m}{\pi^3 D_{mn}} \left[\left(A_{xx} m^2 + A_{ss} n^2 R^2 \right) \left(B_{xs} m^2 + 3 B_{ys} n^2 R^2 \right) \right.$$
$$\left. - n^2 R^2 \left(A_{xy} + A_{ss} \right) \left(3 B_{xs} m^2 + B_{ys} n^2 R^2 \right) \right] \qquad (6.45)$$

$$C_{mn} = q_{mn} \frac{R^4 b^4}{\pi^4 D_{mn}} \left[\left(A_{xx} m^2 + A_{ss} n^2 R^2 \right) \left(A_{xs} m^2 + A_{yy} n^2 R^2 \right) \right.$$
$$\left. - \left(A_{xy} + A_{ss} \right)^2 m^2 n^2 R^2 \right]$$

where

$$D_{mn} = \left\{ \left[\left(A_{xx} m^2 + A_{ss} n^2 R^2 \right) \left(A_{ss} m^2 + A_{yy} n^2 R^2 \right) \right. \right.$$
$$\left. - \left(A_{xy} + A_{ss} \right)^2 m^2 n^2 R^2 \right] \left[D_{xx} m^4 + 2 \left(D_{xy} + 2 D_{ys} \right) m^2 n^2 R^2 + D_{yy} n^4 R^4 \right]$$
$$+ 2 m^2 n^2 R^2 \left(A_{xy} + A_{ss} \right) \left(3 B_{xs} m^2 + B_{ys} n^2 R^2 \right) \left(B_{xs} m^2 + 3 B_{ys} n^2 R^2 \right) \qquad (6.46)$$
$$- n^2 R^2 \left(A_{ss} m^2 + A_{yy} n^2 R^2 \right) \left(B_{xs} m^2 + B_{ys} n^2 R^2 \right)^2$$
$$\left. - m^2 \left(A_{xx} m^2 + A_{ss} n^2 R^2 \right) \left(B_{xs} m^2 + 3 B_{ys} n^2 R^2 \right)^2 \right\}$$

$$R = \frac{a}{b} \qquad (6.47)$$

For the case $B_{xs} = B_{ys} = 0$, we find $u_0 = v_0 = 0$ and

$$w = \frac{R^4 b^4}{\pi^4} \sum_{m=1}^{\infty} \sum_{n=1}^{\infty} q_{mn} \frac{\sin \frac{m \pi x}{a} \sin \frac{n \pi y}{b}}{\left[D_{xx} m^4 + 2 \left(D_{xy} + 2 D_{ss} \right) m^2 n^2 R^2 + D_{yy} n^4 R^4 \right]} \qquad (6.48)$$

Like the anti-symmetric laminate it may be noted that the effect of coupling is very much predominant when only fewer layers are present, but it is substantially reduced with the increase of the number of layers [Fig. 6.5].

6.3 BENDING OF RECTANGULAR PLATES WITH TWO SIMPLY SUPPORTED EDGES

Levy proposed a solution for the bending of rectangular plates having two opposite edges simply supported. The method has been extended for the analysis of similar plates made of composite materials.

A laminate of composite material, which is simply supported along $y = 0, b$ is considered. The laminate is subjected to a transverse load q. The boundary conditions on these edges are

$$w(x, 0) = w(x, b) = 0$$
$$N_y(x, 0) = M_y(x, b) = 0 \qquad (6.49)$$

Levy assumed the solution in the form of a single infinite series as given below

$$w(x, y) = \sum_{n=1}^{\infty} \phi_n(x) \sin \frac{n \pi y}{b} \qquad (6.50)$$

where $\phi_n(x)$ is the unknown function of x which is to be determined.

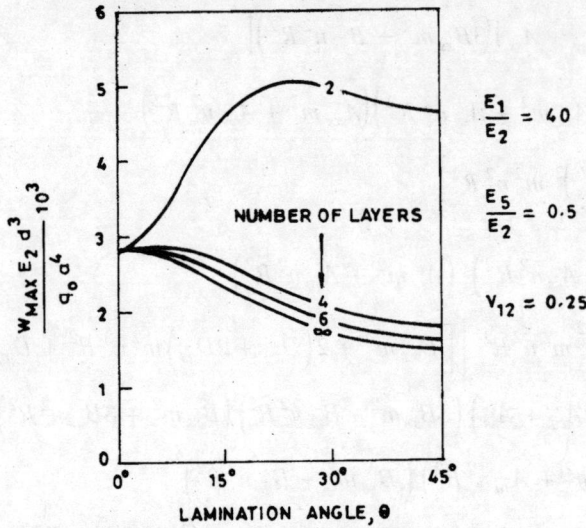

Fig. 6.5 Deflection of a square anti–symmetric angle-ply laminated plate under sinusoidal loading
$q = q_0 \sin \pi x / a \sin \pi y / b$

If the surface load is represented by a double Fourier series given by eqn. (6.21), then the deflection can be expressed as [6.7]

$$w = \sum_{n=1}^{\infty} \phi_n(x) \sin \frac{n\pi y}{b} + \frac{a^4}{\pi^4} \sum_{m=1}^{\infty} \sum_{n=1}^{\infty} \frac{q_{mn}}{D_{mn}} \sin \frac{m\pi x}{a} \sin \frac{n\pi y}{b} \qquad (6.51)$$

where D_{mn} is given by (6.27). This relationship satisfies the boundary conditions for the simply supported edges.

Substituting (6.50), (6.51) into eqn. (6.18) yields

$$\phi_n^{IV}(x) + \frac{(D_{12} + 2D_{66})}{D_{11}} \lambda_n^2 \phi_n''(x) + \frac{D_{22}}{D_{11}} \lambda_x^4 \phi_n(x) = \frac{D_n f_n(x)}{D_{11}} \qquad (6.52)$$

when $\lambda = \dfrac{n\pi}{b}$ \hfill (6.53)

The solution of (6.52) is to be obtained. In order to obtain a homogeneous solution, the right-hand side of (6.52) has to be equal to zero. Therefore,

$$\phi_n^{IV}(x) + \frac{(D_{12} + 2D_{66})}{D_{11}} \lambda_n^2 \phi_n''(x) + \frac{D_{22}}{D_{11}} \lambda_n^4 \phi_n^{(x)} = 0 \qquad (6.54)$$

Solution of (6.52) is given by

$$\phi_n(x) = \exp \frac{n\pi \lambda x}{b} \qquad (6.55)$$

Substituting $\phi_n(x)$ from (6.55) into (6.54), yields

$$D_{11}\lambda^4 - 2(D_{12} + 2D_{66})\lambda^2 + D_{22} = 0 \tag{6.56}$$

Two roots of the characteristic equation are

$$\lambda^2 = \frac{1}{D_{11}}\left[D_{12} + 2D_{66} \pm \sqrt{(D_{12} + 2D_{66})^2 - D_{11}D_{22}}\right] \tag{6.57}$$

The solution of (6.57) will involve four arbitrary constants A_n, B_n, C_n and D_n.

Let us consider a case when roots are real and unequal. The roots are then given by $\pm\lambda_1$ and $\pm\lambda_2$ ($\lambda_1, \lambda_2 > 0$). The solution of (6.52) can then be written as

$$\phi_n(x) = A_n \cosh\frac{n\pi\lambda_1 x}{b} + B_n \sinh\frac{n\pi\lambda_1 x}{b} + C_n \cosh\frac{n\pi\lambda_2 x}{b} + D_n \sinh\frac{n\pi\lambda_2 x}{b} \tag{6.58}$$

Equation of deflection based on (6.51) is given by

$$\sum_{n=1}^{\infty}\left[A_n \cosh\frac{n\pi\lambda_1 x}{b} + B_n \sinh\frac{n\pi\lambda_1 x}{b} + C_n \cosh\frac{n\pi\lambda_2 x}{b}\right.$$

$$\left. + D_n \sinh\frac{n\pi\lambda_2 x}{b} + \frac{a^4}{\pi^4}\sum_{m=1}^{\infty}\frac{q_{mn}}{D_{mn}}\sin\frac{m\pi x}{a}\right]\sin\frac{n\pi y}{b} \tag{6.59}$$

The constants of (6.59) can be determined from two boundary conditions at each edge $x = 0$ and $x = a$. The deflection equation can be simplified for the case of uniform load $q = q_0 = $ constant. Therefore,

$$q(x, y) = \frac{4q_0}{\pi}\sum_{n=1,3,\ldots}^{\infty}\frac{1}{n}\sin\frac{n\pi y}{b} \tag{6.60}$$

The last term of (6.51) is replaced by the particular solution

$$w_p = \frac{4b^4 q_0}{D_{22}\pi^5}\sum_{n=1,3,\ldots}^{5}\frac{1}{n^5}\sin\frac{n\pi y}{b} \tag{6.61}$$

Substituting the deflection function given by (6.59) into simply supported boundary conditions given by (6.19) and solving the algebraic equations yield the values of constants as

$$A_n = -\frac{4\lambda_2^2 b^4 q_0}{D_{12}\pi^5 n^5 (\lambda_2^2 - \lambda_1^2)}$$

$$B_n = -\frac{4\lambda_2^2 b^6 q_0}{D_{22}\pi^5 n^5 (\lambda_2^2 - \lambda_1^2)}\left(\frac{1 - \cosh n\pi\lambda_1 R}{\sinh n\pi\lambda_1 R}\right)$$

$$C_n = \frac{4\lambda_1^2 b^4 q_0}{D_{22}\pi^5 n^5 (\lambda_2^2 - \lambda_1^2)} \tag{6.62}$$

$$D_n = \frac{4\lambda_1^2 b^4 q_0}{D_{22}\pi^4 n^5 (\lambda_2^2 - \lambda_1^2)}\left(\frac{1 - \cosh n\pi\lambda_2 R}{\sinh n\pi\lambda_2 R}\right)$$

$$R = \frac{a}{b}$$

n takes only odd values.

6.4 SHEAR DEFORMATION IN LAMINATED PLATES

Reissner and Mindlin's theories [6.8] [6.9] for homogeneous isotropic plates are extended to laminated plates. This extension was originally due to Yang, Norris and Stavsky [6.10] and was simplified by Whitney and Pagano [6.11]

All assumptions made in Section 6.2.1 of CLPT are valid except assumption 6 which is concerned with the neglect of inter-laminar shear strain ε_{xz} and ε_{yz}. Thus, displacements are now of the form

$$\begin{aligned} u &= u_0(x,y) + z\psi_x \\ v &= v_0(x,y) + z\psi_y \\ w &= w \end{aligned} \quad (6.63)$$

u, v, w, are the displacements at a point (x, y) on the mid-plane, and ψ_x and ψ_y are the rotations of the mid-plane normal about y and x axes.

The following results are based on substituting the above expression in strain–displacement relations

$$\begin{aligned} \varepsilon_x &= \varepsilon_x^0 + z\chi_x \\ \varepsilon_y &= \varepsilon_y^0 + z\chi_y \\ \varepsilon_{xy} &= \varepsilon_{xy}^0 + z\chi_{xy} \end{aligned} \quad (6.64)$$

where $\varepsilon_x^0, \varepsilon_y^0, \varepsilon_{xy}^0$ are mid-plane strains. Further,

$$\chi_x = \frac{\partial \psi_x}{\partial x}, \quad \chi_y = \frac{\partial \psi_y}{\partial y}, \quad \chi_{xy} = \frac{\partial \psi_x}{\partial y} + \frac{\partial \psi_y}{\partial x} \quad (6.65)$$

The inter-laminar shear strains are given by the relationships

$$\begin{aligned} \varepsilon_{xz} &= \frac{\partial u}{\partial z} + \frac{\partial w}{\partial x} = \psi_x + \frac{\partial w}{\partial x} \\ \varepsilon_{yz} &= \frac{\partial v}{\partial z} + \frac{\partial w}{\partial y} = \psi_y + \frac{\partial w}{\partial y} \end{aligned} \quad (6.66)$$

Following steps similar to that for obtaining (5.21), we obtain a constitutive relation exactly in the same form, i.e., in the abbreviated notation,

$$\left\{ \begin{array}{c} \{N\} \\ \{M\} \end{array} \right\} = \left[\begin{array}{c|c} [A] & [B] \\ \hline [B] & [D] \end{array} \right] \left\{ \begin{array}{c} \{\varepsilon\} \\ \{\chi\} \end{array} \right\} \quad (6.67)$$

where $[A], [B]$ etc., are already defined in Chapter 5. Applying the definition of shear force resultants, we include constitutive relations involving transverse shear. Following Reissner and Mindlin, we introduce a parameter k_s in the constitutive relation for transverse shear. Thus, we obtain

$$\begin{Bmatrix} Q_x \\ Q_y \end{Bmatrix} = k_s \begin{bmatrix} A_{xx}^z & A_{xy}^z \\ A_{xy}^z & A_{yy}^z \end{bmatrix} \begin{Bmatrix} \varepsilon_{xz} \\ \varepsilon_{yz} \end{Bmatrix} \qquad (6.68)$$

where $\quad A_{ij}^z = \int_{-h/2}^{h/2} \overline{Q}_{ij}^z \cdot dz \qquad (i, j = x, y) \qquad (6.69)$

where $\overline{Q}_{ij}^z$ terms denote anistropic stiffnesses. $\overline{Q}_{ij}^z$ is given by (8.42). The determination of the factor k_s is discussed later.

The approach for deriving governing equations is similar to that followed in the CLT. The equilibrium equations are

$$\frac{\partial N_x}{\partial x} + \frac{\partial N_{xy}}{\partial y} = 0$$

$$\frac{\partial N_{xy}}{\partial x} + \frac{\partial N_y}{\partial y} = 0$$

$$\frac{\partial M_x}{\partial x} + \frac{\partial M_{xy}}{\partial y} - Q_x = 0 \qquad (6.70)$$

$$\frac{\partial M_{xy}}{\partial x} + \frac{\partial M_y}{\partial y} - Q_y = 0$$

$$\frac{\partial Q_x}{\partial x} + \frac{\partial Q_y}{\partial y} + N_x \frac{\partial^2 w}{\partial x^2} + 2 N_{xy} \frac{\partial^2 w}{\partial x \partial y} + N_y \frac{\partial^2 w}{\partial y^2} + q = 0$$

The above equations given by (6.70) are in terms of force and moment resultants. Combining (6.66) to (6.70) results in following governing equations

$$A_{xx} \frac{\partial^2 u_0}{\partial x^2} + 2 A_{xs} \frac{\partial^2 u_0}{\partial x \partial y} + A_{ss} \frac{\partial^2 u_0}{\partial y^2} + A_{xs} \frac{\partial^2 v_0}{\partial x^2} + (A_{xy} + A_{ss}) \frac{\partial^2 v_0}{\partial x \partial y}$$

$$+ A_{ys} \frac{\partial^2 v_0}{\partial y^2} + B_{xx} \frac{\partial^2 \psi_x}{\partial x^2} + 2 B_{xs} \frac{\partial^2 \psi_x}{\partial x \partial y} + B_{ss} \frac{\partial^2 \psi_x}{\partial y^2} + B_{xs} \frac{\partial^2 \psi_y}{\partial x^2}$$

$$+ (B_{xy} + B_{ss}) \frac{\partial^2 \psi_y}{\partial x \partial y} + B_{ys} \frac{\partial^2 \psi_y}{\partial y^2} = 0 \qquad (6.71)$$

$$A_{xs} \frac{\partial^2 u_0}{\partial x^2} + (A_{xy} + A_{ss}) \frac{\partial^2 u_0}{\partial x \partial y} + A_{ys} \frac{\partial^2 u_0}{\partial y^2} + A_{ss} \frac{\partial^2 v_0}{\partial x^2} + 2 A_{ys} \frac{\partial^2 v_0}{\partial x \partial y}$$

$$+ A_{yy} \frac{\partial^2 v_0}{\partial y^2} + B_{xs} \frac{\partial^2 \psi_x}{\partial x^2} + (B_{xy} + B_{ss}) \frac{\partial^2 \psi_x}{\partial x \partial y} + B_{ys} \frac{\partial^2 \psi_x}{\partial y^2} + B_{ss} \frac{\partial^2 \psi_y}{\partial x^2}$$

$$+ 2B_{ys} \frac{\partial^2 \psi_y}{\partial x \partial y} + B_{yy} \frac{\partial^2 \psi_y}{\partial y^2} = 0 \qquad (6.72)$$

$$B_{xx} \frac{\partial^2 u_0}{\partial x^2} + 2B_{xs} \frac{\partial^2 u_0}{\partial x \partial y} + B_{ss} \frac{\partial^2 u_0}{\partial y^2} + B_{xs} \frac{\partial^2 v_0}{\partial x^2} + \left(B_{xy} + B_{ss}\right) \frac{\partial^2 v_0}{\partial x \partial y}$$

$$+ B_{ys} \frac{\partial^2 v_0}{\partial y^2} + D_{xx} \frac{\partial^2 \psi_x}{\partial x^2} + 2D_{xs} \frac{\partial^2 \psi_x}{\partial x \partial y} + D_{ss} \frac{\partial^2 \psi_x}{\partial y^2} + D_{xs} \frac{\partial^2 \psi_y}{\partial x^2}$$

$$+ \left(D_{xy} + D_{ss}\right) \frac{\partial^2 \psi_y}{\partial x \partial y} + D_{ys} \frac{\partial^2 \psi_y}{\partial y^2} - k_s \left[A_{xx}^z \left(\psi_x + \frac{\partial w}{\partial x} \right) + A_{xy}^z \left(\psi_y + \frac{\partial w}{\partial y} \right) \right] = 0 \qquad (6.73)$$

$$B_{xs} \frac{\partial^2 u_0}{\partial x^2} + \left(B_{xy} + B_{ss}\right) \frac{\partial^2 u_0}{\partial x \partial y} + B_{ys} \frac{\partial^2 u_0}{\partial y^2} + B_{ss} \frac{\partial^2 v_0}{\partial x^2} + 2B_{ys} \frac{\partial^2 v_0}{\partial x \partial y}$$

$$+ B_{yy} \frac{\partial^2 v_0}{\partial y^2} + D_{xs} \frac{\partial^2 \psi_x}{\partial x^2} + \left(D_{xy} + D_{ss}\right) \frac{\partial^2 \psi_x}{\partial x \partial y} + D_{ys} \frac{\partial^2 \psi_x}{\partial y^2} + D_{ys} \frac{\partial^2 \psi_y}{\partial x^2}$$

$$+ 2D_{ys} \frac{\partial^2 \psi_y}{\partial x \partial y} + D_{yy} \frac{\partial^2 \psi_y}{\partial y^2} - k_s \left[A_{xy}^z \left(\psi_x + \frac{\partial w}{\partial x} \right) + A_{yy}^z \left(\psi_y + \frac{\partial w}{\partial y} \right) \right] = 0 \qquad (6.74)$$

$$k_s \left[A_{xx}^z \left(\frac{\partial \psi_x}{\partial x} + \frac{\partial^2 w}{\partial x^2} \right) + A_{xy}^z \left(\frac{\partial \psi_x}{\partial y} + \frac{\partial \psi_y}{\partial x} + 2 \frac{\partial^2 w}{\partial x \partial y} \right) + A_{yy}^z \left(\frac{\partial \psi_y}{\partial y} + \frac{\partial^2 w}{\partial y^2} \right) \right]$$

$$+ q + N_x \frac{\partial^2 w}{\partial x^2} + 2N_{xy} \frac{\partial^2 w}{\partial x \partial y} + N_y \frac{\partial^2 w}{\partial y^2} = 0 \qquad (6.75)$$

For symmetric laminates, (6.71) and (6.72) reduce to plane stress equations as follows:

$$A_{xx} \frac{\partial^2 u_0}{\partial x} + 2A_{xs} \frac{\partial^2 u_0}{\partial x \partial y} + A_{ss} \frac{\partial^2 u_0}{\partial y^2} + A_{xs} \frac{\partial^2 v_0}{\partial x^2}$$

$$+ \left(A_{xy} + A_{ss}\right) \frac{\partial^2 v}{\partial x \partial y} + A_{ys} \frac{\partial^2 v}{\partial y^2} = 0 \qquad (6.76)$$

$$A_{xs} \frac{\partial^2 u_0}{\partial x^2} + \left(A_{xy} + A_{ss}\right) \frac{\partial^2 u_0}{\partial x \partial y} + A_{ys} \frac{\partial^2 u_0}{\partial y^2} + A_{ss} \frac{\partial^2 v_0}{\partial x^2} \qquad (6.77)$$

$$+ 2A_{ys} \frac{\partial^2 v_0}{\partial x \partial y} + A_{yy} \frac{\partial^2 v_0}{\partial y^2} = 0$$

For symmetric laminates, (6.72) and (6.73) reduce to

$$D_{xx} \frac{\partial^2 \psi_x}{\partial x^2} + 2D_{xs} \frac{\partial^2 \psi_x}{\partial x \partial y} + D_{ss} \frac{\partial^2 \psi_x}{\partial y^2} + D_{xs} \frac{\partial^2 \psi_y}{\partial x^2} + \left(D_{xy} + D_{ss}\right) \frac{\partial^2 \psi_y}{\partial x \partial y}$$

$$+ D_{ys}\frac{\partial^2 \psi_y}{\partial y^2} - k_s\left[A_{xx}^z\left(\psi_x + \frac{\partial w}{\partial x}\right) + A_{xy}^z\left(\psi_y + \frac{\partial w}{\partial y}\right)\right] = 0 \qquad (6.78)$$

$$D_{xs}\frac{\partial^2 \psi_x}{\partial x^2} + (D_{xy} + D_{ss})\frac{\partial^2 \psi_x}{\partial x\,\partial y} + D_{ys}\frac{\partial^2 \psi_x}{\partial y^2} + D_{ss}\frac{\partial^2 \psi_y}{\partial x^2} + 2D_{ys}\frac{\partial^2 \psi_y}{\partial x\,\partial y}$$

$$+ D_{yy}\frac{\partial^2 \psi_y}{\partial y^2} - k_s\left[A_{xy}^z\left(\psi_x + \frac{\partial w}{\partial x}\right) + A_{yy}^z\left(\psi_y + \frac{\partial^2 \psi_y}{\partial x\,\partial y}\right)\right] = 0 \qquad (6.79)$$

Equation (6.75) remains unchanged. For homogenous isotropic plates, (6.75), (6.78) and (6.79) reduce to Mindlin's theory.

Ply stresses can be obtained from (5.8) combining with (6.45) and (6.46), yields

$$\begin{Bmatrix}\sigma_x \\ \sigma_y \\ \sigma_{xy}\end{Bmatrix}_k = \begin{bmatrix}Q_{xx} & Q_{xy} & Q_{xs} \\ Q_{yx} & Q_{yy} & Q_{ys} \\ Q_{sx} & Q_{sy} & Q_{ss}\end{bmatrix}_k \begin{Bmatrix}\partial u_0/\partial x \\ \partial v_0/\partial y \\ \dfrac{\partial u_0}{\partial y} + \dfrac{\partial v_0}{\partial x}\end{Bmatrix}$$

$$+ z\begin{bmatrix}Q_{xx} & Q_{xy} & Q_{xs} \\ Q_{yx} & Q_{yy} & Q_{ys} \\ Q_{sx} & Q_{sy} & Q_{ss}\end{bmatrix}_k \begin{Bmatrix}\dfrac{\partial \psi_x}{\partial x} \\ \dfrac{\partial \psi_y}{\partial y} \\ \dfrac{\partial \psi_x}{\partial y} + \dfrac{\partial \psi_y}{\partial x}\end{Bmatrix} \qquad (6.80)$$

As the inter-laminar strains ε_{xz} and ε_{yz} are independent of the z-coordinate, then the inter-laminar shear stresses σ_{xy} and σ_{yz} will also be independent of z,

$$\begin{Bmatrix}\sigma_{xz} \\ \sigma_{yz}\end{Bmatrix} = \begin{bmatrix}Q_{xx}^z & Q_{xy}^z \\ Q_{xy}^z & Q_{yy}^z\end{bmatrix}\begin{Bmatrix}\psi_x + \dfrac{\partial w}{\partial x} \\ \psi_y + \dfrac{\partial w}{\partial y}\end{Bmatrix} \qquad (6.81)$$

Equation (6.81) yields shear stresses, which are uniform throughout the thickness of each ply and discontinuous at ply interfaces. Such a distribution is, however, not realistic.

The following are the boundary conditions

(1) Simply supported

$$N_n = N_{ns} = M_n = \psi_s = w = 0 \qquad (6.82)$$

(2) Hinged – free in the normal direction

$$N_n = u_s^0 = M_n = \psi_s = w = 0 \qquad (6.83)$$

(3) Hinged – free in the tangential direction

$$u_n = N_{ns} = M_n = \psi_s = w = 0 \tag{6.84}$$

(4) Clamped

$$u_n = u_s = \psi_n = \psi_s = w = 0 \tag{6.85}$$

(5) Free

$$N_n = N_{ns} = M_n = M_{ns} = Q_n = 0 \tag{6.86}$$

Here, n and s denote coordinates normal and tangential to the plate edge respectively.

6.5 HIGHER ORDER SHEAR DEFORMATION THEORY

6.5.1 Strain–Displacement Relations

Figure 6.6 (a) – (c) illustrate the assumed cross-sectional behaviour adopted in three plate-bending theories. In the most general case, the displacement fields can be expressed as

$$\begin{aligned} U(x,y,z) &= u_0(x,y) + z\psi_x(x,y) + z^2\xi_x(x,y) + z^3\zeta_x(x,y) \\ V(x,y,z) &= v_0(x,y) + z\psi_y(x,y) + z^2\xi_y(x,y) + z^3\zeta_y(x,y) \\ W(x,y,z) &= w(x,y) \end{aligned} \tag{6.87}$$

in which u_0, v_0 and w denote the displacements of a point (x, y) on the mid-plane, ψ_x and ψ_y are rotations of the mid-plane normal about the y and x axes. Functions ξ_x, ξ_y, ζ_x and ζ_y are used in the higher order theory.

It is assumed that U and V have a cubic variation. The functions ξ_x, ξ_y, ζ_x and ζ_y are determined from the stress conditions of the plate. It is obtained from the condition that the transverse shear stress σ_{xz} and σ_{yz} vanish at the upper and lower surfaces of the plate

$$\begin{aligned} \varepsilon_{yz} &= \frac{\partial U}{\partial z} + \frac{\partial W}{\partial x} = \psi_x + 2z\xi_x + 3z^2\zeta_x + w_{,x} \\ \varepsilon_{xz} &= \frac{\partial V}{\partial z} + \frac{\partial W}{\partial y} = \psi_y + 2z\xi_y + 3z^2\zeta_y + w_{,y} \end{aligned} \tag{6.88}$$

For the condition that $\sigma_{yz}\left(x, y, \pm\frac{h}{2}\right)$ and $\sigma_{xz}\left(x, y, \pm\frac{h}{2}\right)$ are zero, it is possible to write

$$\begin{aligned} \xi_x &= 0, \quad \xi_y = 0 \\ \zeta_x &= -4(w_{,x} + \psi_x)/3h^2, \quad \zeta_y = -4(w_{,y} + \psi_y)/3h^2 \end{aligned} \tag{6.89}$$

The displacement fields of (6.87) are

$$U = u_0 + z\left[\psi_x - \frac{4}{3}\left(\frac{z}{h}\right)^2(\psi_x + w_{,x})\right]$$

$$V = v_0 + z\left[\psi_y - \frac{4}{3}\left(\frac{z}{h}\right)^2\left(\psi_y + w_{,y}\right)\right]$$
$$W = w$$
(6.90)

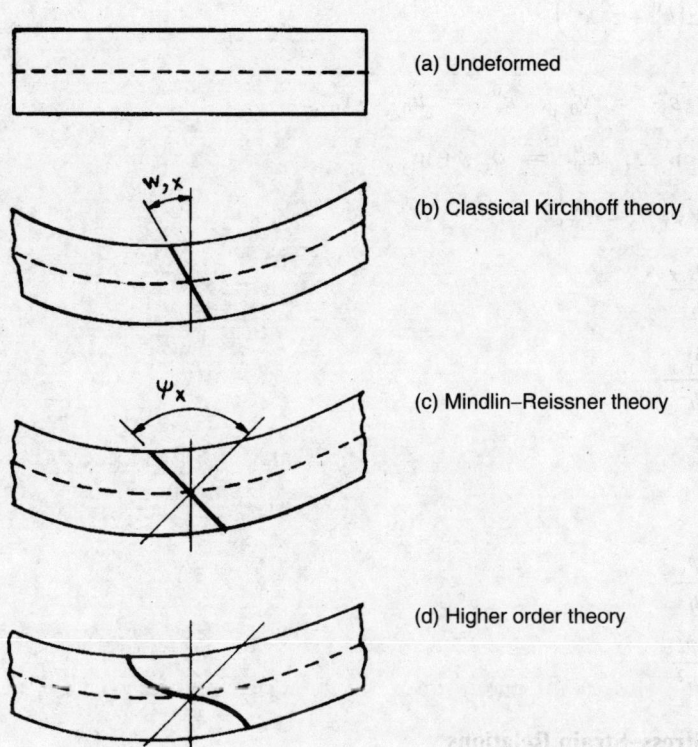

Fig. 6.6 Assumed cross-sectional behaviour of various plate theories

Here,
$$\psi_x = \phi_x - w_{,x} \text{ and } \psi_y = \phi_y - w_{,y} \tag{6.91}$$

Therefore, (6.91) can be written as
$$U = u_0 + z\left[\phi_x - w_{,x} - \frac{4}{3}\left(\frac{z}{h}\right)^2 \phi_x\right]$$
$$V = v_0 + z\left[\phi_y - w_{,y} - \frac{4}{3}\left(\frac{z}{h}\right)^2 \phi_y\right]$$
$$W = w$$
(6.92)

Strains associated with displacements in (6.92) are
$$\varepsilon_x = \varepsilon_x^0 + z\left(\kappa_x^0 + z^2 \kappa_x^2\right)$$
$$\varepsilon_y = \varepsilon_y^0 + z\left(\kappa_y^0 + z^2 \kappa_y^2\right)$$

$$\varepsilon_z = 0$$
$$\varepsilon_{xz} = \varepsilon_{xz}^0 + z^2 \kappa_{xz}^2$$
$$\varepsilon_{yz} = \varepsilon_{yz}^0 + z^2 \kappa_{yz}^2 \tag{6.93}$$
$$\varepsilon_s = \varepsilon_s^0 + z\left(k_s^0 + z^2 \kappa_s^2\right)$$

where

$$\varepsilon_x^0 = u_{0,x}, \quad \varepsilon_y^0 = v_{0,y}, \quad \varepsilon_s^0 = u_{0,y} + v_{0,x}$$
$$\kappa_x^0 = \phi_{x,x} - w_{,xx}, \quad \kappa_y^0 = \phi_{y,y} + w_{,yy}$$
$$\kappa_s = \phi_{x,y} + \phi_{y,x} - 2w_{,xy}$$
$$\kappa_x^2 = -\frac{4\phi_{x,x}}{3h^2}$$
$$\kappa_y^2 = -\frac{4\phi_{y,y}}{3h^2} \tag{6.94}$$
$$\varepsilon_{xy}^0 = \phi_y$$
$$\varepsilon_{yz}^0 = \phi_x$$
$$\kappa_{xz}^2 = -4\frac{\phi_y}{h^2}$$
$$\kappa_{yz}^2 = -4\frac{\phi_x}{h^2}$$

6.5.2 Generalised Stress–Strain Relations

For a plate of constant thickness, which is composed of thin layers of orthotropic material, the constitutive equations are derived as follows.

Stresses in the principal material directions are given by

$$\begin{Bmatrix} \sigma_1 \\ \sigma_2 \\ \sigma_6 \end{Bmatrix} = \begin{bmatrix} Q_{11} & Q_{12} & 0 \\ Q_{12} & Q_{22} & 0 \\ 0 & 0 & Q_{66} \end{bmatrix} \begin{Bmatrix} \varepsilon_1 \\ \varepsilon_2 \\ \varepsilon_6 \end{Bmatrix} \tag{6.95}$$

$$\begin{Bmatrix} \sigma_4 \\ \sigma_5 \end{Bmatrix} = \begin{bmatrix} Q_{44} & 0 \\ 0 & Q_{55} \end{bmatrix} \begin{Bmatrix} \varepsilon_4 \\ \varepsilon_5 \end{Bmatrix} \tag{6.96}$$

Upon transformation, the lamina constitutive equations can be expressed in terms of stresses and strains in plate (laminate) coordinates as

$$\begin{Bmatrix} \sigma_x \\ \sigma_y \\ \sigma_s \end{Bmatrix} = \begin{bmatrix} Q_{xx} & Q_{xy} & Q_{xs} \\ Q_{xy} & Q_{yy} & Q_{ys} \\ Q_{xs} & Q_{ys} & Q_{ss} \end{bmatrix} \begin{Bmatrix} \varepsilon_x \\ \varepsilon_y \\ \varepsilon_s \end{Bmatrix} \tag{6.97}$$

$$\begin{Bmatrix} \sigma_{xz} \\ \sigma_{yz} \end{Bmatrix} = \begin{bmatrix} Q^z_{xx} & Q^z_{xy} \\ Q^z_{xy} & Q^z_{yy} \end{bmatrix} \begin{Bmatrix} \varepsilon_{xz} \\ \varepsilon_{yz} \end{Bmatrix} \tag{6.98}$$

The stress resultants are defined as

$$(N_i, M_i, P_i) = \int_{-h/2}^{h/2} \sigma_i (1, z, z^2) \, dz \qquad i = x, y, s$$

$$(Q_x, R_x) = \int_{-h/2}^{h/2} \sigma_{xz} (1, z^2) \, dz \tag{6.99}$$

$$(Q_y, R_y) = \int_{-h/2}^{h/2} \sigma_{yz} (1, z^2) \, dz$$

The relationship between the stress resultants and the corresponding strain terms can then be written as

$$\begin{Bmatrix} \{N\} \\ \{M\} \\ \{P\} \end{Bmatrix} = \begin{bmatrix} [A] & [B] & [E] \\ [B] & [D] & [F] \\ [E] & [F] & [H] \end{bmatrix} \begin{Bmatrix} \{\varepsilon^0\} \\ \{\kappa^0\} \\ \{\kappa^2\} \end{Bmatrix} \tag{6.100}$$

in which

$$\{N\}^T = \{N_x \ N_y \ N_s\}; \quad \{M\}^T = \{M_x \ M_y \ M_s\}$$

$$\{P\}^T = \{P_x \ P_y \ P_s\}; \quad \{\varepsilon^0\}^T = \{\varepsilon_x \ \varepsilon_y \ \varepsilon_s\} \tag{6.101}$$

$$\{\kappa^0\}^T = \{\kappa^0_x \ \kappa^0_y \ \kappa^0_s\}; \quad \{\kappa^2\}^T = \{\kappa^2_x \ \kappa^2_y \ \kappa^2_s\}$$

$$[A] = \begin{bmatrix} A_{xx} & A_{xy} & A_{xs} \\ A_{yx} & A_{yy} & A_{ys} \\ A_{sx} & A_{sy} & A_{ss} \end{bmatrix}; \quad [B] = \begin{bmatrix} B_{xx} & B_{xy} & B_{xs} \\ B_{yx} & B_{yy} & B_{ys} \\ B_{sx} & B_{sy} & B_{ss} \end{bmatrix}$$

$$[D] = \begin{bmatrix} D_{xx} & D_{xy} & D_{xs} \\ D_{yx} & D_{yy} & D_{ys} \\ D_{sx} & D_{sy} & D_{ss} \end{bmatrix}; \quad [E] = \begin{bmatrix} E_{xx} & E_{xy} & E_{xs} \\ E_{yx} & E_{yy} & E_{ys} \\ E_{sx} & E_{sy} & E_{ss} \end{bmatrix} \tag{6.102}$$

$$[F] = \begin{bmatrix} F_{xx} & F_{xy} & F_{xs} \\ F_{yx} & F_{yy} & F_{ys} \\ F_{sx} & D_{sy} & D_{ss} \end{bmatrix}; \quad [H] = \begin{bmatrix} H_{xx} & H_{xy} & H_{xs} \\ H_{yx} & H_{yy} & H_{ys} \\ H_{sx} & H_{sy} & H_{ss} \end{bmatrix}$$

Furthermore,

$$\begin{Bmatrix} \{Q\} \\ \{R\} \end{Bmatrix} = \begin{bmatrix} [A^z] & [D^z] \\ [D^z] & [F^z] \end{bmatrix} \begin{Bmatrix} \{\varepsilon^0_s\} \\ \{k^2_s\} \end{Bmatrix} \tag{6.103}$$

in which

$$\{Q\}^T = \{Q_x \quad Q_y\}, \quad \{R\}^T = \{R_x \quad R_y\}, \quad \{\varepsilon_s^0\}^T = \{\varepsilon_{xz}^0 \quad \varepsilon_{yz}^0\} \tag{6.104}$$

and $\{k_s^2\} = \{k_{xz}^2 \quad k_{yz}^2\}$ and also

$$[A^z] = \begin{bmatrix} A_{xx}^z & A_{xy}^z \\ A_{xy}^z & A_{yy}^z \end{bmatrix}; \quad [D] = \begin{bmatrix} D_{xx}^z & D_{xy}^z \\ D_{xy}^z & D_{yy}^z \end{bmatrix}; \quad [F] = \begin{bmatrix} F_{xx}^z & F_{xy}^z \\ F_{xy}^z & F_{yy}^z \end{bmatrix} \tag{6.105}$$

where A_{ij}, B_{ij} etc are. the plate stiffnesses defined by

$$\left(A_{ij}, B_{ij}, D_{ij}, E_{ij}, F_{ij}, H_{ij}\right) = \int_{-h/2}^{h/2} Q_{ij}\left(1, z, z^2, z^3, z^4, z^6\right) dz, \quad i, j = x, y, s$$

$$A_{ij}^z, D_{ij}^z, F_{ij}^z = \int_{-h/2}^{h/2} Q_{ij}^z \left(j, z^2, z^4\right) dz, \quad i, j = x, y \tag{6.106}$$

6.6 DETERMINATION OF THE K_S – PARAMETER

From the first of (6.1), we find that for a homogeneous plate

$$\sigma_{xz} = -\frac{\partial}{\partial x} \int_{-h/2}^{z} \sigma_x \, dz - \frac{\partial}{\partial y} \int_{-h/2}^{z} \sigma_{xy} \, dz \tag{6.107}$$

The constitutive relations in (6.80) reduce to the form when the bending of a homogenous, orthotropic plate is considered

$$\begin{Bmatrix} \sigma_x \\ \sigma_y \\ \sigma_{xy} \end{Bmatrix}_k = z \begin{bmatrix} Q_{xx} & Q_{xy} & 0 \\ Q_{xy} & Q_{yy} & 0 \\ 0 & 0 & Q_{ss} \end{bmatrix}_k \begin{Bmatrix} \chi_x \\ \chi_y \\ \chi_{xy} \end{Bmatrix} \tag{6.108}$$

Equation (6.108) is substituted into (6.107) and the integration is then carried out as a result of which we obtain

$$\sigma_{xz} = \frac{h^2}{8} \left[\frac{\partial}{\partial x} \left(Q_{xx} \chi_x + Q_{xy} \chi_y\right) + \frac{\partial}{\partial y} \left(Q_{ss} \chi_{xy}\right) \right] \left(1 - \frac{4z^2}{h^2}\right) \tag{6.109}$$

For a homogeneous orthotropic plate, the following relations are developed based on constitutive relations (6.67)

$$Q_{xx} \chi_x + Q_{xy} \chi_y = \frac{12}{h^3} M_x$$

$$Q_{ss} \chi_{xy} = \frac{12}{h^3} M_{xy} \tag{6.110}$$

Equation (6.85) can now be written as

$$\sigma_{xz} = \frac{3}{2h} \left(\frac{\partial M_x}{\partial x} + \frac{\partial M_{xy}}{\partial y} \right) \left(1 - \frac{4z^2}{h^2}\right) \tag{6.111}$$

Combining (6.11) with (6.111), we obtain the transverse shear stress distribution as

$$\sigma_{xz} = \frac{3}{2h}\left(1 - 4\frac{z^2}{h^2}\right)Q_x \tag{6.112}$$

After following a similar approach as above, we obtain

$$\sigma_{yz} = \frac{3}{2h}\left(1 - 4\frac{z^2}{h^2}\right)Q_y \tag{6.113}$$

There are two existing approaches for determining k_s.

For a homogeneous orthotropic plate, the shear stress–strain relations take the form:

$$\begin{aligned}\sigma_{xz} &= A^z_{xx}\,\varepsilon_{xz} \\ \sigma_{yz} &= A^z_{yy}\,\varepsilon_{yz}\end{aligned} \tag{6.114}$$

Substituting the plate constitutive relations (6.68) into (6.112) and (6.113), we obtain the shear stresses at the plate centre

$$\sigma_{xz} = \frac{3Q_x}{2h}, \quad \sigma_{yz} = \frac{3Q_y}{2h} \tag{6.115}$$

k_s is given by

$$k_s = \frac{2}{3} \tag{6.116}$$

The other method of determination of k_s is based on Reissner's plate theory.

The total strain energy U due to transverse shear only is given by

$$U = \frac{1}{2}\int_{-h/2}^{h/2}\int_0^b\int_0^a \left(\sigma_{xz}\,\varepsilon^0_{xz} + \sigma_{yz}\,\varepsilon^0_{yz}\right)dx\,dy\,dz \tag{6.117}$$

On substitution of (6.68) into (6.117), we obtain

$$U = \frac{1}{2k_s}\int_0^b\int_0^a\left[\int_{-h/2}^{h/2}\left(\frac{Q_x}{A^z_{xx}}\sigma_{xz} + \frac{Q_y}{A^z_{yy}}\sigma_{yz}\right)dz\right]dx\,dy \tag{6.118}$$

or

$$U = \frac{1}{2k_s}\int_0^b\int_0^a\left(\frac{Q_x^2}{A^z_{xx}} + \frac{Q_y^2}{A^z_{yy}}\right)dx\,dy \tag{6.119}$$

Equation (6.117) can be written for a homogeneous orthotropic material as

$$U = \frac{h}{2}\int_0^b\int_0^a\int_{-h/2}^{h/2}\left(\frac{\sigma_{xz}^2}{A^z_{xx}} + \frac{\sigma_{yz}^2}{A^z_{yy}}\right)dz\,dx\,dy \tag{6.120}$$

On substitution of the stress distributions given by (6.112) and (6.113) into (6.120) and integrating with respect to z, we obtain

$$U = \frac{3}{5}\int_0^b\int_0^a\left(\frac{Q_x^2}{A^z_{xx}} + \frac{Q_y^2}{A^z_{yy}}\right)dx\,dy \tag{6.121}$$

Now, if k_s is chosen such that (6.119) and (6.121) are identical, then

$$k_s = \frac{5}{6} \qquad (6.122)$$

The procedure for determining k_s for laminated plates is more complicated than the one presented here, k_s depends to a great content on ply properties and stacking sequences. It is difficult to obtain the solution in a closed form. Investigation on the laminated plates has been reported [6.12]. For ε_{xz} and ε_{yz}, two different shear factors have been obtained.

6.7 FREE VIBRATION OF LAMINATED PLATES (BASED ON CLPT)

The equations of motion for a general laminate is as follows

$$A_{xx}\frac{\partial^2 u_0}{\partial x^2} + 2A_{xs}\frac{\partial^2 u_0}{\partial x \partial y} + A_{ss}\frac{\partial^2 u_0}{\partial y^2} + A_{xs}\frac{\partial^2 v_0}{\partial x^2} + (A_{xy}+A_{ss})\frac{\partial^2 v_0}{\partial x \partial y}$$
$$+ A_{ys}\frac{\partial^2 v_0}{\partial y^2} - B_{xx}\frac{\partial^3 w}{\partial x^3} - 3B_{xs}\frac{\partial^3 w}{\partial x^2 \partial y} - (B_{xy}+2B_{ss})\frac{\partial^3 w}{\partial x \partial y^2} - B_{ys}\frac{\partial^3 w}{\partial y^3} = \rho \frac{\partial^2 u_0}{\partial t^2} \qquad (6.123a)$$

$$A_{xs}\frac{\partial^2 u_0}{\partial x^2} + (A_{xy}+A_{ss})\frac{\partial^2 u_0}{\partial x \partial y} + A_{ys}\frac{\partial^2 u_0}{\partial y^2} + A_{ss}\frac{\partial^2 v_0}{\partial x^2} + 2A_{ys}\frac{\partial^2 v_0}{\partial x \partial y}$$
$$+ A_{yy}\frac{\partial^2 v}{\partial y^2} - B_{xs}\frac{\partial^3 w}{\partial x^3} - (B_{xy}+2B_{ss})\frac{\partial^3 w}{\partial x^2 \partial y} - 3B_{ys}\frac{\partial^3 w}{\partial x \partial y^2} - B_{yy}\frac{\partial^3 w}{\partial y^3} = \rho \frac{\partial^2 v_0}{\partial t^2} \qquad (6.123b)$$

$$D_{xx}\frac{\partial^4 w}{\partial x^4} + 4D_{xs}\frac{\partial^4 w}{\partial x^3 \partial y} + 2(D_{xy}+2D_{ss})\frac{\partial^4 w}{\partial x^2 \partial y^2} + 4D_{ys}\frac{\partial^4 w}{\partial x \partial y^3} + D_{yy}\frac{\partial^4 w}{\partial y^4}$$
$$- B_{xx}\frac{\partial^3 u_0}{\partial x^3} - 3B_{xs}\frac{\partial^3 u_0}{\partial x^2 \partial y} - (B_{xy}+2B_{ss})\frac{\partial^3 u_0}{\partial x \partial y^2} - B_{ys}\frac{\partial^3 u_0}{\partial y^3} - B_{xs}\frac{\partial^3 v_0}{\partial x^3} \qquad (6.123c)$$
$$- (B_{xy}+2B_{ss})\frac{\partial^3 v_0}{\partial x^2 \partial y} - 3B_{ys}\frac{\partial^3 v_0}{\partial x \partial y^2} - B_{yy}\frac{\partial^3 v_0}{\partial y^3} + \rho \frac{\partial^2 w}{\partial t^2} = 0$$

Let us limit our study to unsymmetric angle-ply and cross-ply laminates, that is, all terms with subscript 'xs' and 'ys' are zero. In order to determine natural frequencies, we consider harmonic solutions of the form

$$\begin{aligned} u_0 &= U(x,y)\,e^{i\omega t} \\ v_0 &= V(x,y)\,e^{i\omega t} \\ w &= W(x,y)\,e^{i\omega t} \end{aligned} \qquad (6.124)$$

Substituting (6.124) with (6.123) and neglecting inplane inertia results for a cross-ply laminate

$$A_{xx}\frac{\partial^2 U}{\partial x^2} + A_{ss}\frac{\partial^2 U}{\partial y^2} + (A_{xy}+A_{ss})\frac{\partial^2 V}{\partial x \partial y} - B_{xx}\frac{\partial^3 W}{\partial x^3} = 0$$

$$(A_{xy}+A_{ss})\frac{\partial^2 U}{\partial x \partial y} + A_{ss}\frac{\partial^2 V}{\partial y^2} + A_{xx}\frac{\partial^2 V}{\partial y^2} + B_{xx}\frac{\partial^3 W}{\partial y^3} = 0$$

$$D_{xx}\left(\frac{\partial^4 W}{\partial x^4} + \frac{\partial^4 W}{\partial y^4}\right) + 2\left(D_{xy} + 2D_{ss}\right)\frac{\partial^4 W}{\partial x^2 \partial y^2} - B_{xx}\left(\frac{\partial^3 U}{\partial x^3} - \frac{\partial^3 V}{\partial y^3}\right) - \rho\omega^2 w = 0 \quad (6.125)$$

For hinged supports, free in the normal direction, boundary conditions are
At $x = 0, a$

$$W = 0, \qquad B_{xx}\frac{\partial U}{\partial x} - D_{xx}\frac{\partial^2 W}{\partial x^2} - D_{xy}\frac{\partial^2 W}{\partial y^2} = 0$$

$$V = 0, \qquad A_{xx}\frac{\partial U}{\partial x} + A_{xy}\frac{\partial V}{\partial y} - B_{xx}\frac{\partial^2 W}{\partial x^2} = 0 \quad (6.126)$$

For $y = 0, b$

$$W = 0, \qquad -B_{xx}\frac{\partial V}{\partial y} - D_{xy}\frac{\partial^2 W}{\partial x^2} - D_{xx}\frac{\partial^2 W}{\partial y^2} = 0$$

$$U = 0, \qquad A_{xy}\frac{\partial U}{\partial x} + A_{xx}\frac{\partial W}{\partial y} - B_{xx}\frac{\partial^2 W}{\partial x^2} = 0 \quad (6.127)$$

Equations (6.116) to (6.118) are satisfied by the displacements.
The following solutions are assumed which satisfy boundary conditions:

$$U = U_{mn} \cos\frac{m\pi x}{a} \sin\frac{n\pi y}{b}$$

$$V = V_{mn} \sin\frac{m\pi x}{a} \cos\frac{n\pi y}{b} \quad (6.128)$$

$$W = W_{mn} \sin\frac{m\pi x}{a} \sin\frac{n\pi y}{b}$$

The homogeneous algebraic equations resulting from combining (6.125) and (6.128) are

$$\begin{bmatrix} A_{mn} & B_{mn} & C_{mn} \\ B_{mn} & D_{mn} & E_{mn} \\ C_{mn} & E_{mn} & F_{mn} - \lambda_{mn} \end{bmatrix} \begin{Bmatrix} A \\ B \\ C \end{Bmatrix} = \begin{Bmatrix} 0 \\ 0 \\ 0 \end{Bmatrix} \quad (6.129)$$

where

$$A_{mn} = A_{xx} m^2 + A_{ss} n^2 R^2$$

$$B_{mn} = (A_{xy} + A_{ss})mnR$$

$$C_{mn} = -B_{xx}\frac{m^3 \pi}{Rb}$$

$$D_{mn} = A_{ss} m^2 + A_{xx} n^2 R^2$$

$$E_{mn} = B_{xx}\frac{n^3 R^2 \pi}{R}$$

$$F_{mn} = \frac{\pi^2}{R^2 b^2}\left[D_{xx}\left(m^4 + n^4 R^4\right) + 2\left(D_{xy} + 2D_{ss} m^2 n^2 R^2\right)\right]$$

$$\lambda_{mn} = \rho \frac{\omega_{mn}^2 R^2 b^2}{\pi^2}$$

$$R = \frac{a}{b}$$

In order to obtain a non-trivial solution to (6.129), the determinant of the coefficient matrix must vanish. Thus

$$\begin{vmatrix} A_{mn} & B_{mn} & C_{mn} \\ B_{mn} & D_{mn} & E_{mn} \\ C_{mn} & E_{mn} & F_{mn} - \lambda_{mn} \end{vmatrix} = 0 \qquad (6.130)$$

Solving (6.130) for λ, yields

$$\omega_{mn}^2 = \frac{\pi^4}{\rho R^4 b^4}\left[D_{xx}\left(m^4 + n^4 R^4\right) + 2\left(D_{xy} + 2D_{ss}\right)m^2 n^2 R^2 \right.$$

$$\left. - \frac{B_{xx}^2}{J_3}\left(m^4 J_1 + n^4 R^4 J_2\right)\right] \qquad (6.131)$$

where

$$J_1 = A_{ss} m^4 + A_{xx} m^2 n^2 R^2 + \left(A_{xy} + A_{ss}\right) n^4 R^4$$

$$J_2 = \left(A_{xy} + A_{ss}\right)m^4 + A_{xx} m^2 n^2 R^2 + A_{ss} n^4 R^4$$

$$J_3 = \left(A_{xx} m^2 + A_{ss} n^2 R^2\right)\left(A_{ss} m^2 + A_{xx} n^2 R^2\right) - \left(A_{xy} + A_{ss}\right)^2 m^2 n^2 R^2$$

The variation of the fundamental frequency for the cross-ply laminate with aspect ratio is shown in Fig. 6.7 for $E_1/E_2 = 40$, $E_1/E_2 = 0.5$, $\nu_{12} = 0.25$. It can be seen that the effect of coupling is to reduce the fundamental frequency.

Coupling reduces the fundamental vibration frequency. The fundamental frequency of orthotropic laminates always occurs for $m = n = 1$.

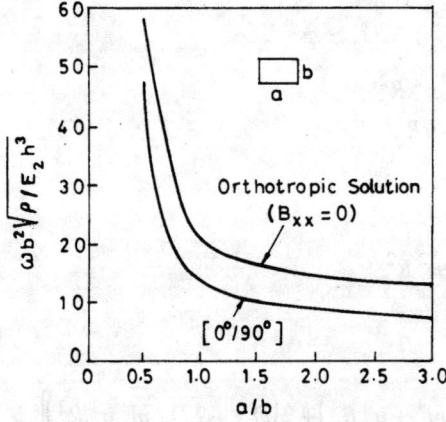

Fig. 6.7 Fundamental vibration frequency for rectangular cross-ply plate

6.7.1 Free Vibration of Unsymmetrical Angle-Ply Laminated Plate (Based on CLPT)

Neglecting the inplane inertia, the equations of motion of the angle-ply laminate is given by

$$A_{xx}\frac{\partial^2 U}{\partial x^2} + A_{ss}\frac{\partial^2 U}{\partial y^2} + \left(A_{xy}+A_{ss}\right)\frac{\partial^2 V}{\partial x \partial y} - 3B_{xs}\frac{\partial^3 W}{\partial x^2 \partial y} - B_{ys}\frac{\partial^3 W}{\partial y^3} = 0$$

$$\left(A_{xy}+A_{ss}\right)\frac{\partial^2 U}{\partial x \partial y} + A_{ss}\frac{\partial^2 V}{\partial x^2} + A_{yy}\frac{\partial^2 W}{\partial y^2} - B_{xs}\frac{\partial^3 W}{\partial x^3} - 3B_{ys}\frac{\partial^3 W}{\partial x \partial y^2} = 0$$

$$D_{xx}\frac{\partial^4 W}{\partial x^4} + 2\left(D_{xy}+2D_{ss}\right)\frac{\partial^4 W}{\partial x^2 \partial y^2} + D_{yy}\frac{\partial^4 W}{\partial y^4}$$

$$- B_{xs}\left(3\frac{\partial^3 U}{\partial x^2 \partial y} + \frac{\partial^3 V}{\partial x^3}\right) - B_{ys}\left(\frac{\partial^4 U}{\partial y^3} + 3\frac{\partial^3 V}{\partial x \partial y^2}\right) - \rho\omega^2 W = 0 \qquad (6.132)$$

For hinged supports, free in the tangential direction, the boundary conditions are
For $x = 0, a$:

$$W = 0, \qquad B_{xs}\left(\frac{\partial U}{\partial y}+\frac{\partial V}{\partial x}\right) - D_{xx}\frac{\partial^2 W}{\partial x^2} - D_{xy}\frac{\partial^2 W}{\partial y^2} = 0$$

$$U = 0, \qquad A_{ss}\left(\frac{\partial U}{\partial y}+\frac{\partial V}{\partial x}\right) - B_{xs}\frac{\partial^2 W}{\partial x^2} - B_{ys}\frac{\partial^2 W}{\partial y^2} = 0 \qquad (6.133)$$

For $y = 0, b$:

$$W = 0, \qquad B_{ys}\left(\frac{\partial U}{\partial y}+\frac{\partial V}{\partial x}\right) - D_{xy}\frac{\partial^2 W}{\partial x^2} - D_{yy}\frac{\partial^2 W}{\partial y^2} = 0$$

$$V = 0, \qquad A_{ss}\left(\frac{\partial U}{\partial y}+\frac{\partial V}{\partial x}\right) - B_{xs}\frac{\partial^2 W}{\partial x^2} - B_{ys}\frac{\partial^2 W}{\partial y^2} = 0 \qquad (6.134)$$

Equations (6.133) and (6.134) are satisfied by assuming the following solution

$$U = U_{mn}\sin\frac{m\pi x}{a}\cos\frac{n\pi y}{b}$$

$$V = V_{mn}\cos\frac{m\pi x}{a}\sin\frac{n\pi y}{b} \qquad (6.135)$$

$$W = W_{mn}\sin\frac{m\pi x}{a}\sin\frac{n\pi y}{b}$$

Combining (6.134) and (6.132) results in (6.130) with C_{mn} and E_{mn} replaced by C'_{mn} and E'_{mn} respectively.

$$C'_{mn} = -\frac{n\pi}{b}\left(3B_{xs}m^2 + B_{ys}n^2 R^2\right)$$

$$E'_{mn} = -\frac{m\pi}{Rb}\left(B_{xs}m^2 + 3B_{ys}n^2 R^2\right) \qquad (6.136)$$

Following steps as for cross-ply laminates, the frequency equation is obtained.

$$\omega_{mn}^2 = \frac{\pi^4}{\rho R^4 b^4}\left\{D_{xx} m^4 + 2(D_{xy} + 2D_{sx})m^2 n^2 R^2 + D_{yy} n^4 R^4 \right. $$
$$\left. - \frac{1}{J_6}\left[m(B_{xs} m^2 + 3B_{ys} R^2 n^2)J_4 + nR(3B_{xs} m^2 + B_{ys} n^2 R^2)J_5\right]\right\} \tag{6.137}$$

where

$$J_4 = \left(A_{xx} m^2 + A_{ss} n^2 R^2\right)\left(B_{xs} m^2 + 3B_{ys} R^2\right) - n^2 R^2 \left(A_{xy} + A_{ss}\right)$$
$$\left(3B_{xs} m^2 + A_{ys} n^2 R^2\right)$$
$$J_5 = \left(A_{ss} m^2 + A_{yy} n^2 R^2\right)\left(3B_{xs} m^2 + B_{ys} n^2 R^2\right) - n^2 R^2 \left(A_{yx} + A_{ss}\right) \tag{6.138}$$
$$\left(B_{xs} m^2 + 3B_{ys} n^2 R^2\right)$$
$$J_6 = \left(A_{xx} m^2 + A_{ss} n^2 R^2\right)\left(A_{ss} m^2 + A_{yy} n^2 R^2\right) - \left(A_{xy} + A_{ss}\right)^2 m^2 n^2 R^2$$

When coupling is neglected $(B_{ij} = 0)$, (6.131) and (6.137) become

$$\omega_{mn}^2 = \frac{\pi^4}{\rho R^4 b^4}\left[D_{xx} m^4 + 2(D_{xy} + 2D_{ss})m^2 n^2 R^2 + D_{yy} n^4 R^4\right] \tag{6.139}$$

Equation (6.139) is the frequency equation for the flexural vibration of a simply supported homogeneous orthotropic plate.

6.8 STABILITY ANALYSIS OF A RECTANGULAR ANGLE-PLY PLATE UNDER UNIFORM BIAXIAL COMPRESSION (BASED ON CLPT)

An angle-ply laminate is considered. It is assumed to have hinged boundary conditions. It is further assumed that the hinged supports are such that uniform displacements along two adjacent boundary conditions are admissible. Thus

For $x = 0$: $\quad u_0 = 0, \quad N_{xy} = 0$ (6.140)

For $x = a$: $\quad u_0 = C_1 = $ constant, $N_{xy} = 0$ (6.141)

For $y = 0$: $\quad v_0 = 0, \quad N_{xy} = 0$ (6.142)

For $y = b$: $\quad v_0 = C_2 = $ constant, $N_{xy} = 0$ (6.143)

For $x = 0, a$: $\quad w = 0, \quad M_x = D_{xs}\left(\dfrac{\partial u_0}{\partial y} + \dfrac{\partial v_0}{\partial x}\right) - D_{xx}\dfrac{\partial^2 w}{\partial x^2} - D_{xy}\dfrac{\partial^2 w}{\partial y^2} = 0$ (6.144)

For $y = 0, b$: $\quad w = 0, \quad M_y = B_{ys}\left(\dfrac{\partial u_0}{\partial y} + \dfrac{\partial v_0}{\partial x}\right) - D_{xy}\dfrac{\partial^2 w}{\partial x^2} - D_{yy}\dfrac{\partial^2 w}{\partial y^2} = 0$ (6.145)

The initial displacement field

$$u_0 = \frac{C_1}{a}x, \quad v_0 = \frac{C_2}{b}y, \quad w = 0 \tag{6.146}$$

satisfies all boundary conditions. The constants C_1 and C_2 are determined such that

$$N_x = -N_0 = \text{constant}, \quad N_y = -kN_0, \quad N_0 > 0 \tag{6.147}$$

where k is constant.

Substituting (6.146) and (6.147) into the expressions of force and moment resultants given below, yields

$$N_x = A_{xx}\frac{\partial u_0}{\partial x} + A_{xs}\left(\frac{\partial u_0}{\partial y} + \frac{\partial v_0}{\partial x}\right) + A_{xy}\frac{\partial v_0}{\partial y}$$
$$- B_{xx}\frac{\partial^2 w}{\partial x^2} - 2B_{xs}\frac{\partial^2 x}{\partial x\,\partial y} - B_{xy}\frac{\partial^2 w}{\partial y^2} \tag{6.148}$$

$$N_y = A_{xy}\frac{\partial u_0}{\partial x} + A_{ys}\left(\frac{\partial u_0}{\partial y} + \frac{\partial v_0}{\partial x}\right) + A_{yy}\frac{\partial v_0}{\partial y}$$
$$- B_{xy}\frac{\partial^2 w}{\partial x^2} - 2B_{ys}\frac{\partial^2 x}{\partial x\,\partial y} - B_{yy}\frac{\partial^2 w}{\partial y^2} \tag{6.149}$$

resulting in

$$C_1 = \frac{(A_{xy}k - A_{yy})N_0}{A_{xx}A_{yy} - A_{xy}^2}, \quad C_2 = \frac{(A_{xy} - kA_{xx})N_0}{A_{xx}A_{yy} - A_{xy}^2} \tag{6.150}$$

For the angle-ply laminate, differential equations of equilibrium are

$$A_{xx}\frac{\partial^2 u_0}{\partial x^2} + A_{ss}\frac{\partial^2 u_0}{\partial y^2} + (A_{xy} + A_{ss})\frac{\partial^2 v_0}{\partial x\,\partial y} - 3B_{xs}\frac{\partial^3 w}{\partial x^2\,\partial y} - B_{ys}\frac{\partial^3 w}{\partial y^3} = 0 \tag{6.151}$$

$$(A_{xy} + A_{ss})\frac{\partial^2 w}{\partial x^2} + A_{ss}\frac{\partial^2 v_0}{\partial x^2} + A_{yy}\frac{\partial^2 v_0}{\partial y^2} - B_{xs}\frac{\partial^3 w}{\partial x^3} - 3B_{ys}\frac{\partial^3 w}{\partial x\,\partial y^2} = 0 \tag{6.152}$$

$$D_{xx}\frac{\partial^4 w}{\partial x^4} + 2(D_{xy} + 2D_{ss})\frac{\partial^4 w}{\partial x^2\,\partial y^2} + D_{yy}\frac{\partial^4 w}{\partial y^4} - B_{xs}\left(3\frac{\partial^3 u_0}{\partial x^2\,\partial y} + \frac{\partial^3 v_0}{\partial x^3}\right)$$
$$- B_{ys}\left(\frac{\partial^3 u_0}{\partial y^3} + 3\frac{\partial^3 v_0}{\partial x\,\partial y^2}\right) + N_0\left(\frac{\partial^2 w}{\partial x^2} + k\frac{\partial^2 w}{\partial y^2}\right) = 0 \tag{6.153}$$

The boundary conditions are

For $x = 0, a$:

$$w = M_x = B_{xs}\left(\frac{\partial v_0}{\partial x} + \frac{\partial u_0}{\partial y}\right) - D_{xx}\frac{\partial^2 w}{\partial x^2} - D_{xy}\frac{\partial^2 w}{\partial y^2} = 0$$
$$u_0 = N_{xy} = A_{ss}\left(\frac{\partial v_0}{\partial x} + \frac{\partial u_0}{\partial y}\right) - B_{xs}\frac{\partial^2 w}{\partial x^2} - B_{ys}\frac{\partial^2 w}{\partial y^2} = 0 \tag{6.154}$$

For $y = a, b$:

$$w = M_y = B_{ys}\left(\frac{\partial u_0}{\partial x} + \frac{\partial u_0}{\partial y}\right) - D_{xy}\frac{\partial^2 w}{\partial x^2} - D_{yy}\frac{\partial^2 w}{\partial y^2} = 0$$

$$v_0 = N_{xy} = A_{ss}\left(\frac{\partial u_0}{\partial x} + \frac{\partial u_0}{\partial y}\right) - B_{xs}\frac{\partial^2 w}{\partial x^2} - B_{ys}\frac{\partial^2 w}{\partial y^2} = 0 \tag{6.155}$$

The following displacement functions satisfy the boundary conditions

$$\begin{aligned} u_0 &= A \sin\frac{m\pi x}{a} \cos\frac{n\pi y}{b} \\ v_0 &= B \cos\frac{m\pi x}{a} \sin\frac{n\pi y}{b} \\ w &= C \sin\frac{m\pi x}{a} \sin\frac{n\pi y}{b} \end{aligned} \tag{6.156}$$

Substituting (6.156) into (6.148) to (6.150) and collecting like coefficients lead to the following homogeneous equations

$$\begin{bmatrix} A_{mn} & B_{mn} & C_{mn} \\ B_{mn} & D_{mn} & E_{mn} \\ C_{mn} & E_{mn} & (F_{mn} - \lambda) \end{bmatrix} \begin{Bmatrix} \{A\} \\ \{B\} \\ \{C\} \end{Bmatrix} = \begin{Bmatrix} \{0\} \\ \{0\} \\ \{0\} \end{Bmatrix} \tag{6.157}$$

where

$$A_{mn} = A_{xx} m^2 + A_{ss} n^2 R^2$$

$$B_{mn} = (A_{xy} + A_{ss}) mnR$$

$$C_{mn} = -\frac{n\pi}{b}\left(3B_{xs} m^2 + B_{ys} n^2 R^2\right)$$

$$D_{mn} = A_{ss} m^2 + A_{yy} n^2 R^2$$

$$E_{mn} = -\frac{m\pi}{Rb}\left(B_{xs} m^2 + 3B_{ys} n^2 R^2\right)$$

$$F_{mn} = \frac{\pi^2}{R^2 b^2}\left[D_{xx} m^4 + 2(D_{xy} + 2D_{ss})m^2 n^2 R^2 + D_{yy} n^4 R^4\right]$$

$$\lambda = N_0(m^2 + kn^2 R^2)$$

$$R = \frac{a}{b}$$

In order to obtain a non-trival solution of (6.157), the determinant of the coefficient matrix must vanish. Thus

$$\begin{vmatrix} A_{mn} & B_{mn} & C_{mn} \\ B_{mn} & D_{mn} & E_{mn} \\ C_{mn} & E_{mn} & (F_{mn} - \lambda) \end{vmatrix} = 0 \tag{6.158}$$

Solving (6.157) for λ, we obtain

$$N_0 = \frac{\pi^2}{R^2 b^2 \left(m^2 + kn^2 R^2\right)} \left[\left\{ D_{xx} m^4 + 2\left(D_{xy} + 2D_{ss}\right) m^2 n^2 R^2 + D_{yy} n^4 R^2 \right\} \right.$$

$$\left. - \frac{1}{J_1} \left[m \left(B_{xs} m^2 + 3B_{ys} n^2 R^2\right) J_2 + nR \left(3B_{xs} m^2 + B_{ys} n^2 R^2\right) J_3 \right\} \right] \quad (6.159)$$

where

$$J_1 = \left(A_{xx} m^2 + A_{ss} n^2 R^2\right)\left(A_{ss} m^2 + A_{yy} n^2 R^2\right) - \left(A_{xy} + A_{ss}\right)^2 m^2 n^2 R^2$$

$$J_2 = \left(A_{xx} m^2 + A_{ss} n^2 R^2\right)\left(B_{xs} m^2 + 3B_{ys} n^2 R^2\right)$$
$$- n^2 R^2 \left(A_{xy} + A_{ss}\right)\left(B_{xs} m^2 + 3B_{ys} n^2 R^2\right)$$

$$J_3 = \left(A_{ss} m^2 + A_{yy} n^2 R^2\right)\left(3B_{xs} m^2 + B_{ys} n^2 R^2\right)$$
$$- n^2 R^2 \left(A_{xy} + A_{ss}\right)\left(B_{xs} m^2 + 3B_{ys} n^2 R^2\right)$$

The buckling load corresponds to N_0. When $B_{xs} = B_{ys} = 0$, (6.141) becomes

$$N_0 = \frac{\pi^2 \left[D_{xx} m^4 + 2\left(D_{xy} + 2D_{ss}\right) m^2 n^2 R^2 + D_{yy} n^4 R^4\right]}{R^2 b^2 \left(m^2 + kn^2 R^2\right)} \quad (6.160)$$

which is the buckling relationship for a simply supported orthotropic plate under uniform bi-axial compression.

6.9 LAYERWISE THEORY

In the analysis of composite laminates with embedded delamination, free edges or regions of 3-D stress fields, one must use a theory based on 3-D kinematics and develop an efficient computational model. In this context, a layerwise theory is presented in this section which appears as an attractive model due to its having some advantages over the conventional one. In this theory, the through-thickness variation of the displacement components is represented by the interpolation functions along the thickness direction and the corresponding quantities for the displacement components at the interfaces between the layers [6.11, 6.12]. In each layer this theory can be treated as a mathematical or physical layer, giving the flexibility to the analyst to treat several physical layers as a sub-laminate wherever necessary (for computational saving) for model delaminations, ply terminations (i.e., ply drop-offs), ply splits and so on.

6.9.1 Displacements and Strains

The three displacements in the three orthogonal directions are given as

$u_1 = u$, $u_2 = v$ and $u_3 = w$

According to the layerwise theory, the ith displacement component is expressed as

$$u_i(x, y, z) = \sum_{J=1}^{N} U_i^J(x, y) \Phi_i^J(z) \quad (6.161)$$

where $i = 1, 2, 3$ (corresponding to the three displacements); N is the number of interfaces between the layers (including the top and bottom surfaces) and Φ_i^J are the interpolation functions based on the thickness coordinate, z which represents the through thickness variation of the displacement component u_i. Summation of repeated indices is implied in (6.161). In general, the same interpolation functions Φ_i^J are used in (6.161) for all the three displacement components for simplicity, but it may be different for different displacement components, especially for w. The functions Φ_i^J are the global interpolation functions associated with layers connected to the J-th interface through the laminate thickness. If a piecewise linear interpolation function is chosen for all the three displacements, the corresponding function $\Phi_i^J(z)$ is

$$\Phi_i^J(z) = \begin{cases} \dfrac{z - z_{J-1}}{z_J - z_{J-1}} & z_{J-1} < z < z_J \\ \dfrac{z_{J+1} - z}{z_{J+1} - z_j} & z_J < z < z_{J+1} \end{cases} \qquad (6.162)$$

where z_J denotes the thickness coordinate of the Jth interface between Jth and $(J + 1)$th layers (Fig. 6.8). U_i^J are nodal values of U_i at the Jth interface (Fig. 6.9). Because of this local nature of Φ_i^J, the displacement components are continuous through the thickness, but their derivatives with respect to z are discontinuous. This implies that the transverse strains will be discontinuous at layer interfaces. Therefore, the inter-laminar transverse stresses computed from layer constitutive equations (i.e., the products of the strains with the elastic stiffness from two layers at the interface) can be continuous. The inplane strains (ε_x, ε_y, ε_{xy}) will be continuous, but the inplane stresses (σ_x, σ_y, σ_{xy}) may be discontinuous at layer interfaces because of the dissimilar material properties at adjacent layers. The resulting theory represents 3-D problems to $3N$ number of 2-D problems. The consideration of aspect ratio in the layerwise models is restricted only to two-dimensions, as opposite to three dimensions in the conventional analysis.

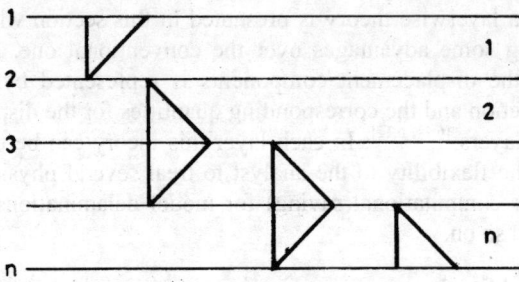

Fig. 6.8 Vibration of Φ_i^J

The value of N in (6.161) can be appropriately selected. When N is chosen such that at least one element per physical layer is used, the inter-laminar stress distributions can be determined accurately. The sub-laminate concept can be used to model several layers as one equivalent single layer, i.e., N is less than the number of physical layers in the laminate. The layerwise laminate theory also yields the conventional, single layer, plate theories as special cases.

Here we consider a laminated composite plate with total thickness h. The linear strain–displacement relations are given by

$$\varepsilon_x = \frac{\partial u}{\partial x} = \sum_{J=1}^{N} \frac{\partial U_1^J}{\partial x} \Phi^J \tag{6.163a}$$

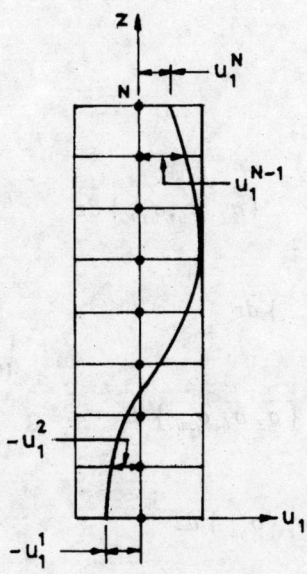

Fig. 6.9 Kinematics of layerwise theory

$$\varepsilon_y = \frac{\partial v}{\partial y} = \sum_{J=1}^{N} \frac{\partial U_2^J}{\partial y} \Phi^J \tag{6.163b}$$

$$\varepsilon_z = \frac{\partial w}{\partial z} = \sum_{J=1}^{N} \frac{\partial U_3^J}{\partial z} \tag{6.163c}$$

$$\varepsilon_{xz} = \frac{\partial v}{\partial z} + \frac{\partial w}{\partial y} = \sum_{J=1}^{N} (U_2^J \frac{d\Phi^J}{dz} + \frac{\partial U_3^J}{\partial y} \Phi^J) \tag{6.164a}$$

$$\varepsilon_{yz} = \frac{\partial u}{\partial z} + \frac{\partial w}{\partial x} = \sum_{J=1}^{N} (U_1^J \frac{d\Phi^J}{dz} + \frac{\partial U_3^J}{\partial x} \Phi^J) \tag{6.164b}$$

$$\varepsilon_s = \frac{\partial u}{\partial y} + \frac{\partial v}{\partial x} = \sum_{J=1}^{N} \left(\frac{\partial U_1^J}{\partial y} + \frac{\partial U_2^J}{\partial x} \right) \Phi^J \tag{6.164c}$$

It may be noted that the subscript i has been omitted from Φ_i^J. The governing equations for the interface variables (U_1^J, U_2^J, U_3^J) can be derived using the principle of virtual displacements. The equations of equilibrium of the layerwise theory are

$$N_{x,x} + N_{xy,y} = 0$$
$$N_{xy,x} + N_{y,y} = 0$$
$$Q_{x,x} + Q_{y,y} + q = 0 \qquad (6.165)$$
$$N_{x,x}^J + N_{xy,y}^J - Q_x^J = 0$$
$$N_{xy,x}^J + N_{y,y}^J - Q_y^J = 0$$

where

$$\{N_x, N_y, N_{xy}\} = \int_{-h/2}^{h/2} \{\sigma_x, \sigma_y, \sigma_{xy}\} \, dz$$

$$\{Q_x, Q_y\} = \int_{-h/2}^{h/2} \{\sigma_{xz}, \sigma_{yz}\} \, dz$$

$$\{N_x^J, N_y^J, N_{xy}^J\} = \int_{-h/2}^{h/2} \{\sigma_x\, \sigma_y\, \sigma_{xy}\} \, dz \qquad (6.166)$$

$$\{Q_x^J, Q_y^J\} = \int_{-h/2}^{h/2} \{\sigma_{xz}, \sigma_{yz}\} \, dz$$

Equation (6.165) consists of $3N$ two-dimensional differential equations in $3N$ variables $(u \ v \ w \ U_1^J \ U_2^J)$, where N denotes the number of interfaces, including the top and bottom faces of the laminate.

For more details the reader may consult Ref. 6.13.

REFERENCES AND SUGGESTED READINGS

6.1 J. N. Reddy, Generalisation of two-dimensional theories of laminated composite plates, *Communication in Applied Numerical Methods*, V.3, 1987, pp. 173–180.

6.2 S. Timoshenko and J. N. Goodier, *Theory of Elasticity*, McGraw Hill Inc., 1959.

6.3 J. T. S. Wang, On the solution of plates of composite materials, *Journal of Composite Materials*, V.3, 1969, pp. 590–592.

6.4 J. E. Ashton, Anisotropic Plate Analysis, General Dynamics Research and Engineering Department, FZM-4099, 12 October, 1967.

6.5 J. M. Whitney and A. W. Leissa, Analysis of heterogeneous anistropic plates, *Journal of Applied Mechanics*, V.36, June 1969, pp. 261–216.

6.6 R. M. Jones, *Mechanics of Composite Materials*, Second Edition, Taylor and Francis Inc., USA, 1999

6.7 J. M. Whitney, *Structural Analysis of Laminated Anisotropic Plates*, Technomic Publishing Co. Inc., USA, 1987.

6.8 E. Reissner, The effect of transverse shear deformation in bending of elastic plates, *Journal of Applied Mechanics*, V.18, 1945, pp. 69–77.

6.9 R. D. Mindlin, Influence of rotary inertia and shear on flexural motions of isotropic elastic plates, *Journal of Applied Mechanics*, V.18, 1951, pp. 326–343.

6.10 P. C. Yang, C. H. Norris and Y. Stavsky, Elastic wave propagation in heterogeneous plates, *International Journal of Solids and Structures*, V.2, 1966, pp. 665-584.

6.11 J. M. Whiney and N. J. Pagano, Shear deflection in heterogeneous anistropic plates, *Journal of Applied Mechanics*, V.37, 1970, pp. 1031–1036.

6.12 J. M. Whitney, Shear correction factors for orthotropic laminate under static loading, *Journal of Applied Mechanics*, V.40, 1973, pp. 301–304.

6.13 O. O. Ocahoa and J. N. Reddy, *Finite Element Analysis of Composite Laminates*, Kluwar Academics, 1996.

EXERCISE 6

6.1 Find the fundamental natural frequency in Hz for an all edges simply supported plate with the following properties

D_{xx} = 188 KN–m D_{xy} = 3 KN–m D_{yy} = 18 KN–m D_{ss} = 4 KN–m

a = 0.75 m b = 0.50 m h = 0.025 m ρ = 5 N–sec^2/mm^4

6.2 Determine the critical buckling load for the plate of Example 6.1

6.3 Derive the basic differential equation of motion of transverse vibration for a symmetric angle-ply laminate.

6.4 For an all edge simply supported square anti-symmetric cross-ply laminate, determine the critical buckling load.

6.5 The following material properties are given for a unidirectional ply laminate, h = 2 mm

$$[A] = \begin{bmatrix} 60 & 19 & 0 \\ 19 & 12 & 0 \\ 0 & 0 & 20 \end{bmatrix} \text{GPa–mm}$$

$$[B] = [0]$$

$$[D] = \begin{bmatrix} 10 & 2 & 0 \\ 2 & 3 & 0 \\ 0 & 0 & 1.5 \end{bmatrix} \text{GPa–mm}^2$$

ρ the mass density is 2 N–s^2/mm^2

If the plate is simply supported on all four edges, what is its fundamental natural frequency?

CHAPTER 7

ANALYSIS OF COMPOSITE BEAMS

7.1 DEFINITION

A beam is a one-dimensional structure. So is a column or a rod. For all of them the cross-sectional dimensions – width b and height h are much smaller in comparison to the length of the structure (Fig. 7.1). If the load acts in the z-direction of the xz plane so as to create bending in the structure, it is referred to as a beam. If a tensile force acts in the x-direction of the structure resulting in its elongation, it is referred to as a rod. If a compressive force acts in the x-direction resulting in its shortening, it is referred to as a column. A combination of these loads is always possible. If a transverse load in the z-direction is accompanied by an inplane compression in the x-direction, the structure is referred to as beam-column. As such it is seen that the same one-dimensional structure is given different names depending on the manner in which the load is applied.

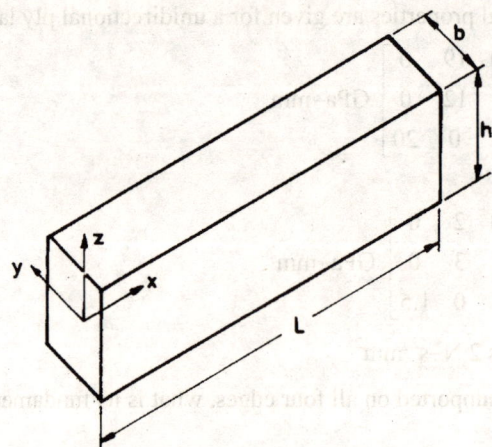

Fig. 7.1 A one-dimensional structure

7.2 BASIC ASSUMPTIONS

The following assumptions are made in the analysis

1) Plane sections normal to the longitudinal axis of the beam bending remain plane and normal to it after bending (Fig. 7.2).

2) Plies of the beam are symmetrically arranged about the neutral surface.
3) Each ply is considered as linearly elastic. There is no shear coupling effect in the ply, as the orientation is either 0° or 90°.
4) Plies are perfectly bonded, so that no slip occurs at the interfaces.
5) The stress components are σ_x and σ_{xz}.
6) The thermal and moisture effects are not considered.
7) All Poisson's ratio effects are ignored

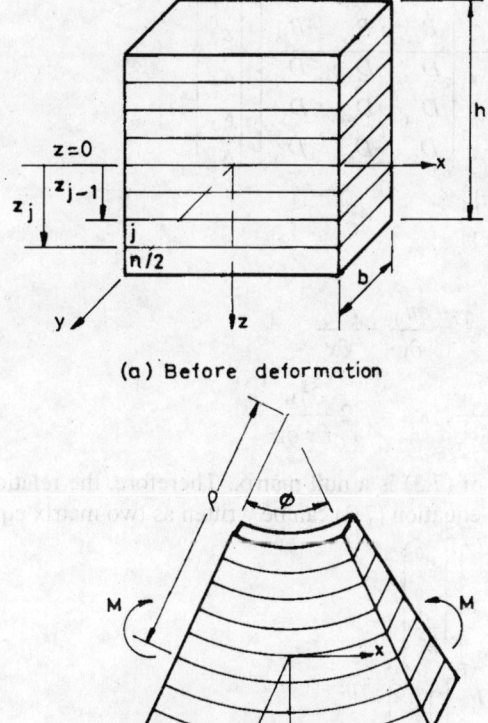

Fig. 7.2 Bending of a laminated beam

7.3 BENDING ANALYSIS OF LAMINATED BEAMS

Classical beam theory can be extended to layered beams provided the principal material axes of each orthotropic ply is parallel to the beam edges and plates are oriented symmetrically about the mid-plane [7.1, 7.2]. In order to do this, equivalent bending stiffness of the laminate is to be considered.

$$E_x^b I = \sum_{k=1}^{n} E_1^k I^k \qquad (7.1)$$

where

E_x^b is the effective bending modulus of the beam
E_1^k is the elastic modulus of the kth layer relative to the beam axis
I^k is the moment of inertia of the kth layer relative to the mid-plane
n is the number of layers of the laminate

Let us reproduce (5.21) below

$$\begin{Bmatrix} N_x \\ N_y \\ N_s \\ \hline M_x \\ M_y \\ M_s \end{Bmatrix} = \left[\begin{array}{ccc|ccc} A_{xx} & A_{xy} & A_{xs} & B_{xx} & B_{xy} & B_{xs} \\ A_{yx} & A_{yy} & A_{ys} & B_{yx} & B_{yy} & B_{ys} \\ A_{sx} & A_{sy} & A_{ss} & B_{sx} & B_{sy} & B_{ss} \\ \hline B_{xx} & B_{xy} & B_{xs} & D_{xx} & D_{xy} & D_{ss} \\ B_{yx} & B_{yy} & B_{ys} & D_{yx} & D_{yy} & D_{ys} \\ B_{sx} & B_{sy} & B_{ss} & D_{sx} & D_{sy} & D_{ss} \end{array} \right] \begin{Bmatrix} \varepsilon_x^0 \\ \varepsilon_y^0 \\ \varepsilon_s^0 \\ \hline \kappa_x \\ \kappa_y \\ \kappa_s \end{Bmatrix} \quad (7.2)$$

or,

$$\begin{Bmatrix} \{N\} \\ \{M\} \end{Bmatrix} = \left[\begin{array}{c|c} [A] & [B] \\ \hline [B] & [D] \end{array} \right] \begin{Bmatrix} \{\varepsilon\} \\ \{\kappa\} \end{Bmatrix} \quad (7.3)$$

where

$$\varepsilon_x^0 = \frac{\partial u_0}{\partial x}, \quad \varepsilon_y^0 = \frac{\partial v_0}{\partial y}, \quad \varepsilon_s^0 = \frac{\partial u_0}{\partial y} + \frac{\partial v_0}{\partial x}$$

$$\kappa_x = -\frac{\partial^2 w}{\partial x^2}, \quad \kappa_y = -\frac{\partial^2 w}{\partial y^2}, \quad \kappa_s = -2\frac{\partial^2 w}{\partial x \, \partial y} \quad (7.4)$$

Due to symmetry, $[B]$ matrix of (7.3) is a null matrix. Therefore, the relationship related to $\{N\}$ and $\{M\}$ becomes uncoupled. The equation (7.2) can be written as two matrix equations given below

$$\begin{Bmatrix} N_x \\ N_y \\ N_s \end{Bmatrix} = \begin{bmatrix} A_{xx} & A_{xy} & A_{xs} \\ A_{yx} & A_{yy} & A_{ys} \\ A_{sx} & A_{sy} & A_{ss} \end{bmatrix} \begin{Bmatrix} \varepsilon_x^0 \\ \varepsilon_y^0 \\ \varepsilon_s^0 \end{Bmatrix} \quad (7.5)$$

and

$$\begin{Bmatrix} M_x \\ M_y \\ M_s \end{Bmatrix} = \begin{bmatrix} D_{xx} & D_{xy} & D_{xs} \\ D_{yx} & D_{yy} & D_{ys} \\ D_{sx} & D_{sy} & D_{ss} \end{bmatrix} \begin{Bmatrix} \kappa_x \\ \kappa_y \\ \kappa_s \end{Bmatrix} \quad (7.6)$$

Equations (7.5) and (7.6) in inverted form (see (5.33)) are

$$\begin{Bmatrix} \varepsilon_x^0 \\ \varepsilon_y^0 \\ \varepsilon_s^0 \end{Bmatrix} = \begin{bmatrix} a_{xx} & a_{xy} & a_{xs} \\ a_{yx} & a_{yy} & a_{ys} \\ a_{sx} & a_{sy} & a_{ss} \end{bmatrix} \begin{Bmatrix} N_x \\ N_y \\ N_s \end{Bmatrix} \quad (7.7)$$

$$\begin{Bmatrix} \kappa_x \\ \kappa_y \\ \kappa_s \end{Bmatrix} = \begin{bmatrix} d_{xx} & d_{xy} & d_{xs} \\ d_{yx} & d_{yy} & d_{ys} \\ d_{sx} & d_{sy} & d_{ss} \end{bmatrix} \begin{Bmatrix} M_x \\ M_y \\ M_s \end{Bmatrix} \quad (7.8)$$

As pure flexure of the composite beam is considered, we write

$$M_y = M_s = 0 \tag{7.9}$$

and concentrate on (7.8)

Equation (7.8) reduces to

$$\kappa_x = -\frac{\partial^2 w}{\partial x^2} = d_{xx} M_x, \quad M_x = -D_{xx}\frac{\partial^2 w}{\partial x^2} = D_{xx}\kappa_x \tag{7.10}$$

It is assumed that

$$w = w(x) \tag{7.11}$$

Based on (7.8) and (7.9), we can write

$$\kappa_y = -\frac{\partial^2 w}{\partial y^2} = d_{yx} M_x, \quad \kappa_s = -2\frac{\partial^2 w}{\partial x\,\partial y} = d_{sx} M_x \tag{7.12}$$

Equation (7.12) clearly reveals that w is dependent on y as well. Due to the effect of Poisson's ratio this one-dimensional assumption (7.11) is strictly speaking not correct even for homogeneous isotropic beam theory. If the length-to-width ratio is reasonably large, then this effect becomes negligible. In (7.12), the shear coupling term 'd_{sx}', may have a more severe effect.

Combining (7.10) and (7.11), yields

$$D_{xx} b \frac{d^2 w}{dx^2} = -M_x b \tag{7.13}$$

Here,

$$D_{xx} = \frac{h^3}{12} E_x^b \tag{7.14}$$

$$M = M_x b, \quad I = \frac{bh^3}{12} \tag{7.15}$$

where b is the width of the beam, h is the depth of the laminated beam.

Combining (7.13) and (7.14), we get

$$\frac{d^2 w}{dx^2} = -\frac{M}{E_x^b I} \tag{7.16}$$

Other equilibrium equations are

$$\frac{dM_x}{dx} - Q_x = 0, \quad \frac{dQ_x}{dx} + p = 0 \tag{7.17}$$

where Q_x is the shear force per unit length.

Converting (7.17) in terms of beam resultants,

$$\frac{dM}{dx} - V = 0, \quad \frac{dV}{dx} + q = 0 \tag{7.18}$$

where $\quad V = Q_x b, \quad q(x) = p(x) b \tag{7.19}$

Combining (7.13) with (7.18), yields

$$bD_{xx}\frac{d^4w}{dx^4} = q(x) \tag{7.20}$$

Equation (7.20) is the governing differential equation for the mid-plane symmetric composite beam. Solution of (7.20) will yield the value of $w(x)$ and the corresponding $M(x)$, $V(x)$ etc. Once the solution of $w(x)$ is obtained, the stresses at any layer of the laminate can be determined. From (5.8) we can write stresses for the kth lamina,

$$\begin{Bmatrix} \sigma_x \\ \sigma_y \\ \sigma_s \end{Bmatrix}_k = z \begin{bmatrix} Q_{xx} & Q_{xy} & Q_{xs} \\ Q_{yx} & Q_{yy} & Q_{ys} \\ Q_{sx} & Q_{sy} & Q_{ss} \end{bmatrix}_k \begin{Bmatrix} \kappa_x \\ \kappa_y \\ \kappa_s \end{Bmatrix} \tag{7.21}$$

Combining (7.8) and (7.21), yields

$$\sigma_x^\kappa = z f_1^\kappa \frac{M}{I} \tag{7.22}$$

$$\sigma_y^\kappa = z f_2^\kappa \frac{M}{I} \tag{7.23}$$

$$\sigma_s^\kappa = z f_3^\kappa \frac{M}{I} \tag{7.24}$$

where

$$f_1^\kappa = \left(Q_{xx}^\kappa d_{xx} + Q_{xy}^\kappa d_{xy} + Q_{xs}^\kappa d_{xs}\right)\frac{h^3}{12} \tag{7.25}$$

$$f_2^\kappa = \left(Q_{xy}^\kappa d_{xx} + Q_{yy}^\kappa d_{xy} + Q_{ys}^\kappa d_{xs}\right)\frac{h^3}{12} \tag{7.26}$$

$$f_3^\kappa = \left(Q_{xs}^\kappa d_{xx} + Q_{ys}^\kappa d_{xy} + Q_{ss}^\kappa d_{ss}\right)\frac{h^3}{12} \tag{7.27}$$

For homogenous beams, $f_1^\kappa = 1$, $f_2^\kappa = f_3^\kappa = 0$ and (7.22) reduces to classical beam theory while (7.23) and (7.24) vanish.

The free edge effect is important and to minimize it, b/h ratio should be pretty high. In fact, the above stresses given by (7.22) to (7.24) will not give correct values in a zone of one laminate thickness, h, away from the free edge, where the state of stress is three-dimensional. [7.3, 7.4].

For the kth layer of laminate, the one basic differential equation of equilibrium is

$$\frac{\partial \sigma_x^k}{\partial x} + \frac{\partial \sigma_{xy}^k}{\partial y} + \frac{\partial \sigma_{xz}^k}{\partial z} = 0 \tag{7.28}$$

Therefore, inter-laminar shear stress σ_{xz} can be determined by integrating all terms of (7.28). Combining (7.22) and (7.28) yields

$$\sigma_{xz}^k = -\frac{1}{I}\int_{-h/2}^{z_k} f_1^k \frac{dM}{dx} z\, dz \tag{7.29}$$

Equation (7.29) can also be written using (7.18),

$$\sigma_{xz}^k = \frac{-V}{I} \int_{-h/2}^{z_k} f_1^k \, z \, dz \tag{7.30}$$

We now consider that N_x acts on the laminate with

$$N_y = N_s = 0 \tag{7.31}$$

If there is no y-direction dependence of any quantity in the differential equation, we get from (7.5) for a symmetric laminate

$$N_x = A_{xx} \varepsilon_x^0 \tag{7.32}$$

where N_x is the force per unit length. So to get beam resultants, we must multiply it by b.

$$P = N_x b \tag{7.33}$$

Therefore combining (7.32) and (7.33), we get

$$P = b A_{xx} \frac{du_0}{dx} \tag{7.34}$$

If P is constant, then u_0, the x-direction displacement of the mid-surface can be obtained from the integration of eqn. (7.34).

$$u_0(x) = \left(\frac{P}{b A_{xx}}\right) x + C_0 \tag{7.35}$$

where C_0 can be obtained from specified end conditions.

If the rod is loaded by tensile load P only, we can write the inplane stresses for the kth lamina as

$$\begin{Bmatrix} \sigma_x \\ \sigma_y \\ \sigma_s \end{Bmatrix}_k = \begin{bmatrix} Q_{xx} & Q_{xy} & Q_{xs} \\ Q_{xy} & Q_{yy} & Q_{ys} \\ Q_{xs} & Q_{ys} & Q_{ss} \end{bmatrix}_k \begin{Bmatrix} \varepsilon_x^0 \\ \varepsilon_y^0 \\ \varepsilon_s^0 \end{Bmatrix} \tag{7.36}$$

Combining (7.7) and (7.36), yields

$$\sigma_x^k = g_1^k \frac{P}{A} \tag{7.37}$$

$$\sigma_y^k = g_2^k \frac{P}{A} \tag{7.38}$$

$$\sigma_s^k = g_3^k \frac{P}{A} \tag{7.39}$$

where

$$g_1^k = (Q_{xx}^k a_{xx} + Q_{xy}^k a_{xy} + Q_{xs}^k a_{xs}) h \tag{7.40}$$

$$g_2^k = (Q_{xy}^k a_{xx} + Q_{yy}^k a_{xy} + Q_{ys}^k a_{xs}) h \tag{7.41}$$

$$g_3^k = (Q_{xs}^k a_{xx} + Q_{ys}^k a_{xy} + Q_{ss}^k a_{xs}) h \tag{7.42}$$

and $A = bh$

If both the inplane and lateral loads occur simultaneously, the stresses in each face can be obtained by summing up (7.22) to (7.37) for σ_x^k and similarly for σ_y^k and σ_z^k.

It may be mentioned that in order that the above theory is valid, the principal material axes and the loading axes are coincident [7.1, 7.2].

$$D_{xx} = D_{11} \quad \text{and} \quad A_{xx} = A_{11} \tag{7.43}$$

7.4 EXAMPLES

7.4.1 Bending of a Laminated Beam under Uniform Load

A beam under uniform loading is shown in Fig. 7.3. The bending moment at a section x is given by

$$M = \frac{p_0 L}{2}x - \frac{p_0 x^2}{2} \quad \text{for} \quad 0 \le x \le L \tag{a}$$

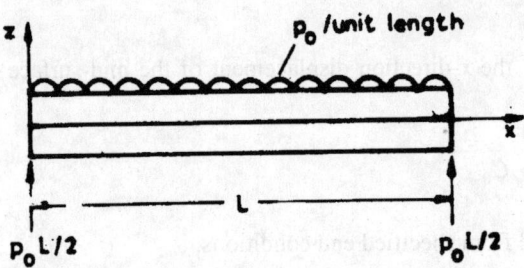

Fig. 7.3 A simply supported beam under uniform load

From (7.16), we get

$$-\frac{d^2 w}{dx^2} = \frac{1}{E_x^b I}\left[\frac{p_0 L}{2}x - \frac{p_0 x^2}{2}\right] \tag{b}$$

The boundary conditions are

at $\quad x = 0, \quad w = 0 \quad$ and $\tag{c}$

at $\quad x = \dfrac{L}{2}, \quad \dfrac{dw}{dx} = 0 \tag{d}$

Integrating (b) twice and substituting boundary conditions from (c) to (d), yields

$$w = -\frac{p_0 L^4}{24 E_x^b I}\left[2\left(\frac{x}{L}\right)^3 - \left(\frac{x}{L}\right)^4 - \left(\frac{x}{L}\right)\right] \tag{e}$$

The effective modulus when expressed in terms of the central deflection w_c is given by

$$E_x^b = \frac{5 p_0 L^4}{384 w_c I} = \frac{5}{32}\frac{p_0 L^4}{bh^3 w_c} \tag{f}$$

The inplane stresses can be determined by substituting (a) in (7.22) to (7.24). The maximum stress occurs at the section of maximum moment, That is, at the centre of the beam (at $x = L/2$). The normal stress in the kth layer is

$$\left(\sigma_x^k\right)_{L/2} = \frac{3}{2} z f_1^k \frac{p_0 L^2}{bh^3} \tag{g}$$

The maximum stress values of σ_y^k and σ_s^k can be evaluated from (7.23) and (7.24) respectively. It may be noted that they are of the same form as eqn. (g). For a laminated beam the maximum stress does not always occur at the outer surface.

7.4.2 A Laminated Cantilever Beam under Tip Loading

A laminated cantilever beam with tip load is considered [Fig. 7.4]. The beam is cantilevered at $x = 0$. The bending moment at a section x is given by

$$M = -P(L-x) \quad \text{for} \quad 0 \le x \le L \tag{a}$$

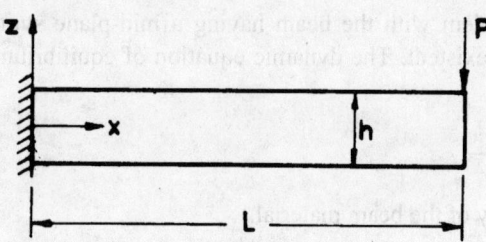

Fig. 7.4 A cantilever beam

From (7.16), we get

$$\frac{d^2 w}{dx^2} = -\frac{1}{E_x^b I}[-P(L-x)] \tag{b}$$

The solution of (b) is

$$w = \frac{1}{E_x^b I}\left[\frac{P(L-x)^3}{6} + C_1 x + C_2\right] \tag{c}$$

The boundary conditions are

at $\quad x = 0 \quad w = 0 \quad , \quad \dfrac{dw}{dx} = 0$ \hfill (d)

which on substitution into (c) gives

$$C_1 = -\frac{PL^2}{2} \quad \text{and} \quad C_2 = \frac{PL^2}{6} \tag{e}$$

Therefore, after replacing C_1 and C_2 from (e) into (c), we get

$$w = -\frac{PL^3}{6 E_x^b I}\left[-3\left(\frac{x}{L}\right)^2 + \left(\frac{x}{L}\right)^3\right] \tag{f}$$

The modulus when expressed in terms of tip deflection is given by

$$E_x^b = \frac{PL^3}{3w_e I} = 4\frac{PL^3}{bh^3 w_e} \tag{g}$$

The normal stress for the kth layer at the fixed end is

$$\sigma_x^k(0) = z f_1^k \frac{12PL}{bh^3} \tag{h}$$

7.5 EIGENVALUE PROBLEMS OF LAMINATED BEAMS

The analysis made so far for the static loading of laminated beams can be extended to free vibration and buckling analyses. Dynamic loads of various types act in the beams and the incorporation of them in the analysis is very important. We limit ourselves in this section to the determination of natural frequencies only, that is, the free vibration of laminated beams. The problem which results is known as the eigenvalue problem. There is another category of eigenvalue problems which shall be taken up in this section, that is, the buckling of laminated beams.

We continue our problem with the beam having a mid-plane symmetry so that the bending-stretching coupling is non–existent. The dynamic equation of equilibrium which includes the inertia force is given by (7.20)

$$bD_{11}\frac{\partial^4 w}{\partial x^4} = -\rho A \frac{\partial^2 w}{\partial t^2} \tag{7.44}$$

where ρ is the mass density of the beam material.

Equation (7.44) is an equation involving time and space and as such, in order to obtain a solution, the variables are to be separated.

Assuming a simply supported laminated beam, a solution for the simply supported boundary conditions can be written in the following form

$$w = W_m \sin\frac{m\pi x}{L} \cos\omega_m t, \quad m = 1, 2, \ldots, r, \ldots \tag{7.45}$$

Combining (7.44) and (7.45), yields

$$\omega_m = \frac{m^2 \pi^2}{L^2}\sqrt{\frac{bD_{11}}{\rho A}} \tag{7.46}$$

Thus, for each value of m, there is a natural frequency ω_m and a mode shape [Fig. 7.5]. As m can have values from 1 to infinity, there are infinite natural frequencies possible.

The same laminated beam is now subjected to an inplane compression P in the x-direction. Inplane loads might result in the lateral deflection and might give rise to a lateral instability problem. The load P that could cause buckling is a very important quantity which needs to be predicted.

The differential equation of a symmetric laminated beam subjected to inplane compression P is given by

$$bD_{11}\frac{d^4 w}{dx^4} = P\frac{d^2 w}{dx^2} \tag{7.47}$$

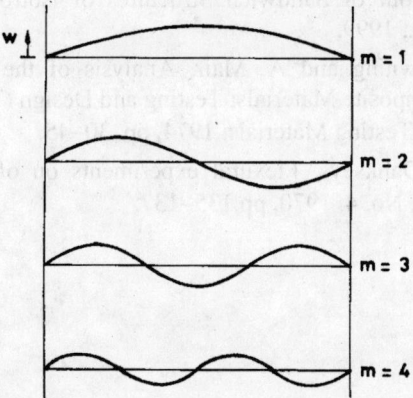

Fig. 7.5 Mode shapes of the beam

For the simply supported beam, the solution assumed is

$$w = A \sin \frac{\pi x}{L} \tag{7.48}$$

Combining (7.48) with (7.46), yields

$$A \left[bD_{11} \frac{\pi^4}{L^4} + P \frac{\pi^2}{L^2} \right] \sin \frac{\pi x}{L} = 0 \tag{7.49}$$

The critical value of P is given by

$$P_{cr} = -\frac{\pi^2}{L^2} bD_{11} \tag{7.50}$$

For the case of an unsymmetric laminate where $B_{11} \neq 0$, it has been shown that for the solution of the eigenvalue problem, D_{11} is to replaced by

$$\frac{A_{11}D_{11} - B_{11}^2}{A_{11}} \tag{7.51}$$

in (7.44) and (7.47) as the case may be.

REFERENCES AND SUGGESTED READINGS

7.1 N.J. Hoff, Strength of laminates and sandwich structural elements, Chapter 1, *Engineering Laminates*, A.G.H. Dietz (Editor), John Wiley and Sons, New York, 1949.

7.2 N.J. Pagano, Analysis of flexural test of bi-directional composites, *Journal of Composite Materials*, V.1, 1967, pp. 336–342.

7.3 N.J. Pagano and R.B. Pipes, Inter-laminar stresses in composite laminates under uniform axial extension, *Journal of Composite Materials*, 4, January, 1971, pp. 538–546.

7.4 N.J. Pagano and R.B. Pipes, Influence of stacking sequence in laminate strength, *Journal of Composite Materials*, V.5, 1971, pp.50–57.

7.5 J.R. Vinson, The Behaviour of Sandwich Structures of Isotropic and Composite Materials, Technomic Publishing Co., 1999.

7.6 J.M. Whitney, C.E. Browning and A. Mair, Analysis of the Flexural Test for Laminated Composite Materials, Composite Materials; Testing and Design (Third Conference), ASTM STP 546, American Society of Testing Materials, 1974, pp. 30–45.

7.7 J.M. Whitney and R.J. Dankseys, Flexural experiments on off-axis composites, *Journal of Composite Materials*, V.4, No. 4, 1970, pp.135–137.

CHAPTER 8

FINITE ELEMENT ANALYSIS OF COMPOSITE STRUCTURES

8.1 INTRODUCTION

The most popular numerical technique for the analysis of composite structures today is the finite element method. Many general purpose finite element softwares are capable of dealing with composite structures. For accurate evaluation of the response of anisotropic structures such as FRP boats, aircraft panels etc., locally or globally, there cannot be a better approach than the application of the finite element method.

8.2 THE FINITE ELEMENT METHOD

In a continuum that is divided into a mesh, two adjacent elements placed side by side will have a common edge. Though to be theoretically correct, we have to satisfy the continuity requirements all along the edge (Fig. 8.1), it has been found that this will result in a more complicated analysis. Therefore, in order to make the analysis simpler, it is assumed that the elements are connected only at nodal points and it is only there that the continuity requirements are to be satisfied. Once the discretisation is made, the analysis follows a rather set procedure. The stiffness matrix of the individual element needs to be formulated. The forces actually distributed in the real structure are transformed to act at the nodal points. In the finite element analysis, the continuum is divided into a 'finite' number of elements, having 'finite' dimensions and reduced from having infinite degrees of freedom to 'finite' degrees of unknowns. The term 'finite element' was coined by Clough [8.1].

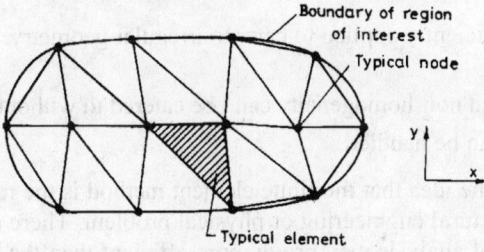

Fig. 8.1 A continuum divided into finite elements

It has been found that the accuracy of the solution in general increases with the increase in the number of elements considered. However, the division of the continuum into more number of elements will result in increased computer cost. A question that will arise in the mind of the analyst is what should be the optimum number of divisions that should give a satisfactory solution. The answer given by the expert to this query even a decade back is that (a) the required subdivision of the continuum should be based on past experience, and (b) if an analysis is attempted for the first time, then the convergence is to be tested based on varied mesh gradings and a suitable choice can be made only after that. Now the adaptive finite element procedure has come to stay in which the mesh divisions are based on an error analysis. For regions where higher concentration of stress is expected such as around the openings, a finer mesh division is to be made (Fig. 8.2).

In the finite element method, the problem is formulated in two stages

a)   the element formulation

b)   the system formulation

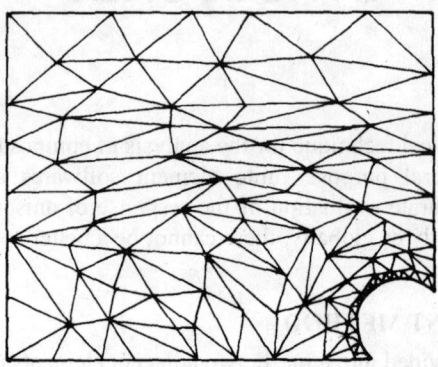

Fig. 8.2 A plate with a hole

The first stage involves the derivation of the element stiffness matrix or element flexibility matrix that yields a relationship between nodal forces and nodal displacements. The next stage is the formulation of the stiffness and loads of the entire structure.

8.3 ADVANTAGES AND DISADVANTAGES OF THE FINITE ELEMENT METHOD

The main advantage of the finite element analysis can be put in one sentence – the physical problems which were so far intractable and complex for any closed-bound solution, can now be analysed by this method. The advantages in relation to the complexity of the problem are stated below

1. The method can be efficiently applied to cater to irregular geometry. It can take care of any type of boundary.
2. Material anisotropy and non-homogeneity can be catered to without much difficulty.
3. Any type of loading can be handled.

One should not form the idea that the finite element method is the most efficient method for the analysis of any type of structural engineering or physical problem. There are many types of problems where some other method of analysis may prove more efficient than the finite element method. One of the main disadvantages of this method is the cost involved in the solution of the problem. In many cases of vibration and stability problems, the cost of analysis of the finite element method may turn out to be very high. It may, therefore be a luxury to undertake vibration and stability analyses of simple structures where application of even simple computer methods such as the finite strip [8.2] or other semi-analytic methods [8.3,8.4] will lead to more economic solution. But these methods work within their own limitations and may not be as versatile as the finite element method [8.5].

There are many approximate methods such as the Rayleigh–Ritz method, the Galerkin's method and others. The finite element method stands superior to all of them. The main advantage of the formulation of the finite element method is that in this method approximations are confined to relatively small sub-domains whereas in other methods, the admissible functions satisfy the boundary conditions of the entire domain which becomes extremely difficult when the domain has an irregular shape. In the finite element method, the admissible functions are valid over the simple domain and have nothing to do with the boundary, however simple or complex it may be.

8.4 BASIC STEPS IN THE FINITE ELEMENT METHOD IN STATIC PROBLEM

The following are the steps adopted for analysing a structural engineering static problem by the finite element method [8.5, 8.6].

Step 1. Discretisation of the domain

The continuum is divided into a number of finite elements by imaginary lines or surfaces. The interconnected elements may have different sizes and shapes. The success of this idealisation lies in how closely this discretised continuum represents the actual continuum. The choice of the simple element or higher order element, straight or curved, its shape and refinements are to be decided before the mathematical formulation starts.

Step 2. Identification of variables

The elements are assumed to be connected at their intersecting points referred to as nodal points. At each node, unknown displacements are to be prescribed. They are dependent on the problem at hand. For example, in a plane stress problems, the minimum number of unknowns are the two linear translations at each node. In a simpler model of the plate bending problem, the unknowns at a node are the deflection and the rotation about two orthogonal axes. In ordinary cases, to every unknown displacement at a node there corresponds a nodal force which can be physically visualised. The problem may be identified in such a way that in addition to the displacement which occurs at the nodes, depending on the physical nature of the problem, certain other quantities such as strain, curvature etc., may need to be specified as nodal unknowns for the element, which however, may not have a corresponding physical quantity in the generalised forces. The value of these quantities can, however, be obtained from variational principles.

Step 3. Choice of approximating functions

Once the variables and the local coordinate system have been chosen, the next step is the choice of the displacement function. In fact, it is the displacement function that is the starting point of the mathematical analysis. This function represents the variation of the displacement within the element. The function can be approximated in a number of ways. The displacement function may be approximated in the form of a linear function or a higher order function. A convenient way to express it is by the use of polynomials. A typical question is posed in the finite element analysis, that is, the solution may converge to the exact solution either by increasing the degrees of the polynomial or by decreasing the element size or by both.

The shape of the element or the geometry may also be approximated. The coordinates of the corner nodes define the element shape accurately if the element is actually made of straight lines or planes. This simplification of geometry may sometimes lead to wrong results, e.g., flat elements in a shell. The analyst will also have to decide on the weightage to be given to the geometry and displacements for a particular problem.

Step 4. Formation of the element stiffness matrix

After the continuum is discretised with desired element shapes, the element stiffness matrix is formulated. This can be done in a number of ways. Basically it is a minimization procedure whatever may be the approach adopted. For certain elements, the form involves a great deal of sophistication.

With the exception of a few simple elements, the element stiffness matrix for the majority of elements is not available in explicit form. As such they require numerical integration for their evaluation.

The geometry of the element is defined with reference to the global frame. In many problems such as those of rectangular plates, the global and local axes systems are parallel and for them no further calculation is needed at the element level beyond the computation of the element stiffness matrix in local coordinates. Coordinate transformation must be done for all elements where it is needed.

Step 5. Formation of the overall stiffness matrix

After the element stiffness matrix in global coordinates is generated, it is assembled to form the overall stiffness matrix. The assembly is done through the nodes which are common to adjacent elements. At the nodes, the continuity of the displacement function and possibly their derivatives are established. The overall stiffness matrix is symmetric and banded.

Step 6. Incorporation of boundary conditions

The boundary restraint conditions are to be imposed in the stiffness matrix. There are various techniques available to satisfy the boundary conditions. In some of these approaches, the size of the stiffness matrix may be reduced or condensed in its final form. To ease the computer programming aspect and to elegantly incorporate the boundary conditions, the size of the overall stiffness matrix may be kept the same by assigning a high value ($= 10^{20}$) to the diagonal element corresponding to the restraint.

Step 7. Formation of the element loading matrix

The loading forms an essential parameter in any structural engineering problem. The loading inside an element is transferred at the nodal points and a consistent element loading matrix is formed. Sometimes, based on the typicality of the problem, the loading matrix may be simplified.

Step 8. Formation of the overall loading matrix

Like the overall stiffness matrix, the element loading matrices are assembled to form the overall loading matrix. This matrix has one column per loading case and it is either a column vector or a rectangular matrix depending on the number of loading conditions.

Step 9. Solution of simultaneous equations

All the equations required for the solution of the problem are now developed. In the displacement method, the unknowns are the nodal displacements. Rarely is the stiffness matrix stored as a solid matrix. Advantages are taken of the symmetric nature of the problem and its banded properties. Gauss elimination and Cholsky's factorisation are the most commonly used procedures for the solution of simultaneous equations. These methods are well-suited to a small or moderate number of equations. For the large-sized problems, the frontal technique is one of the methods of obtaining the solution.

Step 10. Calculation of stresses or stress–resultants

In the previous step, nodal displacements are calculated and these values are utilised for the calculation of stresses or stress–resultants. This may be done for all elements of the continuum or it may be limited only to some predetermined elements.

Results may be obtained by graphical means. It may be desirable to plot the contour of the deformed shape of the continuum. The contour of the principal stresses may be one of the sought-after items for a certain category of problems.

The steps presented above are general in nature and need not be the same for all element types.

8.5 THE BASIC PROBLEMS

The equations of equilibrium for a discretised elastic structural system undergoing small displacements can be expressed in the following matrix form as

1. Static analysis

$$[K]\{\delta\} = \{P\} \tag{8.1}$$

2. Free vibration analysis

$$[K]\{\delta\} + [M]\{\ddot{\delta}\} = \{0\} \tag{8.2}$$

3. Transient vibration analysis

$$[K]\{\delta\} + [C]\{\dot{\delta}\} + [M]\{\ddot{\delta}\} = \{F(t)\} \tag{8.3}$$

4. Buckling analysis

$$([K] - [K]_G)\{\delta\} = \{0\} \tag{8.4}$$

where $[M]$, $[K]$ and $[C]$ are the overall mass, stiffness and damping matrices respectively. $[K]_G$ is the geometric stiffness matrix.

$\{\delta\}$, $\{\dot{\delta}\}$ and $\{\ddot{\delta}\}$ are global nodal displacement, velocity and acceleration vectors respectively with respect to the reference axis system. $\{P\}$ and $\{F(t)\}$ are the static and time-dependent load vectors acting at the nodes.

8.5.1 Mass Matrix

There are two ways of forming the mass matrix of the structure. They are (1) lumped mass matrix and (2) consistent mass matrix.

In a lumped mass matrix, the masses are lumped at nodal points and the mass matrix is diagonal which yields considerable advantages in the computation of different quantities related to the vibration analysis. The consistent mass matrix is much more occupied and requires more storage space, although it does yield more accurate results for certain classes of problems with less mesh division.

8.5.2 Stiffness Matrix

Overall stiffness matrix $[K]$ is formed by assembling the stiffness matrix of the individual elements. $[K]$ is symmetric and banded. These properties may be taken advantage of in economically storing $[K]$. The stiffness matrix is also positive definite.

8.5.3 Damping Matrix

The damping matrix may be considered to be of the viscous damping type where critical damping ratios vary with each mode. Due to the complexity of the analysis it may be considered as a combination of stiffness matrix and mass matrix.

8.6 SOLUTION TECHNIQUES

8.6.1 Static Analysis

Equation (8.1) can be solved by various means. Gauss's method of solving simultaneous equations is one of the popular approaches. Frontal techniques can be resorted to for obtaining the solution [8.7]. For huge size problems, substructure techniques may be tried [8.5].

8.6.2 Free Vibration Analysis

Various methods have been developed for the solution of the eigenvalue problem

$$\omega^2 [M] \{\delta\} = [K] \{\delta\} \qquad (8.5)$$

These methods can be broadly divided into four groups [8.7]. They are the (1) vector iteration method, (2) transformation method, (3) determinant search technique and (4) other methods.

8.7 EIGENVALUE SOLUTION TECHNIQUES

Vector iteration methods are very effectively used for the computation of eigenvalues, as well as the corresponding eigenvectors at the same time. The aim of the method is to directly operate upon (8.5). Various methods come under this category [8.8]. Stodola's method was the earliest one. Investigations of Jennings [8.9], Rutihauser [8.10] and Steward [8.11] have led to the development of modern vector iteration methods. The advanced version applied to the finite element analysis is the subspace iteration technique [8.12].

One of the major advantages of the vector iteration methods is that they may yield a few of the lower eigenvalues and eigenvectors. In most practical problems, only a few of the lower eigenvalues and eigenvectors are of importance. As such, the method avoids the determination of all unknowns in the system.

Transformation methods are preferable when all the eigenvalues and eigenvectors of the problem are required. Transformation methods operate on the following matrices.

$$\left. \begin{array}{l} [\Phi]^T [K] [\Phi] = [\Delta] \\ [\Phi]^T [M] [\Phi] = [I] \end{array} \right\} \qquad (8.6)$$

where $[\Phi]$ is called the modal matrix and formed by modal vectors.

$$[\Phi] = \left[\{\phi^{(1)}\} \{\phi^{(2)}\} \ldots \{\phi^{(n)}\} \right] \qquad (8.7)$$

where $\{\phi^{(1)}\}$ is the mode shape for the first mode and so is the definition of $\{\phi^{(2)}\}$, $\{\phi^{(3)}\}$ etc. It may be noted that $[\Phi]$ is a square matrix of order $n \times n$.

Further, from the orthogonality properties, $[\Delta]$ is a square diagonal matrix containing ω^2-values in the diagonal elements.

Due to transformations of the type given by (8.6), this method of solution requires storage of large matrices.

Since $[\Phi]$, the mode shape matrix is unique, it can be constructed by iteration. Basically, the method involves the reduction of $[K]$ and $[M]$ into the diagonal form, using successive pre-multiplication and post-multiplication of suitable matrices. A sequence of matrix transformations is thus obtained for the final evaluation. In this method, unless all the eigenvalues and eigenvectors are evaluated, the final output cannot be obtained.

Earlier transformation methods are due to Jacobi, Givens and House-holder. It is very difficult to obtain stable methods, which are reliable as well as convergent for unsymmetric eigenvalue problems. The Eberline extension of Jacobi's transformation method is noted for its lack of universal stability. As such, attempts have been made to develop stable forms of the Eberline method. The QR-transformation method is one the methods falling in this group in which the stiffness matrix $[K]$ is transformed into the tri-diagonal form and then rotation matrices are employed.

Techniques which come under the determinant search method, start by assuming a trial frequency and then finding out the value of the determinant $|[D] - \lambda[I]|$. Different methods differ in the process of evaluation of the determinant. Eigenvalues within the prescribed intervals can be obtained by plotting the above determinant against assumed frequency. The zero-crossing points give the natural frequencies of the system [Fig. 8.3]. The earlier methods that come under the category of frequency search methods are the Holzer method and the Myklestad method. The transfer matrix method also belongs to this group.

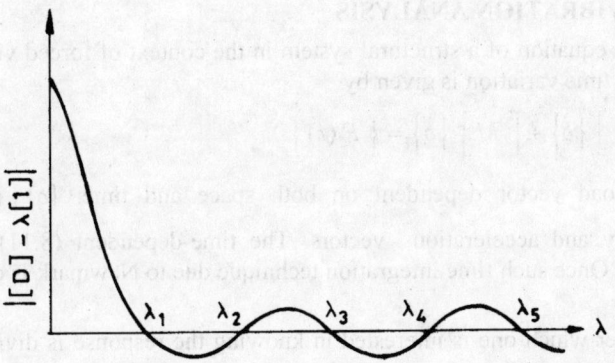

Fig. 8.3 Zero-crossing of frequencies

Methods that do not fall under the above categories are referred to as other methods. The Sturm sequence is one such method.

8.7.1 Simultaneous Iteration Technique

In the simultaneous iteration technique [8.13], $[K]$ is decomposed as

$$[K] = [L][L]^T \tag{8.8}$$

where $[L]$ is the lower triangular matrix.

Using (8.8), (8.5) can be written as

$$\{[L]^{-1}[M][L]^{-T}\}\{[L]^T\{\delta\}\} = \frac{1}{\omega^2}[L]^T\{\delta\} \tag{8.9}$$

or,

$$[A]\{x\} = \lambda\{x\} \tag{8.10}$$

Equation (8.10) represents a standard eigenvalue problem. Any standard procedure can be used to extract the eigenvalues and eigenvectors. The simultaneous iteration method is demonstrated very briefly below

1. Set trial vector $[U]$ and orthonormalize
2. Back substitute $[L]^T[X] = [U]$
3. Multiply $[Y] = [M][X]$
4. Forward substitute $[L]^T[V] = [Y]$
5. Form $[B] = [U]^T[V]$
6. Construct $[T]$ such that $t_{ii} = 1$ and $t_{ij} = 2b_{ij}[b_{ii} - b_{jj} + s(b_{ii} - b_{jj})^2]$

where s = sign of $(b_{ij} - b_{ii})$
7. Multiply $[W] = [V][T]$
8. Perform Schmidt orthonormalization to derive $[\overline{U}]$
9. Check tolerance $[U] = [\overline{U}]$
10. If it is not satisfactory, go to step 2

8.8 TRANSIENT VIBRATION ANALYSIS

The equilibrium equation of a structural system in the context of forced vibration analysis for a load having arbitrary time variation is given by

$$[K]\{\delta\} + [C]\{\dot{\delta}\} + [M]\{\ddot{\delta}\} = \{F(t)\} \tag{8.11}$$

where $\{F(t)\}$ is a load vector dependent on both space and time. $\{\delta\}, \{\dot{\delta}\}$ and $\{\ddot{\delta}\}$ are the displacement, velocity and acceleration vectors. The time-dependent (8.11) is solved by direct integration technique. Once such time integration technique due to Newmark is described very briefly [8.8, 8.12, 8.14].

Total time T over which one is interested in knowing the response is divided into n intervals, each of duration Δt

The following assumptions are used

$$\{\dot{\delta}\}^{t+\Delta t} = \{\dot{\delta}\}^t + \left[(1-\delta_1)\{\ddot{\delta}\}^t + \delta_1\{\ddot{\delta}\}^{t+\Delta t}\right]\Delta t \tag{8.12}$$

$$\{\delta\}^{t+\Delta t} = \{\delta\}^t + \{\dot{\delta}\}^t \Delta t \left[\left(\frac{1}{2} - \alpha_1\right)\{\ddot{\delta}\}^t + \alpha_1\{\ddot{\delta}\}^{t+\Delta t}\right]\Delta t^2 \tag{8.13}$$

where α_1 and δ_1 are parameters that can be determined to obtain integration accuracy and stability. The algorithm for Newmark's method is given below.

A. Initial calculations

1. Compute overall stiffness matrix $[K]$, overall damping matrix $[C]$ and overall mass matrix $[M]$
2. Initialise $\{\delta\}^0$, $\{\dot{\delta}\}^0$ and $\{\ddot{\delta}\}^0$
3. Select time step Δt, parameters α_1 and δ_1 and compute the following integration constants.

$\delta_1 \geq 0.5 \geq 0.25(0.5 + \delta_1)^2$

$$\left.\begin{array}{l} a_0 = \dfrac{1}{\alpha_1 \Delta t^2},\quad a_1 = \dfrac{\delta_1}{\alpha_1 \Delta t},\quad a_2 = \dfrac{1}{\alpha_1 \Delta t},\quad a_3 = \dfrac{1}{2\alpha_1} - 1 \\[1em] a_4 = \dfrac{\delta_1}{\alpha_1} - 1,\quad a_5 = \dfrac{\Delta t}{2}\left(\dfrac{\delta_1}{\alpha_1} - 2\right),\quad a_6 = \Delta t(1-\delta_1),\quad a_7 = \delta_1 \cdot \Delta t \end{array}\right\} \tag{8.14}$$

4. Formation of effective stiffness matrix $[\overline{K}]$

$$[\overline{K}] = [K] + a_0[M] + a_1[C]$$

5. Triangularisation of $[\overline{K}][K] = [L][D][L]^T$ is performed.

For each time step

1. Calculate effective loads at $t + \Delta t$:

$$\{\overline{P}\}^{t+\Delta t} = \{P\}^{t+\Delta t} + [M]\left(a_0\{\delta\}^t + a_2\{\dot{\delta}\}^t + a_3\{\ddot{\delta}\}^t\right) \qquad (8.15)$$

$$+ [C]\left(a_1\{\delta\}^t + a_4\{\dot{\delta}\}^t + a_5\{\ddot{\delta}\}^t\right)$$

2. Solve the displacements at time $t + \Delta t$

$$[L][D][L]^T\{\delta\}^{t+\Delta t} = \{\overline{P}\}^{t+\Delta t} \qquad (8.16)$$

3. Compute accelerations and velocities at time $t + \Delta t$:

$$\{\ddot{\delta}\}^{t+\Delta t} = a_0\left(\{\delta\}^{t+\Delta t} - \{\delta\}^t\right) - a_2\{\dot{\delta}\}^t - a_3\{\ddot{\delta}\}^t \qquad (8.17)$$

$$\{\dot{\delta}\}^{t+\Delta t} = \{\dot{\delta}\}^t + a_6\{\ddot{\delta}\}^t + a_7\{\ddot{\delta}\}^{t+\Delta t} \qquad (8.18)$$

The stresses/stress–resultants can be computed after the calculation of nodal displacements.

8.9 STIFFNESS MATRIX

The degrees of freedom of the structure are first identified. It is referred to as the number of independent joint displacements of the structure.

The intersections or interconnections between the individual elements of the structure are called joints or nodes. Displacements corresponding to each degree of freedom, referred to as nodal displacements, are numbered and positive directions are assumed.

In order to evaluate the elements of the stiffness matrix, the structure is assumed to deform corresponding to a unit value of one of the nodal displacements, while all other displacements are restrained. The forces at all the required points are evaluated. The process is repeated for a unit value for each of the nodal displacements. The formation of the stiffness matrix is explained with the help of an example.

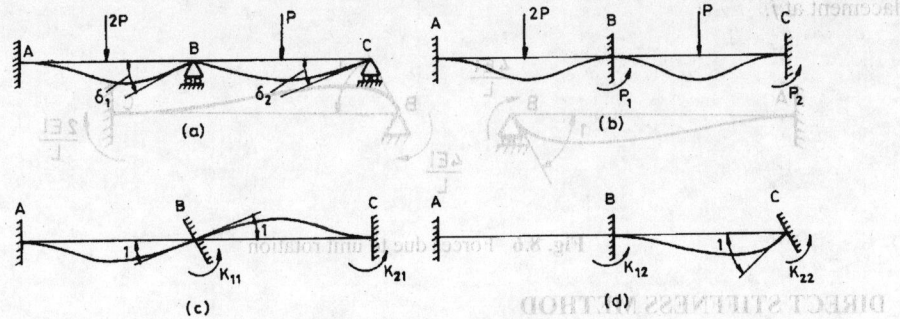

Fig. 8.4 Different steps in the displacement method

The continuous beam of Fig. 8.4, has two degrees of freedom numbered δ_1 and δ_2. Referring to Fig. 8.4 (b), joint B is allowed to rotate by unity in the counter-clockwise direction (i.e., positive direction, see Fig. 8.5) while joint C is kept fixed. The moment developed at B due to unit rotation at B is termed as K_{11} and the moment induced at C due to this displacement is K_{21}. Note that there is a relation of the labels 1 and 2 in K_{11} and K_{21} to joint displacements δ_1 and δ_2.

In order to determine K_{11}, the contribution from both the members AB and BC is to be taken into account. Referring to Fig. 8.6,

$$K_{11} = \frac{4EL}{L} + \frac{4EL}{L} = \frac{8EL}{L} \qquad (8.19a)$$

and

$$K_{21} = \frac{2EL}{L} \qquad (8.19b)$$

Similarly,

$$K_{12} = \frac{2EL}{L} \text{ and } K_{22} = \frac{4EL}{L} \qquad (8.19c)$$

The stiffness matrix for this problem is

$$\begin{bmatrix} K_{11} & K_{12} \\ K_{21} & K_{22} \end{bmatrix} = \frac{2EI}{L} \begin{bmatrix} 4 & 1 \\ 1 & 2 \end{bmatrix} \qquad (8.20)$$

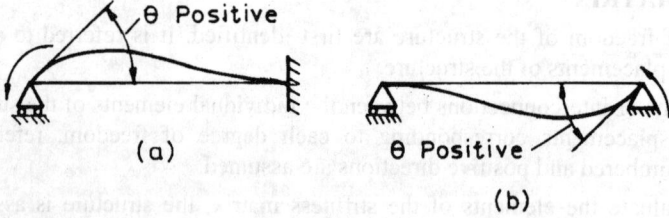

Fig. 8.5 Sign convention

The stiffness coefficient K_{ij} is defined as the force developed at point i corresponding to an unknown displacement at i, due to a unit value of the displacement at j corresponding to an unknown displacement at j.

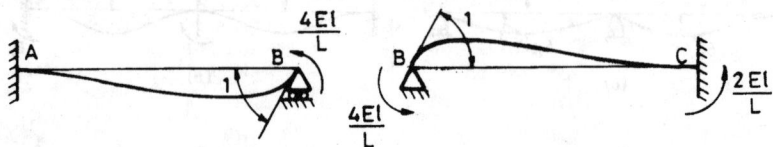

Fig. 8.6 Forces due to unit rotation

8.10 DIRECT STIFFNESS METHOD

Even in the trivial example of Fig. 8.4, it is seen that for determining K_{11}, the contribution from members connected at joint 1 have been taken into account. In this section, we present a more formalized approach to the formation of the stiffness matrix. It will be shown that the overall stiffness matrix of the structure will be generated automatically by the computer, based on the geometric and elastic properties of the individual element.

For this, a local coordinate system is chosen for a particular member. Global axis system is the common coordinate system, which is valid for the entire structure. The two-coordinate systems may be different, depending on the orientation of the member. The difference between the two is explained in Fig. 8.7, for a bar element which can be taken up by axial forces only.

The local and global axes systems are related by the transformation matrix [Fig. 8.7]. If the displacements in the local axis system for a particular member is given by $\{\delta'\}$ and the displacements in the corresponding global axis system are denoted by $\{\delta\}$, then

$$\{\delta'\} = [T]\{\delta\} \tag{8.21}$$

where $[T]$ is the member transformation matrix.

For the truss element of Fig. 8.7, the stiffness matrix of the member in local coordinates is

$$[K]_e = \frac{EA}{L} \begin{bmatrix} 1 & -1 \\ -1 & 1 \end{bmatrix} \tag{8.22}$$

and

$$[T] = \begin{bmatrix} l_1 & m_1 & 0 & 0 \\ 0 & 0 & l_1 & m_1 \end{bmatrix} \tag{8.23}$$

where

$l_1 = \cos\theta$ and $m_1 = \sin\theta$

It can be shown that the stiffness matrix in the global coordinate system is given by

$$[K]_e = [T]^T [k]_e [T] \tag{8.24}$$

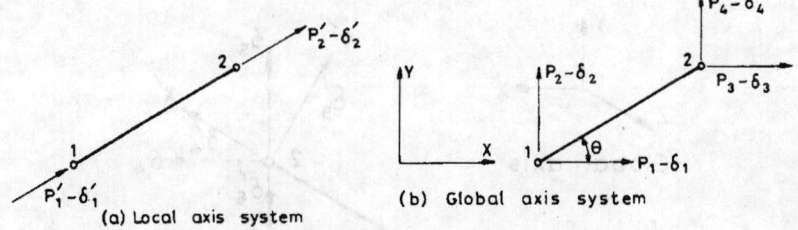

Fig. 8.7 Two sets of axes system

For the truss member

$$[K]_e = \frac{EA}{L} \begin{bmatrix} l_1^2 & & & \text{Symmetrical} \\ l_1 m_1 & m_1^2 & & \\ -l_1^2 & -l_1 m_1 & l_1^2 & \\ -l_1 m_1 & -m_1^2 & l_1 m_1 & m_1^2 \end{bmatrix} \tag{8.25}$$

A typical frame member is shown in Fig. 8.9. The member of the frame has three degrees of freedom at each end – two translations and one rotation. X', Y', Z' is the local axis system and X, Y, Z is the global axis system. The relationship between them is given by

$$\begin{Bmatrix} \delta_1' \\ \delta_2' \\ \delta_3' \\ \hline \delta_4' \\ \delta_5' \\ \delta_6' \end{Bmatrix} = \begin{bmatrix} l_1 & m_1 & 0 & 0 & 0 & 0 \\ -m_1 & l_1 & 0 & 0 & 0 & 0 \\ 0 & 0 & 1 & 0 & 0 & 0 \\ \hline 0 & 0 & 0 & l_1 & m_1 & 0 \\ 0 & 0 & 0 & -m_1 & l_1 & 0 \\ 0 & 0 & 0 & 0 & 0 & 1 \end{bmatrix} \begin{Bmatrix} \delta_1 \\ \delta_2 \\ \delta_3 \\ \hline \delta_4 \\ \delta_5 \\ \delta_6 \end{Bmatrix}$$

(8.26)

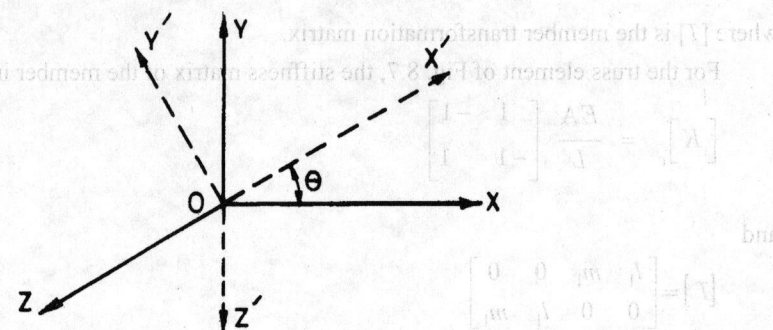

Fig. 8.8 Element of a truss

or, $\{\delta'\} = [T]\{\delta\}$ (8.27)

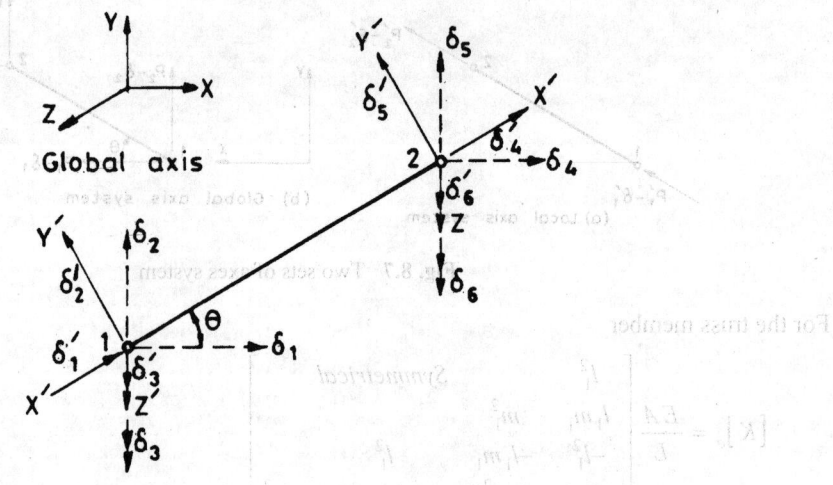

Fig. 8.9 A typical frame member

The stiffness matrix of the member in local axis system is

$$[K]_e = \begin{bmatrix} \dfrac{EA}{L} & 0 & 0 & -\dfrac{EA}{L} & 0 & 0 \\ & \dfrac{12EI}{L^3} & \dfrac{6EI}{L^2} & 0 & -\dfrac{12EI}{L^3} & \dfrac{6EI}{L^2} \\ & & \dfrac{4EI}{L} & 0 & \dfrac{6EI}{L^2} & \dfrac{2Ei}{L} \\ & & & \dfrac{EA}{L} & 0 & 0 \\ & \text{Symmetrical} & & & \dfrac{12EI}{L^3} & -\dfrac{6EI}{L^2} \\ & & & & & \dfrac{4EI}{L} \end{bmatrix} \quad (8.28)$$

The element stiffness matrix in the global coordinate system can be formed by using (8.24).

8.11 OVERALL STIFFNESS MATRIX

The stiffness matrix of an element of the structure forms the basic component. It is now discussed how the element stiffness matrix of the individual element is assembled to form the overall stiffness matrix of the structure.

It might have been noticed earlier that the stiffness at a joint is obtained by adding the stiffness of all the members meeting at the joint.

The degrees of freedom of the structure are first numbered – starting with 1 and ending with NP, where NP is the total degrees of freedom. The restraints are then numbered beyond NP. This numbering is referred to as degrees of freedom corresponding to global numbering or global degrees of freedom. All the restraints can be given the number ($NP + 1$). This procedure will save some storage space.

We have already introduced the concept of local axis system and global axis system. We shall henceforth deal with two different numberings in the global axis system for the displacements at member ends. The first set of numbering is referred to as local numbering, which will remain identical for every member. This numbering refers to the typical displacement labels of the member. The members meet at a joint in the structure. The same member ends will have different displacement labels, when the total structure is considered.

The continuous beam of Fig. 8.10 has four degrees of freedom – rotations at four supports. In Fig. 8.10(a), the numbering from the left-hand side corresponding to global coordinates has been indicated. In Fig. 8.10(b), the local numbering of a typical member and in Fig. 8.10(c), the positive direction of the displacement has been shown. The element stiffness matrix is to be formulated next. This is done for a typical element on the basis of local numbering. For the problem at hand, the size of the element stiffness matrix is 4×4. For the first element, they are, say

$$\begin{array}{c} \text{Global} \quad 5 \quad 1 \quad 6 \quad 2 \\ \text{Local} \quad 1 \quad 2 \quad 3 \quad 4 \\ [k]_e^1 = \begin{array}{c} 5 \\ 1 \\ 6 \\ 2 \end{array} \begin{array}{c} 1 \\ 2 \\ 3 \\ 4 \end{array} \begin{bmatrix} a_{11} & a_{12} & a_{13} & a_{14} \\ a_{21} & a_{22} & a_{23} & a_{24} \\ a_{31} & a_{32} & a_{33} & a_{34} \\ a_{41} & a_{42} & a_{43} & a_{44} \end{bmatrix} \end{array} \quad (8.29)$$

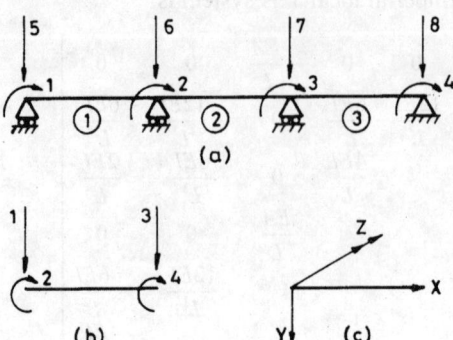

Fig. 8.10 Local and global numbering of a continuous beam

For members which are inclined and where local and global coordinates are not coincident, the element stiffness matrix is to be evaluated on the basis of transformation of (8.24). There is a one-to-one correspondence between the joints of the elements and that of the assembly. The elements of element stiffness matrix of (8.29) should now be put in their proper place in the overall stiffness matrix. Equation (8.29) indicates that the local numbering 1 corresponds to global numbering 5, which means that a_{11} should occupy the fifth row and the fifth column (i.e., K_{55}) of the overall stiffness matrix. Local number 2 corresponds to global number 1. Therefore, a_{12}, a_{21} will occupy positions of K_{51} and K_{11} of the overall stiffness matrix. The proper location of all elements of (8.29) in the overall stiffness matrix is shown in Fig. 8.11.

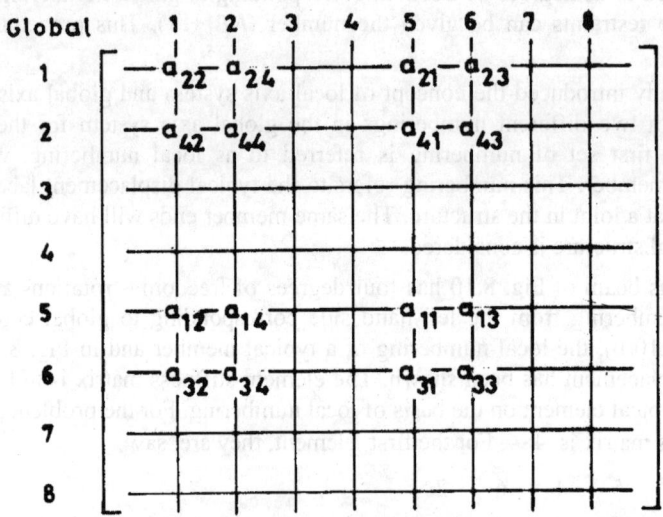

Fig. 8.11 Assembly in global system

The process is to be repeated for all elements. It is to be borne in mind that individual stiffnesses at a particular location in the overall stiffness matrix are to be added. If the element stiffness matrices for the members 2 and 3 are

Finite Element Analysis of Composite Structures 185

$$[k]_e^2 = \begin{array}{c} \text{Global} \\ \text{Local} \\ \\ \begin{matrix} 6 \\ 2 \\ 7 \\ 3 \end{matrix} \end{array} \begin{array}{c} \\ \\ \begin{matrix} 1 \\ 2 \\ 3 \\ 4 \end{matrix} \end{array} \begin{bmatrix} b_{11} & b_{12} & b_{13} & b_{14} \\ b_{21} & b_{22} & b_{23} & b_{24} \\ b_{31} & b_{32} & b_{33} & b_{34} \\ b_{41} & b_{42} & b_{43} & b_{44} \end{bmatrix} \begin{matrix} 6 & 2 & 7 & 3 \\ 1 & 2 & 3 & 4 \end{matrix} \quad (8.30)$$

and

$$[k]_e^3 = \begin{array}{c} \text{Global} \\ \text{Local} \\ \\ \begin{matrix} 7 \\ 3 \\ 8 \\ 4 \end{matrix} \end{array} \begin{array}{c} \\ \\ \begin{matrix} 1 \\ 2 \\ 3 \\ 4 \end{matrix} \end{array} \begin{bmatrix} c_{11} & c_{12} & c_{13} & c_{14} \\ c_{21} & c_{22} & c_{23} & c_{24} \\ c_{31} & c_{32} & c_{33} & c_{34} \\ c_{41} & c_{42} & c_{43} & c_{44} \end{bmatrix} \begin{matrix} 7 & 3 & 8 & 4 \\ 1 & 2 & 3 & 4 \end{matrix} \quad (8.31)$$

then, the overall stiffness matrix is shown in Fig. 8.12. The steps to be followed in the computer program are shown in the flow chart of Fig. 8.13. The overall mass matrix, overall damping matrix and the overall exciting force matrix can similarly be formed from the respective element matrices.

There are various ways, by which labels in the displacements are put. In the scheme that has been discussed, the active degrees of freedom are labelled (their total is NP) first and then the restraints are labelled. The first set of equations NP is operated upon, for obtaining the necessary solution. The procedure of assembly of stiffness matrices for two and three-dimensional problems is similar.

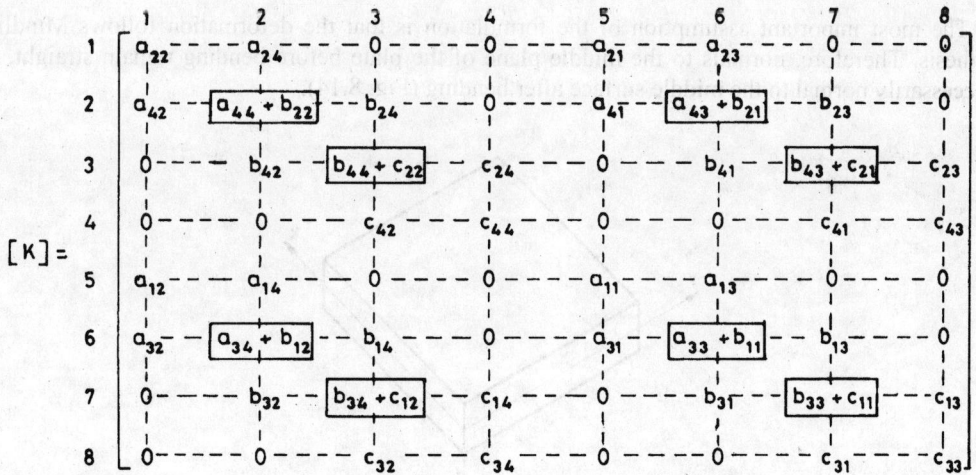

Fig. 8.12 Overall stiffness matrix

8.12 ISOPARAMETRIC ELEMENT FOR THE ANALYSIS OF LAMINATED PLATES

The formulation of an eight-noded isoparametric element for the analysis of the laminated plate is presented in the following [8.15, 8.16]. The element can incorporate transverse shear deformation and is therefore, ideally suited to the analysis of composite plates for which transverse shear deformation is very important. This is due to the large ratio of the longitudinal modulus to transverse shear modulus of the composites.

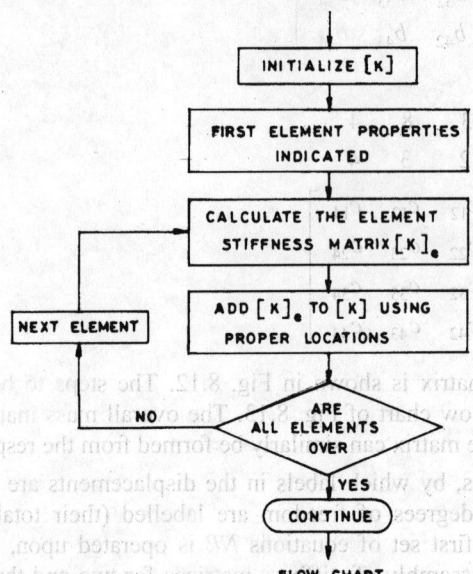

Fig. 8.13 Flow chart for assembly of elements

The most important assumption of the formulation is that the deformation follows Mindlin's hypothesis. Therefore, normals to the middle plane of the plate before bending remain straight, but not necessarily normal to the middle surface after bending (Fig. 8.14).

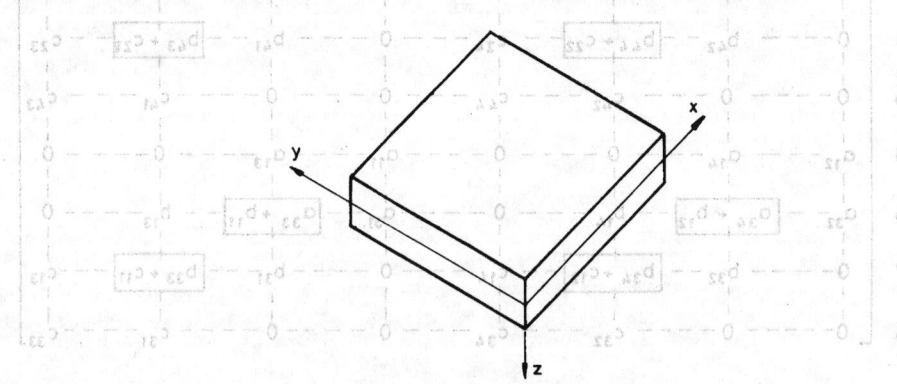

Fig. 8.14 Plate axis system

8.12.1 Constitutive Relations

The displacement field of a laminated plate with respect to the x, y, z coordinate axis system is of the following form [Fig. 8.15].

$$\{\bar{f}\} = \begin{Bmatrix} u \\ v \\ w \end{Bmatrix} = \begin{Bmatrix} u_0 \\ v_0 \\ w \end{Bmatrix} - z \begin{Bmatrix} \theta_x \\ \theta_y \\ 0 \end{Bmatrix} \quad (8.32)$$

where u, v and w are the displacements in the x, y and z-directions at any point and u_0, v_0 and w are those at the middle plane of the plate (Fig. 8.14). θ_x and θ_y are the rotations of the normal to the undeformed mid-plane.

The middle plane of the plate is considered as the reference plane of the plate. This is also a requirement for a general unsymmetrically laminated plate for which the middle plane is not the neutral plane.

The strain components according to the strain–displacement relationship of linear elasticity are

$$\begin{Bmatrix} \dfrac{\partial u}{\partial x} \\ \dfrac{\partial v}{\partial y} \\ \dfrac{\partial u}{\partial y} + \dfrac{\partial v}{\partial x} \\ \dfrac{\partial u}{\partial z} + \dfrac{\partial w}{\partial x} \\ \dfrac{\partial w}{\partial y} + \dfrac{\partial v}{\partial z} \end{Bmatrix} = \begin{Bmatrix} \dfrac{\partial u_0}{\partial x} \\ \dfrac{\partial v_0}{\partial y} \\ \dfrac{\partial u_0}{\partial y} + \dfrac{\partial v_0}{\partial x} \\ \dfrac{\partial w}{\partial x} - \theta_x \\ \dfrac{\partial w}{\partial y} - \theta_y \end{Bmatrix} - z \begin{Bmatrix} \dfrac{\partial \theta_x}{\partial x} \\ \dfrac{\partial \theta_y}{\partial y} \\ \dfrac{\partial \theta_x}{\partial y} + \dfrac{\partial \theta_y}{\partial x} \\ 0 \\ 0 \end{Bmatrix} \quad (8.33)$$

or,

$$\begin{Bmatrix} \varepsilon_x \\ \varepsilon_y \\ \varepsilon_s \\ \varepsilon_{xz} \\ \varepsilon_{yz} \end{Bmatrix} = \begin{Bmatrix} \varepsilon_x^0 \\ \varepsilon_y^0 \\ \varepsilon_s^0 \\ \varepsilon_{xz}^0 \\ \varepsilon_{yz}^0 \end{Bmatrix} + z \begin{Bmatrix} \kappa_x \\ \kappa_y \\ \kappa_s \\ 0 \\ 0 \end{Bmatrix} \quad (8.34)$$

where ε_x^0, ε_y^0 and ε_s^0 are the midplane strains and κ_x, κ_y and κ_s are the curvatures of the laminated plate and from Fig. 8.15.

$$\varepsilon_{xz} = \varepsilon_{xz}^0 = \frac{\partial w}{\partial x} - \theta_x = -\phi_x$$

$$\varepsilon_{yz} = \varepsilon_{yz}^0 = \frac{\partial w}{\partial y} - \theta_y = -\phi_y \quad (8.35)$$

The laminate consists of a number of orthotropic layers oriented arbitrarily with respect to the global coordinate system. The stress-strain relationship of any lamina (kth) with respect to its orthotropic axis (onaxis), i.e., 1, 2, 3 (Fig. 8.16) is [see (4.34)]

$$\begin{Bmatrix} \sigma_1 \\ \sigma_2 \\ \sigma_6 \end{Bmatrix}_k = \begin{bmatrix} \overline{m}E_1 & \overline{m}\nu_{21}E_1 & 0 \\ \overline{m}\nu_{12}E_2 & \overline{m}E_2 & 0 \\ 0 & 0 & E_6 \end{bmatrix}_k \begin{Bmatrix} \varepsilon_1^0 \\ \varepsilon_2^0 \\ \varepsilon_6^0 \end{Bmatrix}_k \quad (8.36)$$

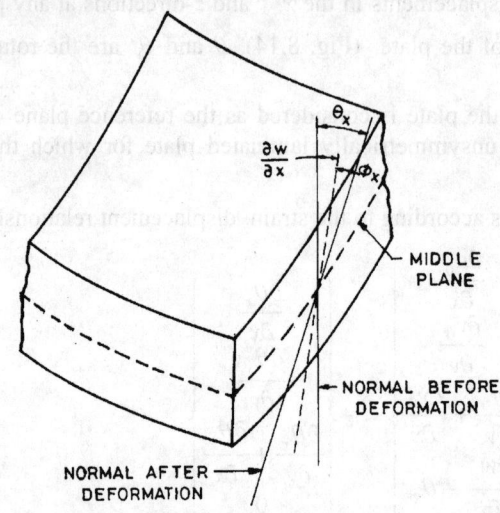

Fig. 8.15 Deformation of the plate in x-z plane

where

$$\overline{m} = \frac{1}{1 - \nu_{12}\nu_{21}} \quad (8.37)$$

and

$$\begin{Bmatrix} \sigma_4 \\ \sigma_5 \end{Bmatrix}_k = \begin{bmatrix} k_s G_{13} & 0 \\ 0 & k_s G_{23} \end{bmatrix} \begin{Bmatrix} \varepsilon_4^0 \\ \varepsilon_5^0 \end{Bmatrix}_k \quad (8.38)$$

in which k_s = shear correction factor = 5/6 for a rectangle

The above relations can be combined as

$$\begin{Bmatrix} \sigma_1 \\ \sigma_2 \\ \sigma_6 \\ \sigma_4 \\ \sigma_5 \end{Bmatrix}_k = \begin{bmatrix} \overline{m}E_1 & \overline{m}\nu_{21}E_1 & 0 & 0 & 0 \\ \overline{m}\nu_{12}E_2 & \overline{m}E_2 & 0 & 0 & 0 \\ 0 & 0 & E_6 & 0 & 0 \\ 0 & 0 & 0 & k_s G_{13} & 0 \\ 0 & 0 & 0 & 0 & k_s G_{23} \end{bmatrix}_k \begin{Bmatrix} \varepsilon_1^0 \\ \varepsilon_2^0 \\ \varepsilon_6^0 \\ \varepsilon_4^0 \\ \varepsilon_5^0 \end{Bmatrix}_k \quad (8.39)$$

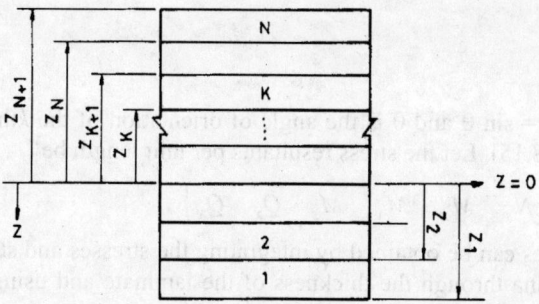

(a) Stacking sequence

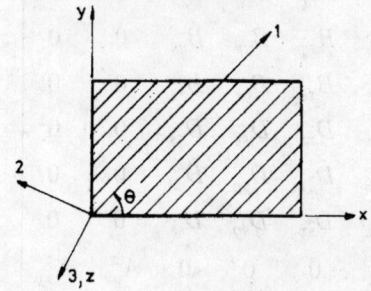

(b) Details of the kth lamina

The stress–strain relationship with respect to global *xyz* axis system (off-axis) is given by

$$\begin{Bmatrix} \sigma_x \\ \sigma_y \\ \sigma_s \\ \sigma_{xz} \\ \sigma_{yz} \end{Bmatrix}_k = \begin{bmatrix} Q_{xx} & Q_{xy} & Q_{xs} & 0 & 0 \\ Q_{yx} & Q_{yy} & Q_{ys} & 0 & 0 \\ Q_{sx} & Q_{sy} & Q_{ss} & 0 & 0 \\ 0 & 0 & 0 & k_s Q^z_{xx} & k_s Q^z_{xy} \\ 0 & 0 & 0 & k_s Q^z_{yx} & k_s Q^z_{yy} \end{bmatrix}_k \begin{Bmatrix} \epsilon_x \\ \epsilon_y \\ \epsilon_s \\ \epsilon_{xz} \\ \epsilon_{yz} \end{Bmatrix}_k \quad (8.40)$$

The reader can refer to (5.7) for part derivations of (8.40). The equation (8.40) is derived using the following transformations.

$$[Q_{ij}]_k = [T_1]^T [Q_{rs}]_k [T_1] \qquad i, j = x, y, s \quad (8.41)$$

$$[Q^z_{ij}]_k = [T_2]^T [Q^s_{pq}]_k [T_2] \qquad i, j = x, y \quad (8.42)$$

in which $[Q_{rs}]_k$ and $[Q^s_{pq}]_k$ are defined in (8.36) and (8.38) respectively. $[T_1]$ and $[T_2]$ are the transformation matrices expressed as

$$[T_1] = \begin{bmatrix} m^2 & n^2 & mn \\ n^2 & m^2 & -mn \\ -2mn & 2mn & m^2 - n^2 \end{bmatrix} \quad (8.43)$$

and

$$[T_2] = \begin{bmatrix} m & -n \\ -n & m \end{bmatrix} \quad (8.44)$$

where $m = \cos\theta$ and $n = \sin\theta$ and θ is the angle of orientation of the kth layer with respect to the global axis system (Fig. 8.15). Let the stress resultants per unit length be

$$\{F\}^T = \{N_x \quad N_y \quad N_s \quad M_x \quad M_y \quad M_s \quad Q_x \quad Q_y\} \quad (8.45)$$

The above quantities can be obtained by integrating the stresses and stress couples derived from the stresses of each lamina through the thickness of the laminate and using (8.34), are expressed as [see (5.65)]

$$\begin{Bmatrix} N_x \\ N_y \\ N_s \\ M_x \\ M_y \\ M_s \\ Q_x \\ Q_y \end{Bmatrix} = \begin{bmatrix} A_{xx} & A_{xy} & A_{xs} & B_{xx} & B_{xy} & B_{xs} & 0 & 0 \\ A_{yx} & A_{yy} & A_{ys} & B_{xy} & B_{yy} & B_{ys} & 0 & 0 \\ A_{xs} & A_{ys} & A_{ss} & B_{xs} & B_{ys} & B_{ss} & 0 & 0 \\ B_{xx} & B_{xy} & B_{xs} & D_{xx} & D_{xy} & D_{xs} & 0 & 0 \\ B_{yx} & B_{yy} & B_{ys} & D_{yx} & D_{yy} & D_{ys} & 0 & 0 \\ B_{sx} & B_{sy} & B_{ss} & D_{sx} & D_{sy} & D_{ss} & 0 & 0 \\ 0 & 0 & 0 & 0 & 0 & 0 & A^z_{xx} & A^z_{xy} \\ 0 & 0 & 0 & 0 & 0 & 0 & A^z_{yx} & A^z_{yy} \end{bmatrix}_k \begin{Bmatrix} \varepsilon^0_x \\ \varepsilon^0_y \\ \varepsilon^0_s \\ \kappa_x \\ \kappa_y \\ \kappa_s \\ \varepsilon^0_{xz} \\ \varepsilon^0_{yz} \end{Bmatrix} \quad (8.46)$$

or, $\quad \{F\} = [D]\{\varepsilon\} \quad (8.47)$

Here,

$$A_{ij}, B_{ij}, D_{ij} = \sum_{k=1}^{N} \int_{z_k}^{z_{k+1}} (Q_{ij})_k (1, z, z^2) dz, \quad i, j = x, y, s \quad (8.48)$$

$$A^z_{ij} = \sum_{k=1}^{N} \int_{z_k}^{z_{k+1}} k_s (Q^z_{ij})_k dz, \quad i, j = x, y \quad (8.49)$$

where N is the number of layers of the laminate.

8.12.2 Plate Element Formulation

The displacement components of node r of the eight-noded isoparametric element are (Fig. 8.17)

$$\{\delta_r\}^T_e = \{u_{0_r} \quad v_{0_r} \quad w_r \quad \theta_{xr} \quad \theta_{yr}\} \quad (8.50)$$

The displacement at any point on the mid-plane of the element can be expressed as

$$\{f\} = \begin{Bmatrix} u_0 \\ v_0 \\ w \\ \theta_x \\ \theta_y \end{Bmatrix} = \sum_{r=1}^{8} N_r [I_5] \begin{Bmatrix} u_{0_r} \\ v_{0_r} \\ w_r \\ \theta_{xr} \\ \theta_{yr} \end{Bmatrix} = [N]\{\delta\}_e \quad (8.51)$$

where $[I_5]$ is a 5×5 identity matrix.

N_r is the shape function expressed in terms of the $\xi - \eta$ coordinates of the element and they are
[8.15, 8.16]

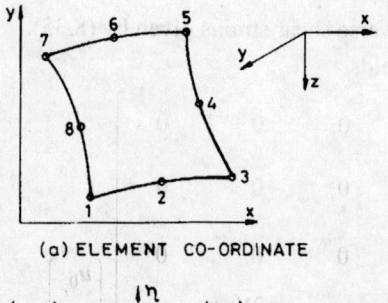

(a) ELEMENT CO-ORDINATE

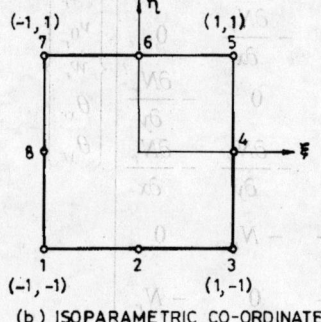

(b) ISOPARAMETRIC CO-ORDINATE

Fig. 8.17 Isoparametric quadratic plate element

At the corner nodes, ($\xi_r = \pm 1$ and $\eta_r = \pm 1$),

$$N_r = \frac{1}{4}\left(1 + \xi_0\right)\left(1 + \eta_0\right)\left(\xi_0 + \eta_0 - 1\right) \tag{8.52}$$

$\xi_0 = \xi \xi_r$ and $\eta_0 = \eta \eta_r$

At middle nodes 2 and 6, ($\xi_r = 0$, $\eta_r = \pm 1$)

$$N_r = \frac{1}{2}\left(1 - \xi^2\right)\left(1 + \eta_0\right) \tag{8.53}$$

At middle nodes 4 and 8, ($\xi_r = \pm 1$, $\eta_r = 0$)

$$N_r = \frac{1}{2}\left(1 + \xi_0\right)\left(1 - \eta^2\right) \tag{8.54}$$

The coordinates within the element are expressed as

$$\begin{Bmatrix} x \\ y \end{Bmatrix} = \sum_{r=1}^{8} N_r \begin{bmatrix} 1 & 0 \\ 0 & 1 \end{bmatrix} \begin{Bmatrix} x_r \\ y_r \end{Bmatrix} \tag{8.55}$$

N_r is the same shape function as discussed above.

x_r, y_r are the coordinates of node r.

The generalised strain components are

$$\{\varepsilon\}^T = \left\{ \frac{\partial u_0}{\partial x} \quad \frac{\partial v_0}{\partial y} \quad \left(\frac{\partial u_0}{\partial y} + \frac{\partial v_0}{\partial x} \right) - \frac{\partial \theta_x}{\partial x} - \frac{\partial \theta_y}{\partial y} - \left(\frac{\partial \theta_x}{\partial y} + \frac{\partial \theta_y}{\partial x} \right) - \phi_x - \phi_y \right\} \quad (8.56)$$

where $-\phi_x$ and $-\phi_y$ are the transverse shear strains given by (8.35).
Combining (8.56) and (8.51), yields,

$$\{\varepsilon\} = \sum_{r=1}^{8} \begin{bmatrix} \frac{\partial N_r}{\partial x} & 0 & 0 & 0 & 0 \\ 0 & \frac{\partial N_r}{\partial y} & 0 & 0 & 0 \\ \frac{\partial N_r}{\partial y} & \frac{\partial N_r}{\partial x} & 0 & 0 & 0 \\ 0 & 0 & 0 & -\frac{\partial N_r}{\partial x} & 0 \\ 0 & 0 & 0 & 0 & -\frac{\partial N_r}{\partial y} \\ 0 & 0 & 0 & -\frac{\partial N_r}{\partial y} & -\frac{\partial N_r}{\partial x} \\ 0 & 0 & \frac{\partial N_r}{\partial x} & -N_r & 0 \\ 0 & 0 & \frac{\partial N_r}{\partial y} & 0 & -N_r \end{bmatrix} \begin{Bmatrix} u_{0_r} \\ v_{0_r} \\ w_r \\ \theta_{xr} \\ \theta_{yr} \end{Bmatrix} \quad (8.57)$$

or,

$$\{\varepsilon\} = \sum_{r=1}^{8} [B]_r \{\delta_r\}_e = [B]\{\delta\}_e \quad (8.58)$$

The shape function derivatives may be calculated by using the chain rule

$$\frac{\partial N_r}{\partial x} = \frac{\partial N_r}{\partial \xi} \cdot \frac{\partial \xi}{\partial x} + \frac{\partial N_r}{\partial \eta} \cdot \frac{\partial \eta}{\partial x} \quad (8.59)$$

$$\frac{\partial N_r}{\partial y} = \frac{\partial N_r}{\partial \xi} \cdot \frac{\partial \xi}{\partial y} + \frac{\partial N_r}{\partial \eta} \cdot \frac{\partial \eta}{\partial y}$$

Equation (8.59) can be written as

$$\begin{Bmatrix} \frac{\partial N_r}{\partial x} \\ \frac{\partial N_r}{\partial y} \end{Bmatrix} = \begin{bmatrix} \frac{\partial \xi}{\partial x} & \frac{\partial \eta}{\partial x} \\ \frac{\partial \xi}{\partial y} & \frac{\partial \eta}{\partial y} \end{bmatrix} \begin{Bmatrix} \frac{\partial N_r}{\partial \xi} \\ \frac{\partial N_r}{\partial \eta} \end{Bmatrix} \quad (8.60)$$

or,

$$\begin{Bmatrix} \frac{\partial N_r}{\partial x} \\ \frac{\partial N_r}{\partial y} \end{Bmatrix} = [J]^{-1} \begin{Bmatrix} \frac{\partial N_r}{\partial \xi} \\ \frac{\partial N_r}{\partial \eta} \end{Bmatrix} \quad (8.61)$$

where

$$[J] = \begin{bmatrix} \dfrac{\partial x}{\partial \xi} & \dfrac{\partial y}{\partial \xi} \\ \dfrac{\partial x}{\partial \eta} & \dfrac{\partial y}{\partial \eta} \end{bmatrix} = \text{Jacobian matrix} \qquad (8.62)$$

Combining eqns. (8.47) and (8.58), yields

$$\{F\} = [D][B]\{\delta\}_e \qquad (8.63)$$

The element stiffness matrix is derived here using the principle of minimum potential energy. The potential energy of the plate element is given by

$$\Phi = \frac{1}{2}\iint \{\varepsilon\}^T \{F\} dx\,dy - \iint \{f\}^T q\, dx\,dy \qquad (8.64)$$

where q is any discrete loading inside the element.

Combining (8.51), (8.58), (8.63) and (8.64), yields

$$\Phi = \frac{1}{2}\iint \{\delta\}_e^T [B]^T [D][B]\{\delta\}_e dx\,dy - \iint \{\delta\}_e^T [N]^T q\, dx\,dy \qquad (8.65)$$

The principle of minimum potential energy requires

$$\left\{\frac{\partial \Phi}{\partial \{\delta\}_e}\right\} = \{0\} \qquad (8.66)$$

Therefore, performing partial differentiation of (8.65), yields

$$\left\{\frac{\partial \Phi}{\partial \{\delta\}_e}\right\} = \iint [B]^T [D][B] dx\, dy\, \{\delta\}_e - \iint [N]^T q\, dx\, dy = \{0\} \qquad (8.67)$$

or, $\quad [K]_e \{\delta\}_e = \{P\}_e \qquad (8.68)$

where

$$[K]_e = \iint [B]^T [D][B] dx\, dy \qquad (8.69)$$

$$\{P\}_e = \iint [N]^T q\, dx\, dy \qquad (8.70)$$

$[B]$ in the element stiffness matrix. $[K]_e$ in (8.69) is expressed in terms of ξ and η. As such, differential area $dx\,dy$ in (8.69) is to be replaced by

$$dx\,dy = |J|\, d\xi\, d\eta \qquad (8.71)$$

where $|J|$ is the determinant of the Jacobian matrix $[J]$. Therefore, the element stiffness matrix is rewritten as

$$[K]_e = \iint [B]^T [D][B] |J|\, d\xi\, d\eta \qquad (8.72)$$

An explicit evaluation of $[K]_e$ is indeed tedious. The integration is carried out by employing the Gaussian quadrature rule. It has been found that even for an isotropic plate, nodes are the worst sampling points for determination of stress resultants. The transverse shear values Q_x and Q_y in particular are reasonable only at the Gauss points and are highly erratic at any other point. This phenomenon is referred to as shear locking. One of the means of avoiding shear locking is to use the reduced integration technique. A 2 × 2 Gauss integration is performed for the evaluation of the stiffness matrix in order to avoid shear locking effect.

8.12.3 Mass Matrix of the Plate

The use of the consistent mass matrix does not afford any special advantage for plate problems; rather the formulation takes more time and computer storage [8.17]. As such, lumped mass matrix is preferred. This is a diagonal matrix with non-zero entries u, v and w only. All the remaining entries including diagonals corresponding to θ_x and θ_y are zero. Hence, the lumped mass matrix for node r is given by

$$[M]_r = \begin{bmatrix} m_{rr} & & & & 0 \\ & m_{rr} & & & \\ & & m_{rr} & & \\ & 0 & & 0 & \\ & & & & 0 \end{bmatrix} \qquad (8.73)$$

where

$$m_{rr} = \frac{\int \rho t \, N_r N_r dA \; \int \rho t \, dA}{\sum_{r=1}^{8} \int \rho t \, N_r N_r dA} \qquad (8.74)$$

Equation (8.74) reveals that the total mass is scaled up.

8.13 FORMULATION OF THE COMPOSITE STIFFENER ELEMENT

A plate structure is strengthened by means of stiffeners. Various attempts have been made to analyse the plate-stiffener combination. The attempt to model the stiffened plates as orthotropic plates or grillages is well-known along with their merits and demerits.

Because of the high ratio of extensional modules to transverse shear modulus, transverse shear deformation effects are more pronounced in the composite laminates and are needed to be incorporated in composite beams.

To include transverse shear deformation in the displacement-based beam element, two basic methods are generally used. In the first method, shear deformation aspects are added to these beam elements. In this case, extra nodal degrees of freedom are generally utilised to represent shear deformation. The second method is the isoparametric formulation in which displacements and rotations are interpolated independently. In this element, shear locking can occur in high slenderness ratio structures.

8.14 FORMULATION OF THE COMPOSITE BEAM ELEMENT

In the following, a laminated composite beam element is presented in which shear deformation is introduced based on classical laminated theory without adding an extra nodal degree of freedom [8.18].

8.14.1 Formulation

To introduce the shear deformation of the first order theory into a laminate based on the classical plate theory, the transverse deflection is considered as a sum of two partial deflections – the deflection due to bending and the deflection due to transverse shear. Then the displacement field can be assumed as

$$u(x, y, z) = u_0(x, y) - z w^b_{,x}$$

$$v(x, y, z) = v_0(x, y) - z w^b_{,y}$$

$$w(x, y, z) = w^b(x, y) + w^s(x, y) \tag{8.75}$$

where u, v, w are in the inplane and transverse displacement components at any point in the laminate along the x, y, z axes respectively.

u_0, v_0 are middle surface displacements along x and y-axes.

w^b, w^s are middle surface transverse displacement components due to bending and shear along the z-axis.

The associated strains are

$$\varepsilon_x = \varepsilon_{x_0} + z k_x$$

$$\varepsilon_y = \varepsilon_{y_0} + z k_y$$

$$\varepsilon_{xy} = \varepsilon_{xy_0} + z k_{xy}$$

$$\varepsilon_{xz} = w^s_{,x}$$

$$\varepsilon_{yz} = w^s_{,y} \tag{8.76}$$

where

$$\varepsilon_{x_0} = u_{0,x}, \quad \varepsilon_{y_0} = v_{0,y}, \quad \varepsilon_{xy_0} = u_{0,y} + v_{0,x}$$

$$k_x = - w^b_{,xx}, \quad k_y = - w^b_{,yy}, \quad k_{xy} = - 2 w^b_{,xy} \tag{8.77}$$

$\varepsilon_{x_0}, \varepsilon_{y_0}, \varepsilon_{xy_0}$ are middle surface strains.

The stress–strain relationship of the kth layer for a laminate for bending deformation is

$$\begin{Bmatrix} \sigma_x \\ \sigma_y \\ \sigma_{xy} \end{Bmatrix}_k = \begin{bmatrix} Q_{xx} & Q_{xy} & Q_{xs} \\ Q_{yx} & Q_{yy} & Q_{ys} \\ Q_{sx} & Q_{sy} & Q_{ss} \end{bmatrix}_k \begin{Bmatrix} \varepsilon_x \\ \varepsilon_y \\ \varepsilon_{xy} \end{Bmatrix}_k \tag{8.78}$$

and the stress–strain relationship for shearing deformation is

$$\begin{Bmatrix} \sigma_{xz} \\ \sigma_{yz} \end{Bmatrix}_k = \begin{bmatrix} k_s Q^z_{xx} & k_s Q^z_{xy} \\ k_s Q^z_{yx} & k_s Q^z_{yy} \end{bmatrix} \begin{bmatrix} \varepsilon_{xz} \\ \varepsilon_{yz} \end{bmatrix} \tag{8.79}$$

The force and moment resultants for a laminated plate are

$$\begin{Bmatrix} N_x \\ N_y \\ N_{xy} \\ M_x \\ M_y \\ M_{xy} \end{Bmatrix} = \begin{bmatrix} A_{xx} & A_{xy} & A_{xs} & B_{xx} & B_{xy} & B_{xs} \\ A_{yx} & A_{yy} & A_{ys} & B_{yx} & B_{yy} & B_{ys} \\ A_{sx} & A_{sy} & A_{ss} & B_{sx} & B_{sy} & B_{ss} \\ B_{xx} & B_{xy} & B_{xs} & D_{xx} & D_{xy} & D_{xs} \\ B_{yx} & B_{yy} & B_{ys} & D_{yx} & D_{yy} & D_{ys} \\ B_{sx} & B_{sy} & B_{ss} & D_{sx} & D_{sy} & D_{ss} \end{bmatrix} \begin{Bmatrix} \varepsilon_{x_0} \\ \varepsilon_{y_0} \\ \varepsilon_{xy_0} \\ k_x \\ k_y \\ k_{xy} \end{Bmatrix} \quad (8.80)$$

and the transverse shear force–strain relationship becomes

$$\begin{Bmatrix} Q_x \\ Q_y \end{Bmatrix} = \begin{bmatrix} A^z_{xx} & A^z_{xy} \\ A^z_{xy} & A^z_{yy} \end{bmatrix} \begin{Bmatrix} \varepsilon_{xz} \\ \varepsilon_{yz} \end{Bmatrix} \quad (8.81)$$

where $A_{ij}, B_{ij}, D_{ij} = \int_{-h/2}^{h/2} Q_{ij}(1, z, z^3)\, dz \quad (i, j = x, y, s)$

$$A^z_{ij} = \int_{-h/2}^{h/2} Q^z_{ij}\, dz \quad (i, j = x, y) \quad (8.82)$$

Now the above constitutive equations of the laminated plate can be reduced to those of a composite laminated beam by equating N_y, M_y and ε_{yz} equal to zero and writing ε_{y_0} and k_y in terms of $\varepsilon_{x_0}, \varepsilon_{xy_0}, k_x$ and k_{xy} in plate constitutive equations.

$$\begin{Bmatrix} N_x \\ N_{xy} \\ M_x \\ M_{xy} \\ Q_x \end{Bmatrix} = \begin{bmatrix} \overline{C}_{xx} & \overline{C}_{xs} & \overline{B}_{xx} & \overline{B}_{xs} & 0 \\ \overline{C}_{xs} & \overline{C}_{ss} & \overline{B}_{xs} & \overline{B}_{ss} & 0 \\ \overline{B}_{xx} & \overline{B}_{xs} & \overline{D}_{xx} & \overline{D}_{xs} & 0 \\ \overline{B}_{xs} & \overline{B}_{ss} & \overline{D}_{xs} & \overline{D}_{ss} & 0 \\ 0 & 0 & 0 & 0 & \overline{C}^z_{xx} \end{bmatrix} \begin{Bmatrix} \varepsilon_{x_0} \\ \varepsilon_{xy_0} \\ k_x \\ k_{xy} \\ \varepsilon_{xz} \end{Bmatrix} \quad (8.83)$$

where

$\overline{C}_{xx} = A_{xx} + (A_{yy} B_{xy}^2 - 2 A_{xy} B_{xy} B_{yy} + A_{xy}^2 D_{yy})/(B_{xy}^2 - A_{yy} D_{yy})$

$\overline{C}_{xs} = A_{xs} + (A_{yy} B_{xy} B_{ys} - A_{ys} B_{xy} B_{yy} - A_{xy} B_{yy} B_{ys} + A_{xy} A_{ys} D_{yy})/(B_{xy}^2 - A_{yy} D_{yy})$

$\overline{B}_{xx} = B_{xx} + (A_{yy} B_{xy} D_{xy} - B_{xy}^2 B_{yy} - A_{xy} B_{yy} - A_{xy} B_{yy} D_{xy} + A_{xy} B_{xy} D_{yy})/(B_{xy}^2 - A_{yy} D_{yy})$

$\overline{B}_{xs} = B_{xs} + (A_{xy} B_{ys} D_{yy} - B_{xy} B_{yy} B_{ys} + A_{yy} B_{xy} D_{ys} - A_{xy} B_{yy} D_{ys})/(B_{xy}^2 - A_{yy} D_{yy})$

$\overline{C}_{ss} = A_{ss} + (A_{yy} B_{ys}^2 - 2 A_{ys} B_{yy} B_{ys} + A_{ys}^2 D_{yy})/(B_{xy}^2 - A_{yy} D_{yy})$

$\overline{B}_{ss} = B_{ss} + (A_{ys} B_{ys} D_{yy} - B_{yy} B_{ys}^2 - A_{ys} B_{yy} D_{ys} + A_{yy} B_{ys} D_{ys})/(B_{xy}^2 - A_{yy} D_{yy})$

$\overline{D}_{xx} = D_{xx} + (A_{yy} D_{xy}^2 - 2 B_{xy} B_{yy} D_{xy} + B_{xy}^2 D_{yy})/(B_{xy}^2 - A_{yy} D_{yy})$

$\overline{D}_{xs} = D_{ss} + (B_{xy} B_{ys} D_{yy} - B_{yy} B_{ys} D_{xy} - B_{xy} B_{yy} D_{ys} + A_{yy} D_{xy} D_{ys})/(B_{xy}^2 - A_{yy} D_{yy})$

$\overline{D}_{ss} = D_{ss} + (B_{ys}^2 D_{yy} - 2 B_{yy} B_{ys} D_{ys} + A_{yy} D_{ys}^2)/(B_{xy}^2 - A_{yy} D_{yy})$

$\overline{C}^z_{xx} = A^z_{xx}$

(8.84)

The principle of virtual displacements can be used to derive the equilibrium equations of the finite element method. Based on the above constitutive equations, the principle of virtual displacements is written as

$$\delta \Pi = \int_L \left\{ N_x \, \delta u_0 + N_{xy} \, \delta v_0 + M_x \, \delta w^b_{,xx} + 2 M_{xy} \, \delta w^b_{,xy} \right.$$
$$\left. + Q_y \, \delta w^s_{,x} \right\} dx - \int_L q_0 \, \delta w \, dx = 0 \qquad (8.85)$$

where

  L indicates the length of the beam

  q_0 is the uniformly distributed transverse load

$$w = w^b + w^s$$

Since (8.85) is the expression for the beam problem, all the variables depending on y must be taken as independent variables. Hence, independent variables in (8.85) are $u_0, w^b, w^s, u_{0,y}$ and $w^b_{,y}$. Among these, the variables, $u_{0,y}$ and $w^b_{,y}$ coming from the coupling effects of the laminated composite beam, represent the inplane shear strain and twist due to bending respectively as a function of x only.

Displacement functions are to be chosen. A linear interpolation is for $u_0, u_{0,y}, w^s$ and $w^b_{,y}$ and a Hermitian interpolation for w^b to consider the characteristics of beam deformation.

We can obtain the discretized internal virtual work by applying the interpolated displacements to (8.85) as follows

$$\delta \Pi_{int} = \delta (\{X\}^b_e)^T [K]^b \{X\}^b_e + \delta (\{X\}^s_e)^T [K]^s \{X\}^s_e \qquad (8.86)$$

where

$$\{X\}^b_e = \{u_1 \quad w^b_1 \quad \theta_1 \quad \phi_1 \quad \beta_1 \quad u_2 \quad w^b_2 \quad \theta_2 \quad \phi_2 \quad \beta_2 \} \qquad (8.87)$$

$$\{X\}^s_e = \{w^s_1 \quad w^s_2\}^T \qquad (8.88)$$

$$u = u_0, \quad \theta = -w^b_{,x}, \quad \phi = w^b_{,y}, \quad \beta = u_{0,y} \qquad (8.89)$$

The stiffness matrices of each deformation mode are obtained as follows

$$[K]^b = \int_L b([B]^b)^T [\overline{C}]^b [B]^b \, dx \qquad (8.90)$$

$$[K]^s = \int_L b([B]^s)^T [\overline{C}]^s [B]^s \, dx \qquad (8.91)$$

where b indicates the width of the beam.

$[\overline{C}]^b$ is the coefficient matrix 4×4 of the constitutive equation for CLT extracted from $[\overline{C}]$ in (8.83) and $[\overline{C}]^s$ is equal to $\overline{C}^z_{xx}$.

Hermitian matrix is given by

$$[H(x)] = \left[2\left(\frac{x}{L}\right)^3 - 3\left(\frac{x}{L}\right)^2 + 1 \quad -\frac{x^2}{L^2} + \frac{2x^2}{L} - x \quad -2\left(\frac{x}{L}\right)^3 + 3\left(\frac{x}{L}\right)^2 \quad -\frac{x^3}{L^2} + \frac{x^2}{L} \right] \quad (8.92)$$

$$[B]^b = \begin{bmatrix} N_{1,x} & 0 & 0 & 0 & 0 & N_{2,x} & 0 & 0 & 0 & 0 \\ 0 & 0 & 0 & 0 & N_1 & 0 & 0 & 0 & 0 & N_2 \\ 0 & -H_{1,xx} & -H_{2,xx} & 0 & 0 & 0 & -H_{3,xx} & -H_{4,xx} & 0 & 0 \\ 0 & 0 & 0 & -2N_{1,x} & 0 & 0 & 0 & 0 & -2N_{2,x} & 0 \end{bmatrix} \quad (8.93)$$

$$[B]^s = \begin{bmatrix} N_{1,x} & N_{2,x} \end{bmatrix} \quad (8.94)$$

$$[N(x)] = \begin{bmatrix} 1 - \frac{x}{L} & \frac{x}{L} \end{bmatrix} \quad (8.95)$$

Since there are two kinds of degrees of freedom (w^b, w^s) related to the deflection in (8.86), it is difficult to apply the boundary conditions and assemble the stiffness matrices in a non-linear beam. Accordingly, these degrees of freedom have to be expressed in terms of essential nodal degrees of freedom (w) to eliminate the extra degrees of freedom in the element stiffness matrix. Shear deflection is to be projected into the bending to obtain the total deflection in (8.75).

To derive the projection matrix of the shear deflection into the bending deflection, we can use equilibrium relations between shear and bending

$$M_{x,x} + 2M_{xy,y} - Q_x = 0 \quad (8.96)$$

By applying (8.77) and (8.83) to (8.96) and interpolating the displacement field, a discretized relation between the bending and the shear is obtained.

Since there exist two shear variables (w_1^s, w_2^s) to be projected into bending variables, an additional constraint equation is required. Such a constraint may be obtained by considering the case of zero applied loads. According to (8.96), a non-trivial solution ($w_1^s = w_2^s = 0$), say, with the following condition

$$w_1^s + w_2^s = 0 \quad (8.97)$$

This equation implies zero shear deflection at the element centre. Similar conditions leading to the non-linear solutions such as $w_1^s = 0$ or $w_2^s = 0$ may be applied instead of the condition above. From (8.96) and (8.97), the projection matrix sought for is applied as follows

$$[R] = ([R]^s)^{-1}[R]^b \quad (8.98)$$

where

$$[R]^s = \begin{bmatrix} -\overline{C}_{xx}^z & \overline{C}_{xx}^z \\ 1 & 1 \end{bmatrix} \quad (8.99)$$

$$[R]^b = \begin{bmatrix} 0 & -12\overline{D}_{xx} & 6\overline{D}_{xx}L & 0 & \overline{C}_0 L^2 & 0 & 12\overline{D}_{xx} & 6\overline{D}_{xx}L & 0 & -\overline{C}_0 L^2 \\ 0 & 0 & 0 & 0 & 0 & 0 & 0 & 0 & 0 & 0 \end{bmatrix} \quad (8.100)$$

$$\overline{C}_0 = \overline{D}_{xx} + 2\overline{B}_{xs} \quad (8.101)$$

We obtain

$$\{X\}_e^s = [R]\{X\}_e^b \quad (8.102)$$

By substituting (8.102) in eqn. (8.86) we can obtain the internal virtual work expressed in terms of $\{X\}_e^b$ such that

$$\delta \pi_{\text{int}} = \delta(\{X\}_e^b)^T ([K]^b + [R]^T [K]^s [R]) \{X\}_e^b \qquad (8.103)$$

Because the total deflection (w) is the sum of w^b and w^s, $\{X\}_e^b$ becomes

$$\{X\}_e^b = ([I] + [\overline{R}])^{-1} \{X\}_e \qquad (8.104)$$

where

$$\{X\}_e = \{u_1 \quad w_1 \quad \theta_1 \quad \phi_1 \quad \beta_1 \quad u_2 \quad w_2 \quad \theta_2 \quad \phi_2 \quad \beta_2\}^T \qquad (8.105)$$

$[I]$ is a identity matrix of size 10×10 and $[\overline{R}]$ consists of $[R]$ augmented by inserting zeros in the proper positions to match the components of $\{X\}_e$. Using (8.104), we obtain

$$\delta \pi_{\text{int}} = \delta \{X\}_e^T [K] \{X\}_e \qquad (8.106)$$

where

$$[K] = [\{([I] + [\overline{R}])^{-1}\}^T ([K]^b + [R]^T [K]^s [R]) ([I] + [\overline{R}])^{-1}] \qquad (8.107)$$

The matrix $[K]$ in the element level is expressed in terms of the essential nodal degrees of freedom and can now be treated in the standard fashion.

8.15 ISOPARAMETRIC STIFFENER STIFFNESS MATRIX

A stiffener is arbitrarily oriented within a plate element [Fig. 8.18(a)]. The figure shows the orientation and stress components of a stiffener placed within the plate element at angle ϕ with respect to the global axis system. The local axis system is x', y', z', whereas the global axis system is x, y, z [Fig. 8.18(b)]. The displacement components at any point of the stiffener in local axis system are (U', V', W')

$$\begin{Bmatrix} U' \\ V' \\ W' \end{Bmatrix} = \begin{Bmatrix} u_0' - z'\theta_{x'} \\ z'\theta_{y'} \\ w' + y''\theta_{y'} \end{Bmatrix} \qquad (8.108)$$

where u' and w' are the reference plane displacements of the stiffener which is a function of x' only. The reference plane is considered as the middle plane of the plate. $\theta_{x'}$ and $\theta_{y'}$ are the rotations of the normals to the underformed middle plane of the plate parallel to the local axis system of the stiffener. y'' is any arbitrary distance along the y'' - axis which is placed at the centroid of the stiffener parallel to the y'-axis [Fig. 8.18(d)].

The strain components are given by

$$\{\varepsilon\}_{x'} = \begin{Bmatrix} \dfrac{\partial u_0'}{\partial x'} - z'\dfrac{\partial \theta_{x'}}{\partial x'} \\ \dfrac{\partial w'}{\partial x'} + y''\dfrac{\partial \theta_{y'}}{\partial x'} - \theta_{x'} \\ -z'\dfrac{\partial \theta_{y'}}{\partial x'} \end{Bmatrix} \qquad (8.109)$$

200 *Mechanics of Composite Materials and Structures*

The generalised strain components of the stiffener in the $x'\,y'$-coordinates are given by

$$\{\bar{\varepsilon}\}_{x'} = \begin{Bmatrix} \dfrac{\partial u'_0}{\partial x'} \\[6pt] -\dfrac{\partial \theta_{x'}}{\partial x'} \\[6pt] \dfrac{\partial w'}{\partial x'} - \theta_{x'} \\[6pt] -\dfrac{\partial \theta_{y'}}{\partial x'} \end{Bmatrix} \tag{8.110}$$

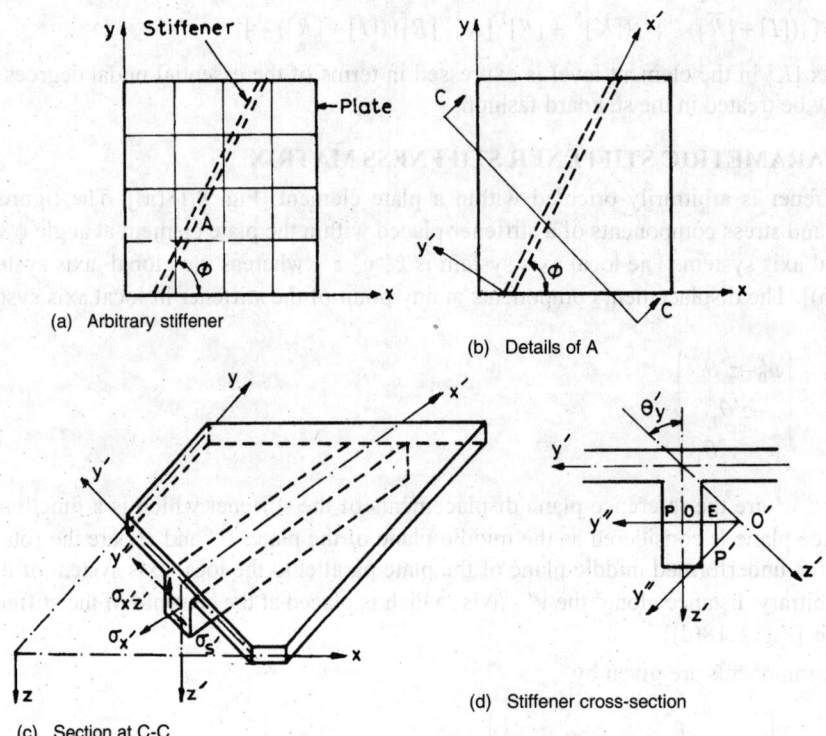

Fig. 8.18 Stiffened composite plate

The relationship between the global x coordinates and the local $x'\,y'$ coordinates is as follows :

$$\{\varepsilon\}_{x'} = \begin{bmatrix} H_s \end{bmatrix} \{\bar{\varepsilon}\}_{x'} \tag{8.111}$$

where

$$[H_s] = \begin{bmatrix} 1 & z' & 0 & 0 \\ 0 & 0 & 1 & -y'' \\ 0 & 0 & 0 & z' \end{bmatrix} \quad (8.112)$$

The relationship between local and global displacements are as follows

$$\begin{Bmatrix} u'_0 \\ \theta_{x'} \\ \theta_{y'} \end{Bmatrix} = \begin{bmatrix} \cos\phi & \sin\phi & 0 & 0 \\ 0 & 0 & \cos\phi & \sin\phi \\ 0 & 0 & -\sin\phi & \cos\phi \end{bmatrix} \begin{Bmatrix} u_0 \\ v_0 \\ \theta_x \\ \theta_y \end{Bmatrix} \quad (8.113)$$

Again,

$$x = x'\cos\phi - y'\sin\phi$$
$$y = x'\sin\phi + y'\cos\phi \quad (8.114)$$

The derivatives of x and y with respect to x' are given by

$$\frac{\partial x}{\partial x'} = \cos\phi \quad (8.115)$$

$$\frac{\partial y}{\partial x'} = \sin\phi$$

Now,

$$\frac{\partial u'_0}{\partial x'} = \frac{\partial u'_0}{\partial x} \cdot \frac{\partial x}{\partial x'} + \frac{\partial u'_0}{\partial y} \cdot \frac{\partial y}{\partial x'} \quad (8.116)$$

Substituting u' from (8.114) in terms of u_0, v_0 and using the relations of (8.115) and (8.116) becomes

$$\frac{\partial u'_0}{\partial x'} = \frac{\partial u_0}{\partial x}\cos^2\phi + \frac{1}{2}\left(\frac{\partial u_0}{\partial y} + \frac{\partial v_0}{\partial x}\right) 2\sin\phi\cos\phi + \frac{\partial v_0}{\partial y}\sin^2\phi \quad (8117)$$

Following similar lines, (8.110) can be expressed in $x\,y$ coordinates as

$$\{\bar{\varepsilon}\}_{x'} = \begin{Bmatrix} \dfrac{\partial u_0}{\partial x}\cos^2\phi + \dfrac{1}{2}\left(\dfrac{\partial u_0}{\partial y} + \dfrac{\partial v_0}{\partial x}\right)\sin 2\phi + \dfrac{\partial v_0}{\partial y}\sin^2\phi \\[6pt] -\dfrac{\partial \theta_x}{\partial x}\cos^2\phi - \dfrac{1}{2}\left(\dfrac{\partial \theta_x}{\partial y} + \dfrac{\partial \theta_y}{\partial x}\right)\sin 2\phi - \dfrac{\partial \theta_y}{\partial y}\sin^2\phi \\[6pt] \left(\dfrac{\partial w}{\partial x} - \theta_x\right)\cos\phi + \left(\dfrac{\partial w}{\partial y} - \theta_y\right)\sin\phi \\[6pt] -\dfrac{1}{2}\left(-\dfrac{\partial \theta_x}{\partial x} + \dfrac{\partial \theta_y}{\partial y}\right)\sin 2\phi + \dfrac{\partial \theta_x}{\partial y}\sin^2\phi - \dfrac{\partial \theta_y}{\partial x}\cos^2\phi \end{Bmatrix} \quad (8.118)$$

The generalised strain components of the stiffener in the $x'\,y'$ coordinates are given by

$$\{\bar{\varepsilon}\}_{x'} = [T]\{\bar{\varepsilon}\}_x \tag{8.119}$$

where $[T]$ is the transformation matrix and is given by

$$[T]^T = \begin{bmatrix} \cos^2\phi & 0 & 0 & 0 \\ \sin^2\phi & 0 & 0 & 0 \\ 0.5\sin 2\phi & 0 & 0 & 0 \\ 0 & \cos^2\phi & 0 & -0.5\sin 2\phi \\ 0 & \sin^2\phi & 0 & 0.5\sin 2\phi \\ 0 & 0.5\sin 2\phi & 0 & \cos^2\phi \\ 0 & 0.5\sin 2\phi & 0 & -\sin^2\phi \\ 0 & 0 & \cos\phi & 0 \\ 0 & 0 & \sin\phi & 0 \end{bmatrix} \tag{8.120}$$

$\{\bar{\varepsilon}\}_x$ is the generalised strain vector in the x, y coordinates and is defined as

$$\{\bar{\varepsilon}\}_x^T = \left\{ \frac{\partial u_0}{\partial x} \quad \frac{\partial v_0}{\partial y} \quad \left(\frac{\partial u_0}{\partial y} + \frac{\partial v_0}{\partial x}\right) \quad -\frac{\partial \theta_x}{\partial x} \right.$$

$$\left. -\frac{\partial \theta_y}{\partial y} \quad -\frac{\partial \theta_y}{\partial x} - \frac{\partial \theta_x}{\partial y} \quad \left(\frac{\partial w}{\partial x} - \theta_x\right) \quad \left(\frac{\partial w}{\partial y} - \theta_y\right) \right\} \tag{8.121}$$

Combining (8.51) and (8.121) yields

$$\{\bar{\varepsilon}\}_x = \sum_{r=1}^{8}[B_s]_r\{\partial_r\}_e = \{B_s\}\{\partial\}_e \tag{8.122}$$

where the expression for $[B_s]_r$ is given in (8.57).

Figure 8.18(c) shows the orientation and stress components of a stiffener placed with the plate element of an angle ϕ with respect to the global axis system. The off-axis stress–strain relationship for the kth layer is given by

$$\begin{Bmatrix} \sigma_{x'} \\ \sigma_{s'} \\ \sigma_{x'z'} \end{Bmatrix}_k = \begin{bmatrix} Q_{x'x'} & Q_{x's'} & 0 \\ Q_{x's'} & Q_{s's'} & 0 \\ 0 & 0 & k_s Q_{x'x'}^z \end{bmatrix}_k \begin{Bmatrix} \varepsilon_{x'} \\ \varepsilon_{s'} \\ \varepsilon_{x's'} \end{Bmatrix}_k \tag{8.123}$$

where the strains components are given by (8.109).
The stress resultants of the stiffener are

$$\{F_s\}^T = \{N_s' \quad M_s' \quad Q_s' \quad T_s'\} \tag{8.124}$$

where

$$N_s' = \sum_{k=1}^{n} \int_{-b/2}^{b/2} \int_{z_k'}^{z_{k+1}'} \sigma_{x'}^{(k)} \, dy'' dz' \tag{8.125}$$

where b is the width of the kth layer and N is the number of layers.

Using (8.123) and (8.109) into (8.125), yields

$$N'_s = \sum_{k=1}^{N} \left[\int_{-b/2}^{b/2} \int_{z'_k}^{z'_{k+1}} Q_{x'x'} \left(\frac{\partial u'_0}{\partial x'} - z' \frac{\partial \theta_{x'}}{\partial x'} \right) + Q_{x's'} \left(-z' \frac{\partial \theta_{y'}}{\partial x'} \right) dy'' \, dz' \right]$$

$$= \sum_{k=1}^{N} \int_{-b/2}^{b/2} \int_{z'_k}^{z'_{k+1}} Q_{x'x'} \left(\frac{\partial u'_0}{\partial x'} \right) dy'' \, dz' +$$

$$\sum_{k=1}^{N} \int_{-b/2}^{b/2} \int_{z'_k}^{z'_{k+1}} Q_{x'x'} \, z' \left(-\frac{\partial \theta_{x'}}{\partial x'} \right) dy'' \, dz' + \sum_{k=1}^{N} \int_{-b/2}^{b/2} \int_{z'_k}^{z'_{k+1}} Q_{x's'} \, z' \left(-\frac{\partial \theta_{y'}}{\partial x'} \right) dy'' \, dz'$$

or,

$$N'_s = A_{x'x'} b \left(\frac{\partial u'_0}{\partial x'} \right) + B_{x'x'} b \left(-\frac{\partial \theta_{x'}}{\partial x'} \right) + B_{x's'} b \left(-\frac{\partial \theta_{y'}}{\partial x'} \right) \quad (8.126)$$

The stiffener laminate properties are obtained for the following relationship

$$A_{ij}, B_{ij}, D_{ij} = \sum_{k=1}^{N} \int_{z'_k}^{z'_{k+1}} (Q_{ij})_K \left(1, z', z'^2 \right) \partial z' \qquad i, j = x', s^r \quad (8.127)$$

In a similar manner, other stress resultants are obtained as given below:

$$M'_s = \sum_{k=1}^{n} \int_{-b/2}^{b/2} \int_{z'_k}^{z'_{k+1}} \sigma_{x'}^{(k)} z' \, dy'' \, dz'$$

or,

$$M'_s = B_{x'x'} b \left(\frac{\partial u'_0}{\partial x'} \right) + D_{x'x'} b \left(-\frac{\partial \theta_{x'}}{\partial x'} \right) + D_{x's'} b \left(-\frac{\partial \theta_{y'}}{\partial x'} \right) \quad (8.128)$$

The transverse shear is

$$Q'_s = \sum_{k=1}^{N} \int_{-b/2}^{b/2} \int_{z'_k}^{z'_{k+1}} \sigma_{x'z'}^{(k)} dy'' \, dz' + \sum_{k=1}^{N} \int_{-b/2}^{b/2} \int_{z'_k}^{z'_{k+1}} k_s Q_{x'x'}^z \left(-\theta_{x'} + \frac{\partial w'}{\partial x'} \right) dy'' \, dz'$$

$$= \sum_{k=1}^{N} \int_{-b/2}^{b/2} \int_{z'_k}^{z'_{k+1}} k_s Q_{x'x'}^z y'' \left(\frac{\partial \theta_{y'}}{\partial x'} \right) dy'' \, dz' \quad (8.129)$$

The second integration term is zero for a symmetric cross-section. However, it is treated as zero for any cross-section.

So, $\quad Q'_s = k_s A^s_{x'x'} b \left(\dfrac{\partial w'}{\partial x'} - \theta_{x'} \right)$ \hfill (8.130)

The torsional moment is given by

$$T'_s = \sum_{k=1}^{N} \int_{-b/2}^{b/2} \int_{z'_k}^{z'_{k+1}} \left(\sigma_{s'} z' - \sigma_{x'x'} y'' \right) dy'' \, dz'$$

$$= \sum_{k=1}^{N} \int_{-b/2}^{b/2} \int_{z'_k}^{z'_{k+1}} Q_{x'x'} \, z' \left(\dfrac{\partial u'}{\partial x'} \right) dy'' \, dz'$$

$$+ \sum_{k=1}^{N} \int_{-b/2}^{b/2} \int_{z'_k}^{z'_{k+1}} Q^z_{x's} \, z'^2 \left(-\dfrac{\partial \theta_{x'}}{\partial x'} \right) dy'' \, dz'$$

$$+ \sum_{k=1}^{N} \int_{-b/2}^{b/2} \int_{z'_k}^{z'_{k+1}} \left(Q_{s's'} z'^2 + k_s Q^z_{x'x'} y''^2 \right) \left(\dfrac{\partial \theta_{y'}}{\partial x'} \right) dy'' \, dz'$$

$$+ \sum_{k=1}^{N} \int_{-b/2}^{b/2} \int_{z'_k}^{z'_{k+1}} k_s Q^z_{x'x'} \, y'' \left(\dfrac{\partial w'}{\partial x'} - \theta_{x'} \right) dy'' \, dz' \hfill (8.131)$$

The last integration term is zero for reasons indicated earlier. The third integration term is interpreted as the torsional rigidity of the stiffener cross-section and is expressed approximately as

$$\dfrac{1}{6} \left(Q_{s's'} + Q^z_{x'x'} \right) hb^3$$

where h is the depth and b is the width of the stiffener. So the expression for torsion T'_s becomes

$$T'_s = B_{x's'} b \left(\dfrac{\partial u'_0}{\partial x'} \right) + D_{x's'} b \left(-\dfrac{\partial \theta_{x'}}{\partial x'} \right) + \dfrac{1}{6} \left(Q_{s's'} + Q^z_{x'x'} \right) hb^3 \left(-\dfrac{\partial \theta_{y'}}{\partial x'} \right) \hfill (8.132)$$

Combining all the stress resultants, results the following matrix equation

$$\begin{Bmatrix} N'_s \\ M'_s \\ Q'_s \\ T'_s \end{Bmatrix} = \begin{bmatrix} A_{x'x'} b & B_{x'x'} b & 0 & B_{x's'} b \\ B_{x'x'} b & D_{x'x'} b & 0 & D_{x's'} b \\ 0 & 0 & k_s A^{z'}_{x'x'} b & 0 \\ B_{x's'} b & D_{x's'} b & 0 & \dfrac{1}{6}\left(Q_{s's'}+Q^z_{x'x'}\right)hb^3 \end{bmatrix} \begin{Bmatrix} \dfrac{\partial u'_0}{\partial x'} \\ -\dfrac{\partial \theta_{x'}}{\partial x'} \\ \dfrac{\partial u'_0}{\partial x'} - \theta_{x'} \\ \dfrac{\partial \theta_{y'}}{\partial x'} \end{Bmatrix} \hfill (8.133)$$

or,

$$\{F_s\} = [D_s]\{\overline{\varepsilon}\}_{x'} \tag{8.134}$$

A_{ij}, B_{ij} and D_{ij} are calculated with respect to the $x'y'$ axis system. To account for the eccentricity of the stiffener with respect to the plate mid-plane, the above properties are to be transformed with respect to the parallel axis theorem [8.18]. Referring to Fig. 5.4(d), the laminate properties of the stiffeners with respect to the oy'' axis system can be transformed in the following manner to the $o'y'$ axis system.

$$[A_{ij}]_{o'y'} = [A_{ij}]_{oy''}$$
$$[D_{ij}]_{o'y'} = [D_{ij}]_{oy''} + 2e[B_{ij}]_{oy''} + e^2[A_{ij}]_{oy''} \tag{8.135}$$
$$[B_{ij}]_{o'y'} = [B_{ij}]_{oy''} + e[A_{ij}]_{oy'}$$

Combining (8.119) and (8.122) yields

$$\{\overline{\varepsilon}\}_{x'} = [T][B_s]\{\delta\}_e = \sum_{r=1}^{8}[B_s]_{x'r}\{\delta\}_e = [B_s]_{x'}\{\delta\}_e \tag{8.136}$$

$$[B_s]_{x'r} = \begin{bmatrix} \frac{\partial N_r}{\partial x}\cos^2\phi + \frac{1}{2}\frac{\partial N_r}{\partial y}\sin 2\phi & \frac{\partial N_r}{\partial y}\sin^2\phi + \frac{1}{2}\frac{\partial N_r}{\partial x}\sin 2\phi & 0 & 0 & 0 \\ 0 & 0 & 0 & -\frac{\partial N_r}{\partial x}\cos^2\phi - \frac{1}{2}\frac{\partial N_r}{\partial y}\sin 2\phi & -\frac{\partial N_r}{\partial y}\sin^2\phi - \frac{1}{2}\frac{\partial N_r}{\partial x}\sin 2\phi \\ 0 & 0 & \frac{\partial N_r}{\partial x}\cos\phi + \frac{\partial N_r}{\partial x}\sin\phi & -N_r\cos\phi & -N_r\sin\phi \\ 0 & 0 & 0 & \frac{1}{2}\frac{\partial N_r}{\partial x}\sin 2\phi + \frac{\partial N_r}{\partial y}\sin^2\phi & -\frac{1}{2}\frac{\partial N_r}{\partial y}\sin 2\phi - \frac{\partial N_r}{\partial x}\cos^2\phi \end{bmatrix} \tag{8.137}$$

Using the principle of virtual work, the element stiffness matrix can be expressed as

$$[K]_e = \iint [B_s]_{x'}^T [D_s][B_s]_{x'}\,dx\,dy \tag{8.138}$$

The integration is to be carried out numerically. The Gaussian integration is adopted in the present formulation. To take into account the shear locking effect, the reduced integration is adopted. A 2×2 integration is carried out for the quadratic isoparameteric element. For the evaluation of the stiffness matrix with respect to the x, y, z axis system, it is to be transformed to the isoparametric ($\xi - \eta$) coordinate system.

According to the present formulation, the location of integration points of an arbitrarily oriented stiffener needs special attention. The procedure is illustrated for a few simple cases of stiffener orientation in Fig. 8.19.

8.15.1 Stiffener Mass Matrix

As in the case of the plate element, a lumped mass matrix has been assumed for the stiffener element. A lumped mass matrix produces a diagonal matrix with non-zero elements of the displacement degrees of freedom only and the stiffener mass is lumped for the inplane and transverse direction in equal proportions. For a stiffener parallel to the x-direction, the mass matrix of node r is

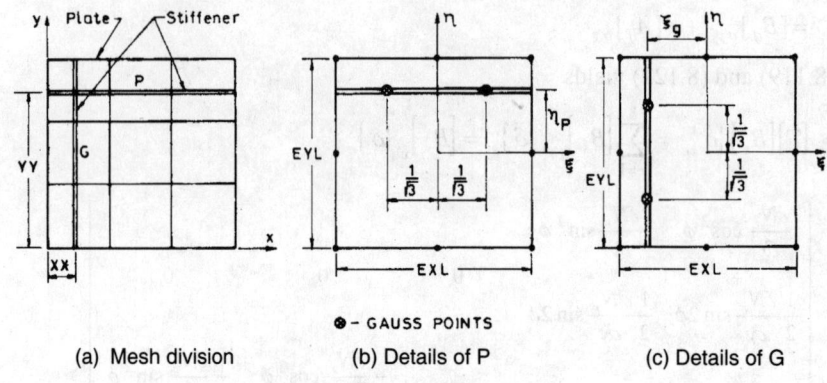

(a) Mesh division   (b) Details of P   (c) Details of G

Fig. 8.19 Gauss points for stiffeners

$$[M_{ss}]_{re} = \begin{bmatrix} m_{rrs} & & & 0 \\ & 0 & & \\ & & m_{rrs} & \\ 0 & & & 0 \\ & & & & 0 \end{bmatrix} \tag{8.139}$$

where

$$m_{rrs} = \frac{\int_{-1}^{1} \rho_s A_s |J| d\xi}{\sum_{r=1}^{8} \int_{-1}^{1} \rho_s A_s N_r N_r |J| d\xi} \cdot \int_{-1}^{1} \rho_s A_s N_r N_r |J| d\xi \tag{8.140}$$

8.16 FINITE ELEMENT ANALYSIS OF THE LAMINATED COMPOSITE SHELL

A composite shell element is presented based as the extension of the Reissner–Mindlin plate element. The formulation is presented for a 9-noded Lagrangian shell element [8.20] (Fig. 8.20(a)). On the surface S of the shell element, two orthogonal intrinsic coordinates s and r are defined [Fig. 8.21(a)]. Consider a laminated shell with a finite number of orthotropic layers of uniform thickness, oriented arbitrarily with respect to shell coordinates (s, r, z). The orthogonal curvilinear coordinate system (s, r, z) is so chosen that the s and r curves are lines of curvature on the mid-surface $z = 0$ and z-curves are straight lines perpendicular to the surface $z = 0$.

A typical shell element and its isoparametric representation are shown in Fig. 8.21(a) and in Fig. 8.21(b). The shell element possesses six degrees of freedom (three translations and three rotations) at each of the nine mid-surface nodes. The stiffness corresponding to the rotation about the normal to the element mid-surface is zero at all nodes. Therefore, a small fictitious rotational spring is introduced at each node to take into account the rotation about the normal.

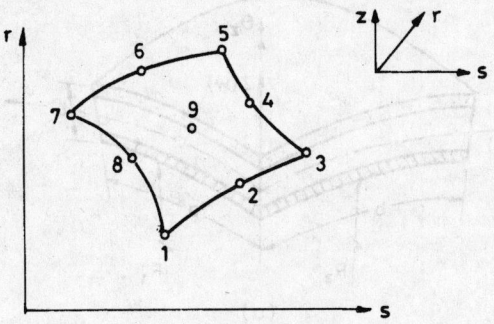

(a) Element co-ordinate

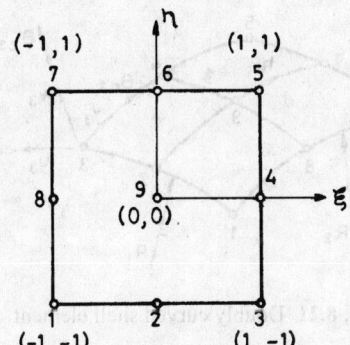

(b) Isoparametric co-ordinate

Fig. 8.20 Isoparametric quadratic shell element

The effective displacements at a node i of the shell element are u_i, v_i, w_i, θ_{xi} and θ_{yi}. The displacements at any point within the shell element can be expressed as [Fig. 8.21(b)].

$$\begin{Bmatrix} u \\ v \\ w \\ \theta_x \\ \theta_y \end{Bmatrix} = \sum N_i [I_5] \begin{Bmatrix} u_i \\ v_i \\ w_i \\ \theta_{xi} \\ \theta_{yi} \end{Bmatrix} \qquad (8.141)$$

where $[I_5]$ is a 5×5 identity matrix.

N_r is a shape function expressed in terms of $\xi-\eta$ coordinates of the element and they are obtained by using Lagrange polynomials. For the kth node and for the ξ- direction, it is

$$N_\xi = \frac{\left(\xi - \xi_1\right)\left(\xi - \xi_2\right)\cdots\left(\xi - \xi_{k-1}\right)\left(\xi - \xi_{k+1}\right)\cdots\left(\xi - \xi_n\right)}{\left(\xi_k - \xi_1\right)\left(\xi_k - \xi_2\right)\cdots\left(\xi_k - \xi_{k-1}\right)\left(\xi_k - \xi_{k+1}\right)\cdots\left(\xi_k - \xi_n\right)} \tag{8.142}$$

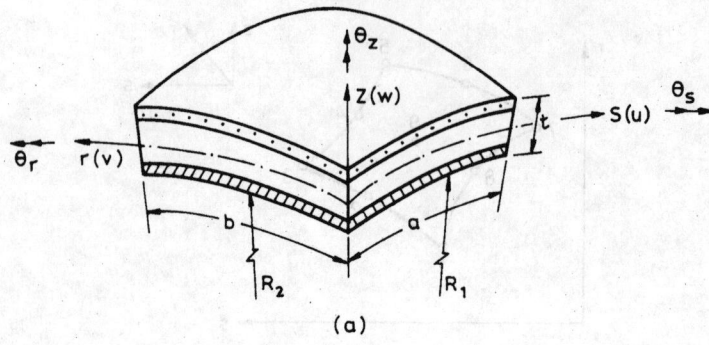

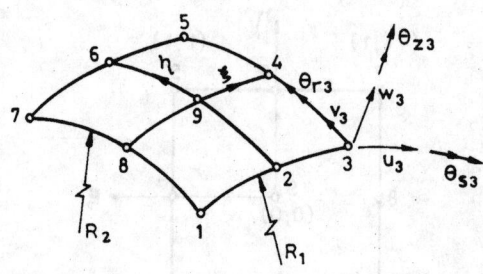

Fig. 8.21 Doubly curved shell element

Similarly, for the η - direction

$$N_\eta = \frac{\left(\eta - \eta_1\right)\left(\eta - \eta_2\right)\cdots\left(\eta - \eta_{k-1}\right)\left(\eta - \eta_{k+1}\right)\cdots\left(\eta - \eta_n\right)}{\left(\eta_k - \eta_1\right)\left(\eta_k - \eta_2\right)\cdots\left(\eta_k - \eta_{k-1}\right)\left(\eta_k - \eta_{k+1}\right)\cdots\left(\eta_k - \eta_n\right)} \tag{8.143}$$

Thus in two directions, the shape function is

$$I = N_\xi \cdot N_\eta \tag{8.144}$$

where N_ξ and N_η are one-dimensional values about the respective axis. These shape functions are indeed very handy.

The first-order shear deformation theory is accounted for using an extension of Sander's shell theory in element stiffness matrix formulation. The shear-displacement equations of the shear deformable theory of doubly curved shells are given by [8.25].

$$\begin{Bmatrix} \varepsilon_s \\ \varepsilon_r \\ \varepsilon_{rz} \\ \varepsilon_{sz} \\ \varepsilon_{rs} \end{Bmatrix} = \begin{Bmatrix} \varepsilon_s^0 \\ \varepsilon_r^0 \\ \varepsilon_{rz}^0 \\ \varepsilon_{sz}^0 \\ \varepsilon_{rs}^0 \end{Bmatrix} + \begin{Bmatrix} \kappa_s \\ \kappa_r \\ 0 \\ 0 \\ \kappa_{rs} \end{Bmatrix} \qquad (8.145)$$

where,

ε_x^0, ε_y^0 and ε_s^0 are the mid-plane strains and κ_r, κ_s and κ_{rs} are the curvatures for the laminated shell and

$$\varepsilon_s^0 = \frac{\partial u}{\partial s} + \frac{w}{R_1}$$

$$\kappa_s = \frac{\partial \theta_r}{\partial s}$$

$$\varepsilon_r^0 = \frac{\partial v}{\partial r} + \frac{w}{R_2} \qquad (8.146)$$

$$\kappa_r = \frac{\partial \theta_s}{\partial r}$$

$$\varepsilon_{rs}^0 = \frac{\partial u}{\partial r} + \frac{\partial v}{\partial s}$$

$$\kappa_{rs} = \frac{\partial \theta_r}{\partial r} + \frac{\partial \theta_s}{\partial s} - C_0 \left(\frac{\partial v}{\partial s} - \frac{\partial u}{\partial r} \right)$$

$$\varepsilon_{sz} = -\theta_r + \frac{\partial w}{\partial s} - \frac{u}{R_1}$$

$$\varepsilon_{rz} = -\theta_s + \frac{\partial w}{\partial r} - \frac{v}{R_2}$$

$$C_0 = \frac{1}{2} \left(\frac{1}{R_1} - \frac{1}{R_2} \right) \qquad (8.147)$$

where u, v, w represent displacements along the s, r and z directions respectively. θ_r and θ_s are the rotations of transverse normals about the r and s axes, respectively. The C_0 term accounts for the conditions of zero strain for the rigid body motion in Sander's shell theory. R_1 and R_2 are the radii of curvature in the R_1 and R_2 directions.

The stress–strain relation is same as that of (8.46) except that x and y indices are to be changed to s and r.

The following relationship is valid for Mindlin's theory

$$\begin{Bmatrix} \phi_s \\ \phi_r \end{Bmatrix} = \begin{bmatrix} \dfrac{\partial w}{\partial s} + \theta_r \\ \dfrac{\partial w}{\partial r} + \theta_s \end{bmatrix} \tag{8.148}$$

The generalised strain components are given by

$$\{\varepsilon\}^T = \left\{ \dfrac{\partial u}{\partial s}\, \dfrac{\partial v}{\partial r} \left(\dfrac{\partial u}{\partial r} + \dfrac{\partial v}{\partial s} \right) - \dfrac{\partial \theta_r}{\partial s} - \dfrac{\partial \theta_s}{\partial r} - \left(\dfrac{\partial \theta_r}{\partial r} + \dfrac{\partial \theta_s}{\partial s} \right) \right.$$
$$\left. \left(\dfrac{\partial w}{\partial s} - \phi_r \right) \left(\dfrac{\partial w}{\partial r} - \phi_s \right) \right\} \tag{8.149}$$

The last two terms are related to transverse shear.

Combining (8.141) and (8.149), yields

$$\{\varepsilon\} = \sum_{r=1}^{9} \begin{bmatrix} \dfrac{\partial N_r}{\partial s} & 0 & 0 & 0 & 0 \\ 0 & \dfrac{\partial N_r}{\partial r} & 0 & 0 & 0 \\ \dfrac{\partial N_r}{\partial r} & \dfrac{\partial N_r}{\partial s} & 0 & 0 & 0 \\ 0 & 0 & 0 & -\dfrac{\partial N_r}{\partial s} & 0 \\ 0 & 0 & 0 & 0 & -\dfrac{\partial N_r}{\partial r} \\ 0 & 0 & 0 & -\dfrac{\partial N_r}{\partial r} & \dfrac{\partial N_r}{\partial s} \\ 0 & 0 & \dfrac{\partial N_r}{\partial s} & -N_r & 0 \\ 0 & 0 & \dfrac{\partial N_s}{\partial r} & 0 & -N_r \end{bmatrix} \tag{8.150}$$

or,

$$\{\varepsilon\} = \sum_{r=1}^{9} [B]_r \{\delta_r\}_e = [B]\{\delta\}_e \tag{8.151}$$

Following steps similar to that mentioned in the previous section, the element stiffness matrix of the laminated shell is given by

$$[k]_e = \iint [B]^T [D][B]\, ds\, dr \tag{8.152}$$

If $[k]_e$ is to be evaluated numerically, (8.119) is written as

$$[k]_e = \int_{-1}^{1}\int_{-1}^{1} [B]^T [D][B]\, |J|\, d\xi\, d\eta \tag{8.153}$$

The element formulation is different than that of the conventional isoparametric element. On the surface S of the shell element [Fig. 8.21], two orthogonal intrinsic coordinates s and r can be defined. They are known as curvilinear coordinates and are defined in the following way

$$s(\xi) = \int_0^\xi \sqrt{\left(\frac{\partial x}{\partial \xi}\right)^2 + \left(\frac{\partial y}{\partial \xi}\right)^2 + \left(\frac{\partial z}{\partial \xi}\right)^2} \qquad (8.154)$$

$$r(\eta) = \int_0^\eta \sqrt{\left(\frac{\partial x}{\partial \eta}\right)^2 + \left(\frac{\partial y}{\partial \eta}\right)^2 + \left(\frac{\partial z}{\partial \eta}\right)^2}$$

and ξ and η are valid in the region $[-1, +1]$.

The Jacobian matrix of transformation is given by

$$[J] = \begin{bmatrix} \frac{\partial x}{\partial \xi} & \frac{\partial y}{\partial \xi} & \frac{\partial z}{\partial \xi} \\ \frac{\partial x}{\partial \eta} & \frac{\partial y}{\partial \eta} & \frac{\partial z}{\partial \eta} \end{bmatrix} \qquad (8.155)$$

and is calculated as

$$[J] = \begin{bmatrix} \sum_{i=1}^m \frac{\partial N_i}{\partial \xi} x_i & \sum_{i=1}^m \frac{\partial N_i}{\partial \xi} y_i & \sum_{i=1}^m \frac{\partial N_i}{\partial \xi} z_i \\ \sum_{i=1}^m \frac{\partial N_i}{\partial \eta} x_i & \sum_{i=1}^m \frac{\partial N_i}{\partial \eta} y_i & \sum_{i=1}^m \frac{\partial N_i}{\partial \eta} z_i \end{bmatrix} \qquad (8.156)$$

where x, y, z are the global coordinates for the calculation of nodal coordinates, m is the total number of nodes and N is defined by (8.144).

The Jacobian matrix given by (8.156) is rectangular and the procedure for obtaining its determinant is as follows :

$$J_1 = \det \begin{bmatrix} \sum_{i=1}^m \frac{\partial N_i}{\partial \xi} y_i & \sum_{i=1}^m \frac{\partial N_i}{\partial \xi} z_i \\ \sum_{i=1}^m \frac{\partial N_i}{\partial \eta} y_i & \sum_{i=1}^m \frac{\partial N_i}{\partial \eta} z_i \end{bmatrix}$$

$$J_2 = -\det \begin{bmatrix} \sum_{i=1}^m \frac{\partial N_i}{\partial \xi} x_i & \sum_{i=1}^m \frac{\partial N_i}{\partial \xi} z_i \\ \sum_{i=1}^m \frac{\partial N_i}{\partial \eta} x_i & \sum_{i=1}^m \frac{\partial N_i}{\partial \eta} z_i \end{bmatrix} \qquad (8.157)$$

$$J_3 = \det \begin{bmatrix} \sum_{i=1}^{m} \frac{\partial N_i}{\partial \xi} x_i & \sum_{i=1}^{m} \frac{\partial N_i}{\partial \xi} y_i \\ \sum_{i=1}^{m} \frac{\partial N_i}{\partial \eta} x_i & \sum_{i=1}^{m} \frac{\partial N_i}{\partial \eta} y_i \end{bmatrix}$$

det $[J]$ can be obtained by the following expression

$$|J| = \det[J] = \sqrt{J_1^2 + J_2^2 + J_3^2} \tag{8.158}$$

The derivatives of r and s are obtained as follows

$$\frac{\partial N_i}{\partial s} = \frac{\partial N_i}{\partial \xi} \cdot \frac{\partial \xi}{\partial s} = \frac{\partial N_i}{\partial \xi} \cdot \frac{1}{(\partial s / \partial \xi)}$$

or,

$$\frac{\partial N_i}{\partial s} = \frac{\partial N_i}{\partial \xi} \frac{1}{\sqrt{\left(\frac{\partial x}{\partial \xi}\right)^2 + \left(\frac{\partial y}{\partial \xi}\right)^2 + \left(\frac{\partial z}{\partial \xi}\right)^2}} \tag{8.159}$$

Similarly,

$$\frac{\partial N_i}{\partial r} = \frac{\partial N_i}{\partial \eta} \frac{1}{\sqrt{\left(\frac{\partial x}{\partial \eta}\right)^2 + \left(\frac{\partial y}{\partial \eta}\right)^2 + \left(\frac{\partial z}{\partial \eta}\right)^2}} \tag{8.160}$$

We know,

$$\frac{\partial x}{\partial \xi} = \sum_{i=1}^{m} \frac{\partial N_i}{\partial \xi} x_i \quad \text{etc.} \tag{8.161}$$

Curvatures of the curvilinear coordinates are calculated as follows:

$$C_1 = \left[\left\{ \left(\frac{\partial y}{\partial \xi} \cdot \frac{\partial^2 z}{\partial \xi^2} - \frac{\partial z}{\partial \xi} \cdot \frac{\partial^2 y}{\partial \xi^2} \right)^2 + \left(\frac{\partial z}{\partial \xi} \cdot \frac{\partial^2 x}{\partial \xi^2} - \frac{\partial x}{\partial \xi} \cdot \frac{\partial^2 z}{\partial \xi^2} \right)^2 \right. \right.$$

$$\left. \left. + \left(\frac{\partial x}{\partial \xi} \cdot \frac{\partial^2 y}{\partial \xi^2} - \frac{\partial y}{\partial \xi} \cdot \frac{\partial^2 x}{\partial \xi^2} \right)^2 \right\} \right]^{\frac{1}{2}} \tag{8.162a}$$

$$C_2 = \left[\left\{ \left(\frac{\partial y}{\partial \eta} \cdot \frac{\partial^2 z}{\partial \eta^2} - \frac{\partial z}{\partial \eta} \cdot \frac{\partial^2 y}{\partial \eta^2} \right)^2 + \left(\frac{\partial z}{\partial \eta} \cdot \frac{\partial^2 x}{\partial \eta^2} - \frac{\partial x}{\partial \eta} \cdot \frac{\partial^2 z}{\partial \eta^2} \right)^2 \right. \right.$$

$$+\left(\frac{\partial x}{\partial \eta} \cdot \frac{\partial^2 y}{\partial \eta^2} - \frac{\partial y}{\partial \eta} \cdot \frac{\partial^2 x}{\partial \eta^2}\right)^2\right]^{\frac{1}{2}} \qquad (8.162b)$$

Likewise for y and z, and the same for the derivatives of x, y, z with respect to η.

The curvature $1/R_1$, as a function of ξ and $1/R_2$ as a function of η have the following expressions.

$$\frac{1}{R_1} = \frac{1}{\sqrt{\left[\left(\frac{\partial x}{\partial \xi}\right)^2 + \left(\frac{\partial y}{\partial \xi}\right)^2 + \left(\frac{\partial z}{\partial \xi}\right)^2\right]^3}} C_1^*$$

$$\frac{1}{R_2} = \frac{1}{\sqrt{\left[\left(\frac{\partial x}{\partial \eta}\right)^2 + \left(\frac{\partial y}{\partial \eta}\right)^2 + \left(\frac{\partial z}{\partial \eta}\right)^2\right]^3}} C_2 \qquad (8.163)$$

8.17 FEM OF LAMINATED PLATES BASED ON HIGHER ORDER PLATE THEORY

The basic equations for higher order plate theory for laminated plates have been presented in Chapter 6. The finite element formulation of the higher order theory is presented here [8.21] for symmetrically laminated plates.

A 4-noded rectangular element is considered. It has five degrees of freedom per node – w, $w_{,x}$, $w_{,y}$, ϕ_x and ϕ_y [Fig. 8.22]. The displacement and the shear rotation parameters at any point (ξ, η) in the 4-noded element can be expressed as

$$\begin{Bmatrix} w \\ \phi_x \\ \phi_y \end{Bmatrix} = \sum_{i=1}^{4} \begin{bmatrix} N_i & N_{ix} & N_{iy} & 0 & 0 \\ 0 & 0 & 0 & N_{i\phi x} & 0 \\ 0 & 0 & 0 & 0 & N_{i\phi y} \end{bmatrix} \begin{Bmatrix} w_i \\ (w,_x)_i \\ (w,_y)_i \\ \phi_{xi} \\ \phi_{yi} \end{Bmatrix}$$

$$= \sum_{i=1}^{4} [N]_i \{\delta\}_e \qquad (8.164)$$

where

$$\{\delta_i\}_e^T = \{w_i \; (w,_x)_i \; (w,_y)_i \; \phi_{x_i} \; \phi_{y_i}\}$$

$$N_i = \frac{1}{8}\left(1 + \xi_0\right)\left(1 + \eta_0\right)\left(2 + \xi_0 + \eta_0 - \xi^2 - \eta^2\right)$$

$$N_{ix} = \frac{a}{8}\left(1 + \xi_0\right)^2\left(1 - \xi_0\right)\left(1 + \eta_0\right)$$

$$N_{iy} = \frac{b}{8}\left(1 + \eta_0\right)^2 \left(1 - \eta_0\right)\left(1 + \xi_0\right) \tag{8.165}$$

$$N_{i\phi x} = N_{i\phi y} = \frac{1}{4}\left(1 + \eta_0\right)\left(1 + \xi_0\right)$$

$$\xi_0 = \xi\xi_i, \quad \eta_0 = \eta\eta_i, \quad \xi = \frac{x}{a}, \quad \eta = \frac{y}{b}$$

The relationships between bending strains and nodal displacement parameters are written as

$$\{\varepsilon_b\} = \sum_{i=1}^{4} [B_b]_i \{\delta_i\}_e \tag{8.166}$$

where

$$\{\varepsilon_b\} = \begin{Bmatrix} \phi_{x,x} - w_{,xx} \\ \phi_{y,y} - w_{,yy} \\ \phi_{x,y} + \phi_{y,x} - 2w_{,xy} \\ -4\phi_{x,x}/3h^2 \\ -4\phi_{y,y}/3h^2 \\ -4(\phi_{x,y} + \phi_{y,x})/3h^2 \end{Bmatrix} \tag{8.167}$$

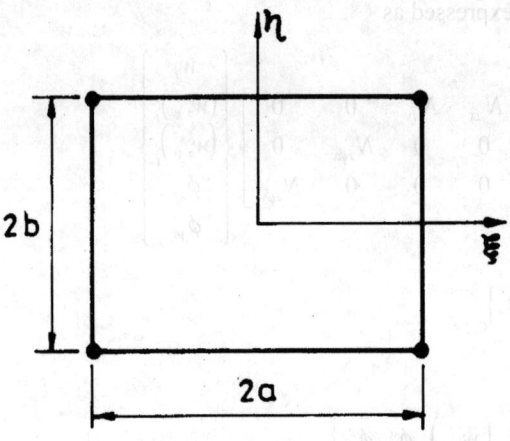

Fig. 8.22 Rectangular finite element based on higher order plate theory

and

$$[B_b]_i = \begin{bmatrix} -N_{i,xx} & -N_{ix,xx} & -N_{iy,xx} & N_{i\phi x,x} & 0 \\ -N_{i,yy} & -N_{ix,yy} & -N_{iy,yy} & 0 & N_{i\phi y,y} \\ -2N_{i,xy} & -2N_{ix,xy} & -2N_{iy,xy} & N_{i\phi x,y} & N_{i\phi y,x} \\ 0 & 0 & 0 & -4\dfrac{N_{i\phi x,x}}{3h^2} & 0 \\ 0 & 0 & 0 & 0 & -4\dfrac{N_{i\phi y,y}}{3h^2} \\ 0 & 0 & 0 & -4\dfrac{N_{i\phi x,y}}{3h^2} & -4\dfrac{N_{i\phi y,x}}{3h^2} \end{bmatrix} \qquad (8.168)$$

The relationship between the shear strain terms and nodal displacement parameters may be written as

$$\{\varepsilon_s\} = \sum_{i=1}^{4} [B_s]_i \{\delta_i\}_e \qquad (8.169)$$

in which

$$\{\varepsilon_s\} = \begin{Bmatrix} \phi_x \\ \phi_y \\ -4\phi_x/h^2 \\ -4\phi_y/h^2 \end{Bmatrix} \qquad (8.170)$$

and

$$[B_s]_i = \begin{bmatrix} 0 & 0 & 0 & 0 & N_{i\phi x} \\ 0 & 0 & 0 & N_{i\phi y} & 0 \\ 0 & 0 & 0 & 0 & -4N_{i\phi y}/h^2 \\ 0 & 0 & 0 & -4N_{i\phi x}/h^2 & 0 \end{bmatrix} \qquad (8.171)$$

Total potential energy of the plate is written as

$$\Pi = \frac{1}{2} \int_A \{\varepsilon_b\}^T [D_b] \{\varepsilon_b\} dA + \frac{1}{2} \int_A \{\varepsilon_s\}^T [D_s] \{\varepsilon_s\} dA - \frac{1}{2} \int_A qw \, dA \qquad (8.172)$$

where q is the applied transverse load.

Substituting (8.168) and (8.171) into (8.172) and upon subsequent minimisation of the total potential energy with respect to nodal variables, it is possible to obtain the discretized equations in matrix form as

$$[K]_e \{\delta\}_e = \{f\}_e \qquad (8.173)$$

in which

$$[K]_e = \int [B_b]^T [D_b][B_b] dA + \int [B_s]^T [D_s][B_s] dA \qquad (8.174)$$

A 4×4 Gauss–Legendre rule is used to evaluate $[K]_e$.

8.18 FEM OF LAMINATED PLATES BASED ON HIGHER ORDER THEORY: ANOTHER APPROACH

A large number of elements exist for laminated plates using higher order theory [8.23]. Another approach is presented in this section. As this treatment of the unsymmetric laminated plate is more general, the theoretical formulation is based on this case only.

The displacement fields are expressed in Taylor's series expansion

$$u(x, y, z) = u(x, y, 0) + z\left(\frac{\partial u}{\partial z}\right)_0 + \frac{1}{2!}z^2\left(\frac{\partial^2 u}{\partial z^2}\right)_0 + \cdots + \infty$$

$$v(x, y, z) = v(x, y, 0) + z\left(\frac{\partial v}{\partial z}\right)_0 + \frac{1}{2!}z^2\left(\frac{\partial^2 v}{\partial z^2}\right)_0 + \cdots + \infty \quad (8.175)$$

$$w(x, y, z) = w(x, y, 0) + z\left(\frac{\partial v}{\partial z}\right)_0 + \frac{1}{2!}z^2\left(\frac{\partial^2 w}{\partial z^2}\right)_0 + \cdots + \infty$$

After neglecting higher order strain terms, the appropriate displacement field can be expressed as

$$u(x, y, z) = u_0(x, y) + z\,\theta_x(x, y) + z^3\theta_x^*(x, y)$$
$$v(x, y, z) = v_0(x, y) + z\,\theta_y(x, y) + z^3\theta_y^*(x, y) \quad (8.176)$$
$$w(x, y, z) = w_0(x, y) + z^2 w_3^*(x, y)$$

where

u_0, v_0 are inplane displacements at a point in the middle plane

w_0 is the transverse displacement at a point in the middle plane

θ_x and θ_y are the rotations of the normal to the middle plane about the y and x-axis respectively.

w_0^*, θ_x^* and θ_y^* are higher order terms of Taylor's series expression.

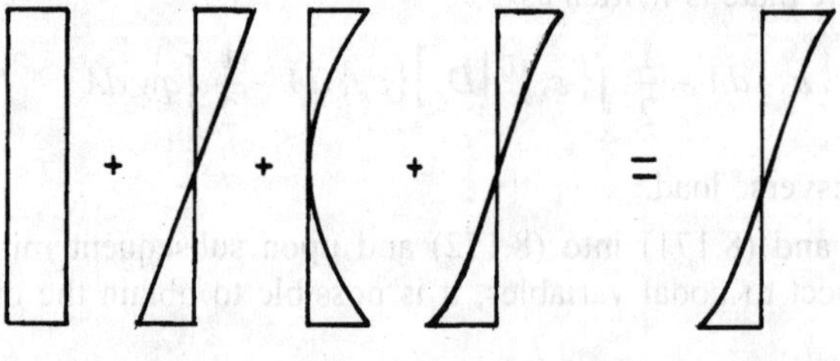

Fig. 8.23 Variation of various displacement components through the thickness

It may be noted that (8.176) is based on the consideration of the existence of the transverse normal strain condition through the laminate thickness. Those different terms mentioned above are shown in Fig. 8.23

8.18.1 Strain–Displacement Relationship

The strain-displacement relations are given by

$$\varepsilon_x = \frac{\partial u}{\partial x} \qquad \varepsilon_s = \frac{\partial u}{\partial y} + \frac{\partial v}{\partial x}$$

$$\varepsilon_y = \frac{\partial v}{\partial y} \qquad \varepsilon_{yz} = \frac{\partial v}{\partial z} + \frac{\partial w}{\partial y} \qquad (8.177)$$

$$\varepsilon_z = \frac{\partial w}{\partial z} \qquad \varepsilon_{zx} = \frac{\partial w}{\partial x} + \frac{\partial u}{\partial z}$$

Substituting the expressions for u, v, w from (8.176) into (8.177), yields

$$\varepsilon_x = \varepsilon_{x_0} + z\,\kappa_x + z^3\,\kappa_x^*$$

$$\varepsilon_y = \varepsilon_{y_0} + z\,\kappa_y + z^3\,\kappa_y^*$$

$$\varepsilon_z = z\,\kappa_z^* \qquad (8.178)$$

$$\varepsilon_s = \varepsilon_{s_0} + z\,\kappa_s + z^3\,\kappa_s^*$$

$$\varepsilon_{yz} = \phi_y + z^2\,\phi_y^*$$

$$\varepsilon_{zx} = \phi_y + z^2\,\phi_x^*$$

where

$$(\varepsilon_{x_0}, \varepsilon_{y_0}, \varepsilon_{s_0}) = \left(\frac{\partial u_0}{\partial x}, \frac{\partial v_0}{\partial y}, \frac{\partial u_0}{\partial y} + \frac{\partial v_0}{\partial x}\right)$$

$$(\kappa_x, \kappa_y, \kappa_s, \kappa_x^*, \kappa_y^*, \kappa_z^*) = \left(\frac{\partial \theta_x}{\partial x}, \frac{\partial \theta_y}{\partial y}, \frac{\partial \theta_x}{\partial y} + \frac{\partial \theta_y}{\partial x}, \frac{\partial \theta_x^*}{\partial x}, \frac{\partial \theta_y^*}{\partial y}, 2w_0^*\right)$$

$$\varepsilon_{s_0} = \frac{\partial u_0}{\partial y} + \frac{\partial v_0}{\partial x}, \qquad \kappa_s^* = \frac{\partial \theta_x^*}{\partial y} + \frac{\partial \theta_y^*}{\partial x} \qquad (8.179)$$

$$(\phi_x, \phi_y, \phi_x^*, \phi_y^*) = \left(\theta_x + \frac{\partial w_0}{\partial x}, \theta_y + \frac{\partial w_0}{\partial y}, 3\theta_x^* + \frac{\partial w_0^*}{\partial x}, 3\theta_y^* + \frac{\partial w_0}{\partial y}\right)$$

The flexural strains in the kth layer of the laminate at a distance z from the middle plane can be written separately as

$$\{\varepsilon_b\}^k = \begin{Bmatrix} \varepsilon_x \\ \varepsilon_y \\ \varepsilon_z \\ \varepsilon_s \end{Bmatrix}_k = \begin{Bmatrix} \varepsilon_{x_0} \\ \varepsilon_{y_0} \\ 0 \\ \varepsilon_{s_0} \end{Bmatrix} + z \begin{Bmatrix} \kappa_x \\ \kappa_y \\ \kappa_z^* \\ \kappa_s \end{Bmatrix}_k + z^3 \begin{Bmatrix} \kappa_x^* \\ \kappa_y^* \\ 0 \\ \kappa_s^* \end{Bmatrix}_k \qquad (8.180)$$

or, $\quad \{\varepsilon_b\}^k = \{\varepsilon_0\} + z\,\{\kappa\}_k + z^3\,\{\kappa^*\}_k \qquad (8.181)$

and the transverse shear strain for the kth layer is given by

$$\{\varepsilon_s\}^k = \begin{Bmatrix} \varepsilon_{yz} \\ \varepsilon_{zx} \end{Bmatrix}^k = \begin{Bmatrix} \phi_y \\ \phi_x \end{Bmatrix}^k + z^2 \begin{Bmatrix} \phi_y^* \\ \phi_x^* \end{Bmatrix} \tag{8.182}$$

or, $\quad \{\varepsilon_s\}^k = \{\phi\}^k + z^2 \{\phi^*\}$ \hfill (8.183)

8.18.2 Stress–Strain Relationship

Total potential energy is given by

$$\pi = U - W \tag{8.184}$$

or, $\quad \pi = \dfrac{1}{2} \displaystyle\int_v \{\varepsilon\}^T \{\sigma\} \, dv - \int_v \{\delta\}^T \{P\} \, dv$ \hfill (8.185)

where

U is the strain energy stored in the laminate
V is the work done by the external load
$\{P\}$ is the load vector
$\{\delta\}$ is the generated displacement vector at the middle plane of the plate

$$\begin{aligned}
\{\sigma\}^T &= \{\sigma_x \quad \sigma_y \quad \sigma_z \quad \sigma_s \quad \sigma_{yz} \quad \sigma_{zx}\} \\
\{\varepsilon\}^T &= \{\varepsilon_x \quad \varepsilon_y \quad \varepsilon_z \quad \varepsilon_s \quad \varepsilon_{yz} \quad \varepsilon_{zx}\} \\
\{\delta\}^T &= \{u \quad v \quad w\} \\
\{P\}^T &= \{P_x \quad P_y \quad P_z\}
\end{aligned} \tag{8.186}$$

The expression of energy assumes the following form after the strain components are substituted in (8.185) [8.24].

$$\pi = \dfrac{1}{2} \int_P \{\bar{\varepsilon}\}^T \{\bar{\sigma}\} \, dA - \int_A \{\bar{\delta}\}^T \{\bar{P}\} \, dA \tag{8.187}$$

where

$$\begin{aligned}
\{\bar{\sigma}\}^T &= \{N_x \quad N_y \quad N_s \quad M_x \quad M_y \quad M_s \quad M_x^* \quad M_y^* \quad M_s^* \quad Q_x \quad Q_y \quad Q_x^* \quad Q_y^*\} \\
\{\bar{\varepsilon}\}^T &= \{\varepsilon_{x0} \quad \varepsilon_{y0} \quad \varepsilon_{s0} \quad \kappa_x \quad \kappa_y \quad \kappa_s \quad \kappa_x^* \quad \kappa_y^* \quad \kappa_s^* \quad \kappa_z \quad \phi_x \quad \phi_y \quad \phi_x^* \quad \phi_y^*\} \\
\{\bar{P}\} &= \{P_{x0} \quad P_{y0} \quad P_{z0} \quad M_{x0} \quad M_{y0} \quad P_{z0}^* \quad M_{x0}^* \quad M_{y0}^*\} \\
\{\bar{\delta}\}^T &= \{u_0 \quad u_0 \quad w_0 \quad \theta_x \quad \theta_y \quad w_0^* \quad \theta_x^* \quad \theta_y^*\}
\end{aligned} \tag{8.188}$$

Then the resultants defined above are given by

$$\begin{Bmatrix} N_x \\ N_y \\ N_s \end{Bmatrix} = \sum_{k=1}^{N} \int_{z_k}^{z_{k+1}} \begin{Bmatrix} \sigma_x \\ \sigma_y \\ \sigma_s \end{Bmatrix} dz \tag{8.189}$$

$$\begin{bmatrix} M_x & M_x^* \\ M_y & M_y^* \\ M_z^* & 0 \\ M_s & M_s^* \end{bmatrix} = \sum_{k=1}^{N} \int_{z_k}^{z_{k+1}} \begin{Bmatrix} \sigma_x \\ \sigma_y \\ \sigma_z \\ \sigma_s \end{Bmatrix} (z, z^3) \, dz \tag{8.190}$$

$$\begin{bmatrix} Q_x & Q_x^* \\ Q_y & Q_y^* \end{bmatrix} = \sum_{k=1}^{N} \int_{z_k}^{z_{k+1}} \begin{Bmatrix} \sigma_{xz} \\ \sigma_{yz} \end{Bmatrix} (1, z^2) \, dz \tag{8.191}$$

where N is the total number of layers of the laminate.

After integrating through the thickness, (8.189) to (8.191) can be combined and written as

$$\begin{Bmatrix} \{N\} \\ \{M\} \\ \{Q\} \end{Bmatrix} = \begin{bmatrix} [A] & [B] & [0] \\ [B] & [D]_b & [0] \\ [0] & [0] & [D]_s \end{bmatrix} \begin{Bmatrix} \{\varepsilon_0\} \\ \{\kappa\} \\ \{\phi\} \end{Bmatrix} \tag{8.192}$$

or, $\quad \{\overline{\sigma}\} = [D]\{\overline{\varepsilon}\} \tag{8.193}$

where

$$\{N\}^T = \{N_x \quad N_y \quad N_s\}$$

$$\{M\}^T = \{M_x \quad M_y \quad M_s \quad M_x^* \quad M_y^* \quad M_s^*\} \tag{8.194}$$

$$\{Q\}^T = \{Q_x \quad Q_y \quad Q_x^* \quad Q_y^*\}$$

$$\{\varepsilon_0\} = \{\varepsilon_{x_0} \quad \varepsilon_{y_0} \quad \varepsilon_{s_0}\}$$

$$\{\kappa\} = \{\kappa_x \quad \kappa_y \quad \kappa_s \quad \kappa_x^* \quad \kappa_y^* \quad \kappa_s^* \quad \kappa_z^*\} \tag{8.195}$$

$$\{\phi\}^T = \{\phi_x \quad \phi_y \quad \phi_x^* \quad \phi_y^*\}$$

$\{\overline{\sigma}\}$ and $\{\overline{\varepsilon}\}$ are explained in (8.178).

Carrying out the necessary integration, the following relationship is obtained

$$[D] = \begin{bmatrix} A_{xx} & A_{xy} & A_{xs} \\ A_{yx} & A_{yy} & A_{ys} \\ A_{sx} & A_{sy} & A_{ss} \end{bmatrix} \tag{8.196}$$

$$[B] = \begin{bmatrix} B_{xx} & B_{xy} & B_{xs} \\ B_{yx} & B_{yy} & B_{ys} \\ B_{sx} & B_{sy} & B_{ss} \end{bmatrix} \tag{8.197}$$

where

$$A_{ij}, B_{ij} = \sum_{k=1}^{N} \int_{z_k}^{z_{k+1}} (Q_{ij})_k (1, z) \, dz \qquad i, j = x, y, s \tag{8.198}$$

$$[D]_B = \begin{bmatrix} D_{xx} & & & & & \\ D_{yx} & D_{yy} & & \text{symmetrical} & & \\ D_{sx} & D_{sy} & D_{ss} & & & \\ \overline{D}_{xx}^* & \overline{D}_{xy}^* & \overline{D}_{xs}^* & \overline{D}_{xx}^* & & \\ \overline{D}_{yx}^* & \overline{D}_{yy}^* & \overline{D}_{ys}^* & D_{ys}^* & D_{yy}^* & \\ \overline{D}_{sx}^* & \overline{D}_{sy}^* & \overline{D}_{ss}^* & D_{sx}^* & D_{sy}^* & D_{ss}^* \end{bmatrix} \tag{8.199}$$

where

$$D_{ij} = \sum_{k=1}^{N} \int_{z_k}^{z_{k+1}} (Q_{ij})_k z^2 \, dz \qquad i, j = x, y, s$$

$$D_{ij}^* = \sum_{k=1}^{N} \int_{z_k}^{z_{k+1}} (Q_{ij})_k z^6 \, dz \qquad i, j = x, y$$

$$D_{is}^* = \sum_{k=1}^{N} \int_{z_k}^{z_{k+1}} (Q_{is})_k z^4 \, dz \qquad i, j = x, y \tag{8.200}$$

$$\overline{D}_{ij}^* = \sum_{k=1}^{N} \int_{z_k}^{z_{k+1}} (Q_{ij})_k z^4 \, dz \qquad i, j = x, y$$

$$\overline{D}_{is}^* = \sum_{k=1}^{N} \int_{z_k}^{z_{k+1}} (Q_{ij})_k z^2 \, dz \qquad i, j = x, y$$

Similarly,

$$[D]_s = \begin{bmatrix} A_{xx}^z & A_{xy}^z & \overline{A}_{xx}^* & \overline{A}_{xy}^* \\ A_{xy}^z & A_{yy}^* & \overline{A}_{xy}^* & \overline{A}_{yy}^* \\ \overline{A}_{xx}^* & \overline{A}_{xy}^* & A_{xx}^* & A_{xy}^* \\ \overline{A}_{xy}^* & \overline{A}_{yy}^* & A_{yx}^* & A_{yy}^* \end{bmatrix} \tag{8.201}$$

$$A_{ij}^z = \sum_{k=1}^{N} \int_{z_k}^{z_{k+1}} k_s (Q_{ij}^z)_k \, dz \qquad i, j = x, y$$

$$A_{ij}^{*z} = \sum_{k=1}^{N} \int_{z_k}^{z_{k+1}} k_s (Q_{ij}^z)_k z^2 \, dz \qquad i, j = x, y \tag{8.202}$$

$$\overline{A}_{ij}^{*z} = \sum_{k=1}^{N} \int_{z_k}^{z_{k+1}} k_s (Q_{ij}^z)_k \, z^4 \, dz \qquad i,j = x, y$$

where k_s is the shear coefficient.

It may be noted that the formulation given so far results in non-vanishing transverse shear stress on the bounding planes of the laminate. The zero shear conditions have been incorporated, that is, ε_{xz} and ε_{yz} equal to zero at $z = \pm t/2$. The shear rigidity given by (8.191) is modified as follows.

If ε_{xz} and ε_{yz} are made zero at $z = \pm t/2$, then from (8.178) we get

$$\theta_x = -\frac{t^2}{4} \theta_x^*, \qquad \theta_y = -\frac{t^2}{4} \theta_y^*$$

$$\theta_x^* = -\frac{4}{t^2} \theta_x, \qquad \theta_y^* = -\frac{4}{t^2} \theta_y \qquad (8.203)$$

Then (8.201) becomes

$$[D]_{sh} = \begin{bmatrix} \overline{A}_{xx}^z & \overline{A}_{xy}^z & 0 & 0 \\ \overline{A}_{xy}^z & \overline{A}_{yy}^z & 0 & 0 \\ 0 & 0 & \overline{A}_{xx}^* & \overline{A}_{xy}^* \\ 0 & 0 & \overline{A}_{yx}^* & \overline{A}_{yy}^* \end{bmatrix} \qquad (8.204)$$

where

$$\overline{A}_{ij}^z = \sum_{i=1}^{N} k_s (Q_{ij})_k \left(H_1 - \frac{4}{t^2} H_3 \right) \qquad i,j = x, y$$

$$\overline{A}_{ij}^z = \sum_{i=1}^{N} k_s (Q_{ij})_k \left(H_5 - \frac{t^2}{4} H_3 \right) \qquad i,j = x, y \qquad (8.205)$$

$$H_i = \left(t_{k+1}^i - t_k^i \right) / i$$

8.18.3 Derivation of the Stiffness Matrix

The derivation of the stiffness matrix follows the routine procedure. The displacement vector within an element is expressed as

$$\{\delta\} = [N]\{X\}_e \qquad (8.206)$$

At each mode, the unknown displacements are

$$\{X\}_e^i = \{u_{0i} \quad v_{0i} \quad w_{oi} \quad \theta_{xi} \quad \theta_{yi} \quad w_{0i}^* \quad \theta_{xi}^* \quad \theta_{yi}^*\}^T \qquad (8.207)$$

The strain vector corresponding to the membrane part is given by

$$\left\{\begin{array}{c}\varepsilon_x\\ \varepsilon_y\\ \varepsilon_s\end{array}\right\}=\sum_{i=1}^{NN}\begin{bmatrix}\dfrac{\partial N_i}{\partial x} & 0 & 0 & 0 & 0 & 0 & 0 & 0\\ 0 & \dfrac{\partial N_i}{\partial y} & 0 & 0 & 0 & 0 & 0 & 0\\ \dfrac{\partial N_i}{\partial y} & \dfrac{\partial N_i}{\partial x} & 0 & 0 & 0 & 0 & 0 & 0\end{bmatrix}\left\{\begin{array}{c}u_{0i}\\ v_{0i}\\ w_{0i}\\ \theta_{xi}\\ \theta_{yi}\\ w_{0i}^*\\ \theta_{xi}^*\\ \theta_{yi}^*\end{array}\right\} \qquad (8.208)$$

where NN is the total number of nodes.

The strain component corresponding to the bending part is given by

$$\left\{\begin{array}{c}\kappa_x\\ \kappa_y\\ \kappa_s\\ \kappa_x^*\\ \kappa_y^*\\ \kappa_s^*\\ \kappa_z\end{array}\right\}=\sum_{i=1}^{NN}\begin{bmatrix}0 & 0 & 0 & \dfrac{\partial N_i}{\partial x} & 0 & 0 & 0 & 0\\ 0 & 0 & 0 & 0 & \dfrac{\partial N_i}{\partial y} & 0 & 0 & 0\\ 0 & 0 & 0 & \dfrac{\partial N_i}{\partial y} & \dfrac{\partial N_i}{\partial x} & 0 & 0 & 0\\ 0 & 0 & 0 & 0 & 0 & 0 & \dfrac{\partial N_i}{\partial x} & 0\\ 0 & 0 & 0 & 0 & 0 & 0 & 0 & \dfrac{\partial N_i}{\partial y}\\ 0 & 0 & 0 & 0 & 0 & 0 & \dfrac{\partial N_i}{\partial y} & \dfrac{\partial N_i}{\partial x}\\ 0 & 0 & 0 & 0 & 0 & 2N_i & 0 & 0\end{bmatrix}\left\{\begin{array}{c}u_{0i}\\ v_{0i}\\ w_{0i}\\ \theta_{xi}\\ \theta_{yi}\\ w_{0i}^*\\ \theta_{xi}^*\\ \theta_{yi}^*\end{array}\right\} \qquad (8.209)$$

The strain component corresponding to the shear part is given by

$$\left\{\begin{array}{c}\phi_x\\ \phi_y\\ \phi_x^*\\ \phi_y^*\end{array}\right\}=\sum_{i=1}^{NN}\begin{bmatrix}0 & 0 & \dfrac{\partial N_i}{\partial x} & N_i & 0 & 0 & 0 & 0\\ 0 & 0 & \dfrac{\partial N_i}{\partial y} & 0 & N_i & 0 & 0 & 0\\ 0 & 0 & 0 & 0 & 0 & \dfrac{\partial N_i}{\partial x} & 3N_i & 0\\ 0 & 0 & 0 & 0 & 0 & \dfrac{\partial N_i}{\partial y} & 0 & 3N_i\end{bmatrix}\{X\}_e^i \qquad (8.210)$$

Combining (8.208) to (8.210), yields

$$\{\varepsilon\}=[B]\{X\}_e \qquad (8.211)$$

Thus the stiffness matrix is derived as

$$[K]_e = \int_A [B]^T [D][B] \, dA \qquad (8.212)$$

8.18.4 Boundary Conditions

The boundary conditions for higher order elements may appear a bit confusing. In the displacement based finite element approach only the displacement boundary conditions are to be specified. For a simply supported plate for the element under consideration, the boundary conditions are as follows

$$w_0 = w_0^* = \theta_y = \theta_y^* = 0 \quad \text{at the edge} \quad x = a \qquad (8.213)$$

For a clamped plate for the element under consideration, boundary conditions are as follows

$$\theta_x = \theta_x^* = \theta_y = \theta_y^* = 0$$

$$w_0 = w_0^* = 0 \qquad (8.214)$$

at the edge $x = a$

8.19 NUMERICAL EXAMPLES

Examples are presented for bare composite plates and shells in static loading and free vibration.

8.19.1 Bare Composite Plates Under Transverse Load

Symmetric laminated plates have been analysed by Chen and Liu [8.25] using an elasticity approach. All the plates are of square shape. Two edges of the plate are simply supported and the other two are free. The entire plate is subjected to a uniform load of intensity q_0. These laminated plates are also analysed by Chattopadhyay et al. [8.15] using a 4×4 mesh division of the quarter plate for varying number of layers and specified boundary conditions. Results of both the investigators are presented in Table 8.1 in the following non-dimensional form for deflection and bending moments

$$\overline{w} = \frac{wh^3 E_2}{q_0 a^4}, \qquad \overline{M}_x \text{ or } \overline{M}_y = \frac{M_x \text{ or } M_y}{q_0 a^2} \qquad (8.215)$$

using usual notations.

It is interesting to note that the deflection $\overline{w}$ and bending moments $\overline{M}_x$ and $\overline{M}_y$ obtained by both the methods are close enough.

Table 8.1 Deflections and bending moments of square cross-ply laminated plates with two opposite edges simply supported and the others free under uniform loading.

No. of layers	$\overline{w}$ at free edge mid-point		$\overline{M}_x$ at free edge mi-dpoint		$\overline{M}_y$ at centre	
	Chen & Liu [8.25]	Chattopadhyay et al. [8.15]	Chen & Liu [8.25]	Chattopadhyay et al. [8.15]	Chen & Liu [8.25]	Chattopadhyay et al. [8.15]
3	0.06691	0.6698	0.1253	0.1267	0.00077	0.00077
5	0.01995	0.02000	0.1254	0.1267	0.00034	0.00035

7	0.01554	0.1537	0.1254	0.1268	0.00031	0.00032
9	0.01390	0.01393	0.1254	0.1268	0.00030	0.00031
11	0.01305	0.01308	0.1254	0.1268	0.00030	0.00030

8.19.2 Bare Composite Plate in Free Vibration

Anti-symmetric angle-ply plates having two different orientations of fibres have been analysed in free vibration. Results of Reddy [8.26] and Chattopadhyay et al. [8.16] using 4x4 mesh division for the whole plate have been presented in Table 8.2 which indicates only the fundamental frequency. The plates have all edges simply supported. Results of both the investigators are quite close.

Table 8.2 Fundamental frequency parameter $\lambda = \overline{\omega} a^2 \left(\rho h^2 / E_2\right)^{\frac{1}{2}}$ for a square simply supported plate laminate considered: Anti-symmetric angle-ply $(\theta/-\theta/\theta/-\theta \cdots)$

No. of layers	$\theta = 30^0$		$\theta = 45^0$	
	Reddy [8.26]	Chattopadhyay et al. [8.16]	Reddy [8.26]	Chattopadhyay et al. [8.16]
2	15.00	14.59	15.71	15.31
4	17.69	17.61	18.61	18.53
6	18.00	18.23	18.29	18.85
8	18.10	18.42	18.63	18.95
16	18.19	18.60	19.12	19.02

8.19.3 Composite Cylindrical Shell under Transverse Load

A cross-ply (0°/90°) cylindrical shell of length equal to 4 times the radius and subjected to sinusoidal loading is analysed. The cylinder is subjected to internal sinusoidal pressure of intensity $q = q_0 \sin\left(\frac{\pi x}{l}\right) \cos\left[\frac{4y}{R}\right]$. This composite shell is analysed by the theory of elasticity approach which is considered as exact [8.26]. A finite element analysis using isoparametric element has also been carried out using a 2×2 mesh discretization adopting 1/8 along the circumference and 1/2 along longitudinals [8.27]. The finite element results are those obtained by using higher order shear deformation theory (HOST). Results have been presented in Table 8.3 in terms of the following non-dimentional quantities.

$$\overline{w}_0 = \left(\frac{h^3 E_2}{q_0 R^4} \times 10\right) \frac{w_0}{\sin\left(\frac{\pi x}{R}\right)\cos\left(\frac{4y}{R}\right)} \tag{8.216}$$

and

$$\sigma_x \text{ or } \sigma_y = \left(\frac{10h^3}{q_0 R^2}\right)\left(\sin\frac{\pi x}{l}\right)\cos\left(\frac{4y}{R}\right) \tag{8.217}$$

It may be noted from Table 8.3 that at higher values of *R/h* ratio, the results obtained by the theory of elasticity approach is closer to that of the finite element analysis.

Table 8.3 Maximum displacement and extreme fibre stresses of an unsymmetric cross-ply (0°/90°) cylindrical shell of finite length subjected to sinusoidal transverse load.

$\dfrac{R}{h}$	Theory	$\overline{w}_0$	$\sigma_{x_{top}}$	$\sigma_{x_{bot}}$	$\sigma_{y_{top}}$	$\sigma_{y_{bot}}$
2	Exact	14.0340	2.6600	0.2510	3.046	9.775
	HOST	16.3658	1.8720	0.2848	2.485	13.544
4	Exact	6.1000	0.9610	0.2120	1.7800	10.3100
	HOST	6.1923	0.8516	0.2146	1.6458	11.4738
10	Exact	3.3300	0.1689	0.1930	1.3430	10.590
	HOST	3.2447	0.1651	0.1769	1.2404	9.284

Exact – Ref. 8.26 and HOST – Ref. 8.27

8.19.4 Free Vibration of a Composite Shallow Shell

A spherical shell made of graphite shown in Fig. 8.24 has the following material properties.

$E_1 = 60.7$ GPa, $E_2 = 24.8$ GPa, $G_{12} = 12.0$ GPa, $\nu_{12} = 0.23$

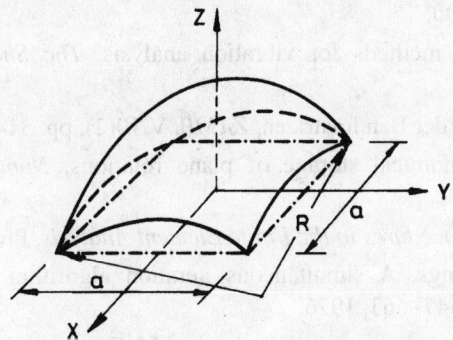

Fig. 8.24 Free vibration of a spherical shell

The shallow laminated cross-ply shell is completely free along all edges. In Table 8.4 are given the first five non-dimensional frequency parameters $\Omega = \omega a^2 \sqrt{\rho/E_1 h^2}$. The cross-ply laminate has 3 layers - 0°/90°/0°: $R_x/R_y = 1$ and two ratios of a/R_x have been considered [8.28].

Table 8.4 Frequency parameters of completely free, cross-ply laminated shallow shells

$\dfrac{a}{b} = 1$, $\dfrac{a}{h} = 100$, Laminate: $0^o/90^o/0^o$

$\dfrac{R_x}{R_y}$	$\dfrac{a}{R_x}$	Ω				
		1	2	3	4	5
1	0.1	1.5073	3.0713	3.6816	5.6224	6.3610
	0.5	1.5205	4.3732	5.0317	6.9544	7.9116

REFERENCES & SUGGESTED READINGS

8.1 R.W. Clough, The finite element method in plane stress analysis, *Proc. of the Second Conference on Electronic Computation*, ASCE, Pittsburg, USA, 1960.

8.2 Y.K. Cheung, *Finite Strip Method in Structural Analysis*, Pergamon Press, 1976.

8.3 M. Mukhopadhyay, A semi-analytic solution for the free vibration of rectangular plates, *Journal of Sound and Vibration*, V.60, No. 1, pp.71–85, 1978.

8.4 M. Mukhopadhyay, A semi-analytic solution for radially supported curved plates in bending, *Forschung im Ingenieurwesen*, V. 44. No. 6, pp.187–196, 1978.

8.5 M. Mukhopadhyay, *Structures: Matrix and Finite Element*, 3rd Edn. A.A. Balkema & Co., The Netherlands, 1994.

8.6 O.C. Zienkiewicz and R.L. Taylor, *The Finite Element Method*, V.1 and 2, McGraw-Hill Book Co., 1989.

8.7 B.M. Irons, A frontal solution program for the finite element analysis, *IJNME*, V 2, pp. 5–32, 1970.

8.8 M. Mukhopadhyay, *Vibration, Dynamics and Structural Systems*, Oxford & IBH Publishing Co., India, Second Edition, 2000.

8.9 A. Jennings, Eigenvalue methods for vibration analysis, *The Shock and Vibration Digest*, V.12(2), pp. 3–16, 1980.

8.10 H. Rutihauser, Deflection der Bandmatrizen, *ZAMP*, V.10(2), pp. 314–319, 1959.

8.11 G.W. Steward, The economical storage of plane rotations, *Numerical Mathematics*, V.24, pp.137–138, 1976.

8.12 K.J. Bathe, *Numerical Procedures in the Finite Element Analysis*, Prentice Hall, 1995.

8.13 R.B. Corr and A. Jennings, A simultaneous iteration algorithm for symmetric eigenvalue problems, *IJNME*, V.10, 647–663, 1976.

8.14 R.W. Clough and J. Penzien, *Dynamics of Structures*, McGraw Hill Inc., 2nd Edition, 1994

8.15 B. Chattopadhyay, P.K. Sinha and M. Mukhopadhyay, Finite element analysis of blade stiffened composite plates under transverse loads, *Journal of Reinforced Plastics and Composites*, V. 12, No. 1, pp. 76–10, 19930.

8.16 B. Chattopadhyay, P. K. Sinha and M. Mukhopadhyay, Free vibration analysis of eccentrically stiffened composite plates, *Journal of Reinforced Plastics and Composites*, V. 11, No. 9., pp. 1003–1034,1992.

8.17 A. Mukherjee and M. Mukhopadhyay, Finite element free vibration of eccentrically stiffened plates, *Computers and Structures*, V. 30, No. 6, pp. 1303–1317, 1988.

8.18 Jeong Su Kuo and Byung Man Kwak, A laminated composite beam element separately interpolated for bending and shear deflections without increase of nodal DOF, *Computers and Structures*, V.53, No. 5, pp. 1091–1098, 1994.

8.19 C. C. Chao and J. C. Lee, Vibration of eccentrically stiffened laminates, *Journal of Composite Materials*, V. 15, No. 1, pp. 233–255, 1980.

8.20 S. Goswami and M. Mukhopadhyay, Finite element analysis of laminated composite stiffened shells, *Journal of Reinforced Plastics and Composites*, V. 13, No. 7, pp. 574–617, 1994.

8.21 K. Chandrasekhara, Free vibrations of anisotropic laminated doubly curved shells, *Computers and Structures*, V. 33, pp. 435–440, 1989.

8.22 J. G. Ren and E. Hinton, The finite element analysis of homogeneous and laminated composite plates using a simple higher order theory, *Communications in Applied Numerical Methods*, V.2, pp. 217–228, 1986.

8.23 Mallikarjuna and T. Kant, A critical review and some results of recently refined theories of fibre-reinforced laminated composites and sandwiches, *Composite Structures*, V.23, 1993, pp. 293–312

8.24 Mallikarjuna and T. Kant, Finite element transient response of composite and sandwich plates with a refined theory, *Journal of Applied Mechanics, Transactions ASME*, V.57, 1995, pp. 1084–1086.

8.25 W.C. Chen and W.H. Liu, Deflections and free vibrations of laminated plates - Levy type solutions, *International Journal of Mechanical Sciences*, V. 32, pp. 779–793, 1990.

8.26 J.N. Reddy, Free vibration of anti-symmetric angle-ply laminated plates including transverse shear deformation by the finite element method, *Journal of Sound and Vibration*, V. 66, pp. 565–576, 1979.

8.27 T. Kant and J. R. Kommeni, Geometrically nonlinear analysis of doubly curved laminated and fibre reinforced composite shells with higher order theory and C^0 finite elements, *Journal of Reinforced Plastics and Composites*, V. 11, No. 9, pp. 1048–1076, 1992.

8.28 T. K. Varadan and K. Bhaskar, Bending of laminated orthotropic cylindrical shells - an elasticity approach, *Composite Science and Technology*, V. 17, pp. 141–156, 1991.

8.29 M.S. Qatu and A. W. Leissa, Free vibrations of completely free doubly curved laminated composite shallow shells, *Journal of Sound and Vibration*, V. 151, pp. 9–29, 1991.

CHAPTER 9

HYGROTHERMAL EFFECTS IN LAMINATES

9.1 INTRODUCTION

The analysis of FRP laminates has so far been based on externally applied loads which is referred to as mechanical loading. Nothing has so far been mentioned about environmental effects. But during the initial fabrication and final use, composites are subjected to changing environmental conditions. Out of various environmental loads that may exist, the ones that are of interest to us arise from two considerations. The first is the effect of temperature known as thermal effects. The second is the effect of moisture absorption from the atmosphere known as hygroscopic effects. The combined effect of temperature and moisture is known as hygrothermal effects.

Hygrothermal effects induce a dimensional change of laminate. But due to the mismatch of the properties of the constituents of the laminate, its free movement is inhibited. As a result of which deformations and corresponding stresses are set up. The net result is the occurrence of residual stress and warpage.

There are two principal effects of changes in hygrothermal environment on the mechanical behaviour of polymer composites

1. Matrix dominated properties such as stiffness and strength under transverse, off-axis and shear loading are altered. Increased temperature causes a gradual softening of the polymer matrix material upto a point. If the temperature is increased beyond the so-called "glass transition" region (indicating a transition from glassy behaviour to rubbery behaviour), however, the polymer becomes too soft for use as a structural material (Fig. 9.1). Plasticization of the polymer by absorbed moisture causes a reduction of the glass transition temperature T_g and a corresponding degradation of composite properties (Fig. 9.1).

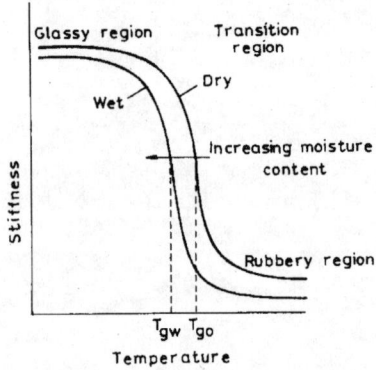

T_g - glass transition temperature T_{go} - dry T_g T_{gw} - wet T_g

Fig. 9.1 Variartion of stiffness with temperature for a typical polymer

2. Hygrothermal expansions or contractions change the stress or strain distribution of the composite. Increased temperature and/or moisture content causing swelling or contraction of the matrix is resisted by the fibres and residual stresses developed in the composite. Similarly, at the laminate level differential contractions or expansions of the laminae may take place.

9.2 EFFECT OF HYGROTHERMAL FORCES ON MECHANICAL BEHAVIOUR

Hygrothermal effects are most pronounced in matrix dominated properties, as fibres are least influenced by environment. The effect of temperature on matrix dominated properties of composites such as transverse strength and inplane shear are shown in Fig. 9.2 and 9.3 respectively for carbon/epoxy composites. Increase of temperature decreases the transverse modulus at a uniform rate and the ultimate strain at failure is the same at all three temperatures (Fig. 9.2). Influence of temperature on inplane shear stress–strain curves follows similar patterns, but increase in temperature reduces the shear strength (Fig. 9.3).

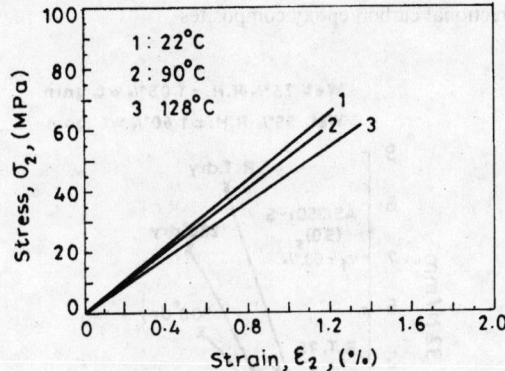

Fig. 9.2 Transverse tensile stress–strain curve for carbon/epoxy composites at different temperature

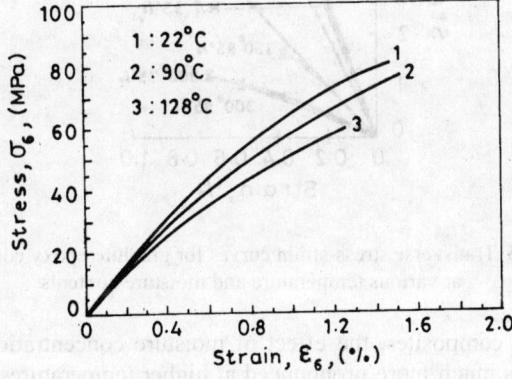

Fig. 9.3 Inplane shear stress–strain curves for unidirectional carbon/epoxy composites at varying temperature

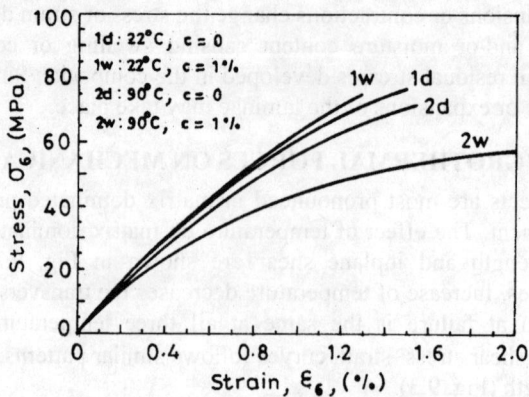

Fig. 9.4 Combined effects of temperature and moisture on inplane shear stress–strain curves for unidirectional carbon/epoxy composites

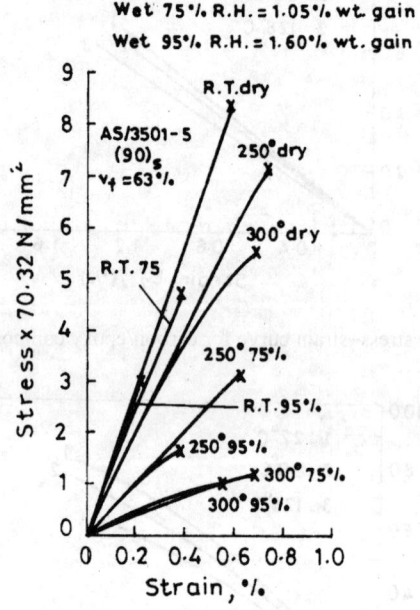

Fig. 9.5 Transverse stress-strain curves for graphite/epoxy composites at various temperature and moisture contents

For polymer matrix composites, the effect of moisture concentration is similar to that of the temperature. This effect is much more pronounced at higher temperatures. In the shear stress versus shear strain curve of Fig. 9.4, the difference in shear stress for 1 percent moisture content seem to be marginal at room temperature, but the difference is significant at a temperature of 90°C. Another result of the transverse stress–strain curve is shown in Fig. 9.5. Strength and stiffness are reduced considerably for both the values of relative humidity. Other investigations have revealed that the most deleterious effects on stiffness and strength are produced by a combination of elevated temperature and high moisture concentration [9.1, 9.2].

Composition and type of matrix resin have significant influence on strength and stiffness due to hygrothermal effects.

9.3 MICROMECHANICS OF HYGROTHERMAL PROPERTIES

Two basic hygrothermal properties for a unidirectional lamina are the two principal coefficients of thermal expansion (CTE), α_1 and α_2 and the two principal coefficients of moisture expansion (CME), β_1 and β_2. These properties are related to the corresponding properties of the constituent materials.

For isotropic constituents, the approximate rule of mixtures for the longitudinal coefficient of thermal expansion is given as [9.3]

$$\alpha_1 = \frac{E_{1f}\alpha_f V_f + E_m \alpha_m V_m}{E_{1f} V_f + E_m V_m} \tag{9.1}$$

where E_{1f}, E_m = Elastic modulus of the fibre in the longitudinal direction and that of matrix respectively

α_f, α_m = Coefficients of thermal expansion of the fibre and the matrix respectively

V_f, V_m = Fibre and matrix volume fractions respectively

Equation (9.1) has been derived by Schapery using energy principles [9.3]. Same equations can be derived by using the mechanics of materials approach [9.4]. Interestingly, the simple expression given by (9.1) yields sufficient accurate results.

Transverse coefficients of thermal expansion for composites with isotropic constituents have been proposed by Schapery [9.3] as

$$\alpha_2 = \alpha_f V_f (1+\nu_f) + \alpha_m V_m (1+\nu_m) - \nu_{12}\alpha_1 \tag{9.2}$$

where

α_1 = Longitudinal coefficient of thermal expansion given by (9.1)

ν_f, ν_m = Poisson's ratio for the fibre and the matrix respectively

ν_{12} = Major Poisson's ratio and is given by (3.52).

Hashin [9.5] has derived α_2 for composites with orthotropic fibres as

$$\alpha_2 = \alpha_{2f} \cdot V_f \left(1+\nu_{12f} \cdot \frac{\alpha_{1f}}{\alpha_{2f}}\right) + \alpha_m V_m (1+\nu_m)$$
$$- \left(\nu_{12f} \cdot V_f + \nu_m V_m\right)\frac{(E\alpha)_1}{E_1} \tag{9.3}$$

where

α_{1f}, α_{2f} = longitudinal and transverse CTEs of fibres respectively

α_m = longitudinal and transverse CTE of the matrix

ν_{12f}, ν_m = longitudinal Poisson's ratio for fibres and matrix respectively

$$(E\alpha)_1 = E_{1f} \cdot \alpha_{1f} \cdot V_f + E_m \alpha_m V_m \tag{9.4}$$

$$E_1 = E_{1f} \cdot V_f + E_m \cdot V_m \tag{9.5}$$

It may be noted that carbon and aramid fibres are orthotropic. The matrix however is considered as isotropic and matrix properties are related to a single value as given by (9.3) to (9.5).

The off-axis values of the coefficient of thermal expansion can be obtained from the transformation relation. Thus, the coefficients in an arbitrary orthogonal xy axis system with x-axis orientation θ to 1–axis direction are related to those in the principal material axis as

$$\begin{aligned}
\alpha_x &= \alpha_1 m^2 + \alpha_2 n^2 \\
\alpha_y &= \alpha_1 n^2 + \alpha_2 m^2 \\
\alpha_{xy} &= \alpha_s = 2(\alpha_1 - \alpha_2) mn
\end{aligned} \tag{9.6}$$

where $m = \cos\theta$ and $n = \sin\theta$

The micromechanical approach for the determination of the coefficient of moisture expansion is identical to the above approach given for the CTE. Fibres, however, usually do not absorb any moisture from the surrounding so that $\beta_f = \beta_{1f} = \beta_{2f} = 0$

Longitudinal and transverse coefficients of moisture expansion for orthotropic constituents are

$$\beta_1 = \beta_m \frac{E_m V_m}{E_{1f} V_f + E_m V_m} \tag{9.7}$$

and

$$\beta_2 = \beta_{2m} \cdot \frac{V_m}{E_1} \left[\left(1 + v_{12m} \cdot \frac{\beta_{1m}}{\beta_{2m}}\right) E_1 - \left(v_{12f} V_f + v_{12m} \cdot V_m\right) E_{1m} \cdot \frac{\beta_{1m}}{\beta_{2m}} \right] \tag{9.8}$$

For composites with orthotropic fibres and an isotropic matrix, β_1 and β_2 are given by

$$\beta_1 = \beta_{1m} \cdot \frac{E_m V_m}{E_{1f} \cdot V_f + E_m V_m} \tag{9.9}$$

$$\beta_2 = \beta_{2m} \cdot \frac{V_m}{E_{1f} V_f + E_m V_m} \left[(1 + v_m) E_{1f} \cdot V_f + (V_m - v_{12f} \cdot V_f) E_m \right] \tag{9.10}$$

For determining $\beta_x, \beta_y, \beta_{xy}$ the same transformation law applies as given by (9.6)

$$\begin{aligned}
\beta_x &= \beta_1 m^2 + \beta_2 n^2 \\
\beta_y &= \beta_1 n^2 + \beta_2 m^2 \\
\beta_{xy} &= \beta_s = 2(\beta_1 - \beta_2) mn
\end{aligned} \tag{9.11}$$

where $m = \cos\theta$, $n = \sin\theta$

9.4 UNIDIRECTIONAL LAMINA : HYGROTHERMAL STRAINS

The coefficients of the thermal expansion have been found to vary with the temperature. However, it is assumed that the coefficient of thermal expansion (CTE) and the coefficient of moisture expansion (CME) remain constant over the range of temperature and moisture conditions which a laminate is subjected to. The CTEs and CMEs of fibres are distinctly different from that of matrix materials. CTEs and CMEs are indicated for typical materials in Table 9.1.

Let (T_0, C_0) be the reference hygrothermal state. Let the uniform change in temperature be $\Delta T = T - T_0$ and the change in moisture concentration $\Delta C = C - C_0$. A lamina undergoes hygrothermal deformation due to the above changes. It is further assumed that thermal deformation and moisture deformation are uncoupled.

Hygrothermal strains in the principal material axes direction of the lamina are

$$e_1 = \alpha_1 \Delta T + \beta_1 \Delta C$$
$$e_2 = \alpha_2 \Delta T + \beta_2 \Delta C \qquad (9.12)$$
$$e_6 = 0$$

Referring to Fig. 9.6, the hygrothermal strains in the arbitrary xy-axis system is

$$e_x = e_1 m^2 + e_2 n^2$$
$$e_y = e_1 n^2 + e_2 m^2 \qquad (9.13)$$
$$e_{xy} = e_s = 2(e_1 = e_2) mn$$

where $m = \cos\theta, n = \sin\theta$

Table 9.1 CTEs and CMEs of typical composites

Material	Thermal expansion coefficient (10^{-6} m/m)		Coefficient of moisture expansion m/m	
	α_1	α_2	β_1	β_2
A. S. graphite/epoxy	0.88	31.0	0.090	0.30
E-glass/epoxy	6.30	20.0	0.014	0.29
AF-126-2 adhesive	29.00	29.0	0.200	0.20
1020 steel	12.00	12.0	–	–
Carbon/epoxy* (AS4/3501-6)	– 0.90	24.0	–	–
Kevlar/epoxy*	– 4.00	57.6	–	–
Boron/epoxy*	6.10	30.3	–	–

* Values indicated are at 24° C

Combining (9.6), (9.12) and (9.13) yields

$$e_x = \alpha_x \Delta T + \beta_x \cdot \Delta C$$
$$e_y = \alpha_y \Delta T + \beta_y \cdot \Delta C \qquad (9.14)$$
$$e_s = \alpha_s \Delta T + \beta_s \cdot \Delta C$$

9.5 FREE THERMAL STRAINS

For a fibre-reinforced material, the coefficients of thermal expansion are different in each of the three principal material directions.

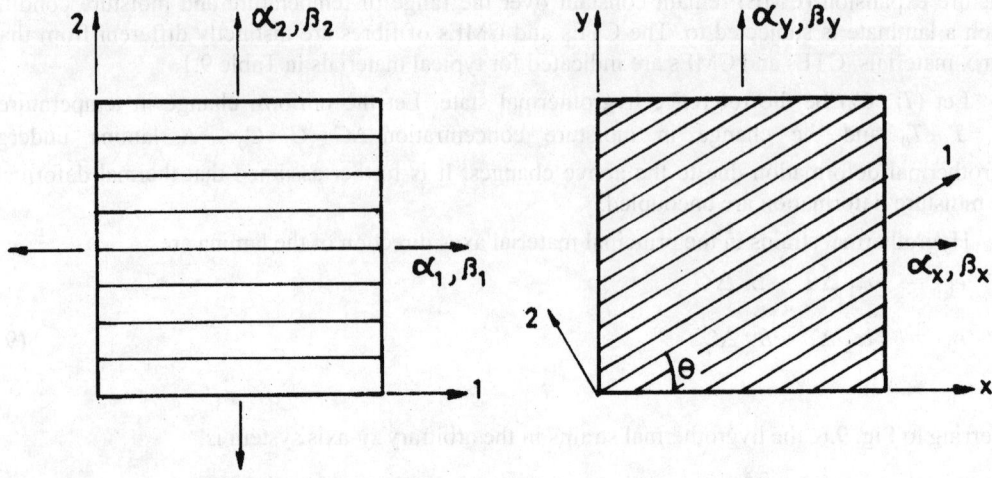

Fig. 9.6 α and β values of a unidirectional lamina

Let us consider a graphite-polymer composite and heat it. Graphite fibres will contract when heated while the polymer matrix will expand. As such the deformation of the composite in the fibre direction is dependent on the relative proportion of the fibre and the matrix – the composite may expand or contract or exhibit no change in length. But as the matrix properties are dominant in the other two directions, the composite will expand in these directions.

If the element has no externally applied load acting on it, the strain is termed as free thermal strain. The free thermal strains can be related to three coordinates in terms of the difference in temperature (already shown in (9.12)).

$$e_1 = \alpha_1 \Delta T, \qquad e_2 = \alpha_2 \Delta T \quad \text{and} \quad e_3 = \alpha_3 \Delta T \qquad (9.15)$$

where ΔT is the temperature difference.

It is interesting to note that thermal strains do not induce shear deformations. The strains that occur due to temperature variation are dilatational (change of volume) rather than distortional (change of shape). Further, as the material is allowed to deform freely, no stresses result due to this.

However, a point is to be made at this stage. In the graphite-polymer composite mentioned, if both the fibre and the matrix are to be viewed separately, then stresses will exist in both of them due to temperature change. But when they are treated as smeared, the stresses when integrated over the entire volume will yield a null value.

Similarly, if ΔC is the difference in moisture content, then free moisture strains related to the three coordinates are

$$e_1 = \beta_1 \Delta C, \qquad e_2 = \beta_2 \Delta C \quad \text{and} \quad e_3 = \beta_3 \Delta C \qquad (9.16)$$

9.6 STRESS-STRAIN RELATIONSHIP OF A LAMINA WHEN FREE THERMAL STRAINS ARE INCLUDED

It is the mechanical strains that play a key role in stress-strain relationship. Total strains consist of mechanical strains plus free thermal strains. As such

$$\begin{Bmatrix} \varepsilon_{1e} \\ \varepsilon_{2e} \\ \varepsilon_{3e} \\ \varepsilon_{4e} \\ \varepsilon_{5e} \\ \varepsilon_{6e} \end{Bmatrix} = \begin{Bmatrix} \varepsilon_1 - e_1 \\ \varepsilon_2 - e_2 \\ \varepsilon_3 - e_3 \\ \varepsilon_4 \\ \varepsilon_5 \\ \varepsilon_6 \end{Bmatrix} = \begin{Bmatrix} \varepsilon_1 - \alpha_1 \Delta T \\ \varepsilon_2 - \alpha_2 \Delta T \\ \varepsilon_3 - \alpha_3 \Delta T \\ \varepsilon_4 \\ \varepsilon_5 \\ \varepsilon_6 \end{Bmatrix} \tag{9.17}$$

in (9.17), $\varepsilon_{1e}, \varepsilon_{2e}$ etc. are mechanical strains, $\varepsilon_1, \varepsilon_2$ etc., are total strains and e_1, e_2 etc., are free thermal strains.

Therefore, the stress–strain relationship in terms of stiffnesses and compliances can be written as

$$\begin{Bmatrix} \varepsilon_1 - \alpha_1 \Delta T \\ \varepsilon_2 - \alpha_2 \Delta T \\ \varepsilon_3 - \alpha_3 \Delta T \\ \varepsilon_4 \\ \varepsilon_5 \\ \varepsilon_6 \end{Bmatrix} = \begin{bmatrix} S_{11} & S_{12} & S_{13} & 0 & 0 & 0 \\ S_{12} & S_{22} & S_{23} & 0 & 0 & 0 \\ S_{13} & S_{23} & S_{33} & 0 & 0 & 0 \\ 0 & 0 & 0 & S_{44} & 0 & 0 \\ 0 & 0 & 0 & 0 & S_{55} & 0 \\ 0 & 0 & 0 & 0 & 0 & S_{66} \end{bmatrix} \begin{Bmatrix} \sigma_1 \\ \sigma_2 \\ \sigma_3 \\ \sigma_4 \\ \sigma_5 \\ \sigma_6 \end{Bmatrix} \tag{9.18}$$

and

$$\begin{Bmatrix} \sigma_1 \\ \sigma_2 \\ \sigma_3 \\ \sigma_4 \\ \sigma_5 \\ \sigma_6 \end{Bmatrix} = \begin{bmatrix} C_{11} & C_{12} & C_{13} & 0 & 0 & 0 \\ C_{21} & C_{22} & C_{23} & 0 & 0 & 0 \\ C_{31} & C_{32} & C_{33} & 0 & 0 & 0 \\ 0 & 0 & 0 & C_{44} & 0 & 0 \\ 0 & 0 & 0 & 0 & C_{55} & 0 \\ 0 & 0 & 0 & 0 & 0 & C_{66} \end{bmatrix} \begin{Bmatrix} \varepsilon_1 - \alpha_1 \Delta T \\ \varepsilon_2 - \alpha_2 \Delta T \\ \varepsilon_3 - \alpha_3 \Delta T \\ \varepsilon_4 \\ \varepsilon_5 \\ \varepsilon_6 \end{Bmatrix} \tag{9.19}$$

If due to restraints in all directions, no deformation results due to temperature changes, then we get

$$\varepsilon_1 = \varepsilon_2 = \varepsilon_3 = \varepsilon_4 = \varepsilon_5 = \varepsilon_6 = 0 \tag{9.20}$$

Substitution of the values from (9.20) into (9.17), yields

$$\begin{Bmatrix} \varepsilon_{1e} \\ \varepsilon_{2e} \\ \varepsilon_{3e} \\ \varepsilon_{4e} \\ \varepsilon_{5e} \\ \varepsilon_{6e} \end{Bmatrix} = \begin{Bmatrix} -\alpha_1 \Delta T \\ -\alpha_2 \Delta T \\ -\alpha_3 \Delta T \\ 0 \\ 0 \\ 0 \end{Bmatrix} \tag{9.21}$$

Equation (9.19) reduces to

$$\begin{Bmatrix} \sigma_1 \\ \sigma_2 \\ \sigma_3 \\ \sigma_4 \\ \sigma_5 \\ \sigma_6 \end{Bmatrix} = \begin{bmatrix} C_{11} & C_{12} & C_{13} & 0 & 0 & 0 \\ C_{21} & C_{22} & C_{23} & 0 & 0 & 0 \\ C_{31} & C_{32} & C_{33} & 0 & 0 & 0 \\ 0 & 0 & 0 & C_{44} & 0 & 0 \\ 0 & 0 & 0 & 0 & C_{55} & 0 \\ 0 & 0 & 0 & 0 & 0 & C_{66} \end{bmatrix} \begin{Bmatrix} -\alpha_1 \Delta T \\ -\alpha_2 \Delta T \\ -\alpha_3 \Delta T \\ 0 \\ 0 \\ 0 \end{Bmatrix} \qquad (9.22)$$

Example 9.1

The stiffness matrix for a glass-polymer composite is given by

$$[C] = \begin{bmatrix} 160 & 5.5 & 5.5 & 0 & 0 & 0 \\ 5.5 & 15 & 7.2 & 0 & 0 & 0 \\ 5.5 & 7.2 & 15 & 0 & 0 & 0 \\ 0 & 0 & 0 & 3.5 & 0 & 0 \\ 0 & 0 & 0 & 0 & 4.5 & 0 \\ 0 & 0 & 0 & 0 & 0 & 4.5 \end{bmatrix} \text{ GPa}$$

Given

$\alpha_1 = 6.34 \times 10^{-6}\,/°\text{C}, \quad \alpha_2 = 23.3 \times 10^{-6}\,/°\text{C} \quad \text{and} \quad \alpha_3 = 23.3 \times 10^{-6}\,/°\text{C}$

$\Delta T = 100°\text{C}$

The composite material is fully restrained.

Calculate the change in the stresses.

$\sigma_1 = -(\alpha_1 C_{11} + \alpha_2 C_{12} + \alpha_3 C_{13})\Delta T$
$\quad = -[6.34 \times 10^{-6} \times 160 \times 10^3 + 23.3 \times 10^{-6} \times 5.5 \times 10^3 + 23.3 \times 10^{-6} \times 5.5 \times 10^3] \times 100$ MPa
$\quad = -12.707$ MPa

$\sigma_2 = -[\alpha_1 C_{12} + \alpha_2 C_{22} + \alpha_3 C_{23}]\Delta T$
$\quad = -[6.34 \times 10^{-6} \times 5.5 \times 10^3 + 23.3 \times 10^{-6} \times 15 \times 10^3 + 23.3 \times 10^{-6} \times 7.2 \times 10^3] \times 100$ MPa
$\quad = -5.5213$ MPa

$\sigma_3 = -[\alpha_1 C_{13} + \alpha_2 C_{23} + \alpha_3 C_{33}]\Delta T$
$\quad = -[6.34 \times 10^{-6} \times 5.5 \times 10^3 + 23.3 \times 10^{-6} \times 7.2 \times 10^3 + 23.3 \times 10^{-6} \times 15 \times 10^3] \times 100$ MPa
$\quad = -5.5213$ MPa

$\sigma_4 = \sigma_5 = \sigma_6 = 0$

9.7 HYGROTHERMOELASTIC STRESS–STRAIN RELATIONS

Following the steps of the earlier section, the plane stress–strain relations including free thermal and moisture strain effects for the kth lamina in the principal material directions are given by

$$\begin{Bmatrix} \varepsilon_1 - \alpha_1 \Delta T - \beta_1 \Delta C \\ \varepsilon_2 - \alpha_2 \Delta T - \beta_2 \Delta C \\ \varepsilon_6 \end{Bmatrix}_k = \begin{bmatrix} S_{11} & S_{12} & 0 \\ S_{12} & S_{22} & 0 \\ 0 & 0 & S_{66} \end{bmatrix}_k \begin{Bmatrix} \sigma_1 \\ \sigma_2 \\ \sigma_6 \end{Bmatrix}_k \qquad (9.23)$$

and

$$\begin{Bmatrix} \sigma_1 \\ \sigma_2 \\ \sigma_6 \end{Bmatrix}_k = \begin{bmatrix} Q_{11} & Q_{12} & 0 \\ Q_{12} & Q_{22} & 0 \\ 0 & 0 & Q_{66} \end{bmatrix}_k \begin{Bmatrix} \varepsilon_1 - \alpha_1 \Delta T - \beta_1 \Delta C \\ \varepsilon_2 - \alpha_2 \Delta T - \beta_2 \Delta C \\ \varepsilon_6 \end{Bmatrix}_k \qquad (9.24)$$

The mechanical strains are defined as

$$\begin{Bmatrix} \varepsilon_{1e} \\ \varepsilon_{2e} \\ \varepsilon_{6e} \end{Bmatrix}_k = \begin{Bmatrix} \varepsilon_1 - \alpha_1 \Delta T - \beta_1 \Delta C \\ \varepsilon_2 - \alpha_2 \Delta T - \beta_2 \Delta C \\ \varepsilon_6 \end{Bmatrix}_k \qquad (9.25)$$

Equation (9.6) can be written as

$$\begin{Bmatrix} \alpha_x \Delta T \\ \alpha_y \Delta T \\ \dfrac{\alpha_s}{2} \Delta T \end{Bmatrix}_k = \begin{bmatrix} \cos^2 \theta & \sin^2 \theta & -2\sin\theta\cos\theta \\ \sin^2 \theta & \cos^2 \theta & 2\sin\theta\cos\theta \\ \sin\theta\cos\theta & -\sin\theta\cos\theta & \cos^2\theta - \sin^2\theta \end{bmatrix} \begin{Bmatrix} \alpha_1 \Delta T \\ \alpha_2 \Delta T \\ 0 \end{Bmatrix}_k \qquad (9.26)$$

or,

$$\begin{Bmatrix} \alpha_1 \Delta T \\ \alpha_2 \Delta T \\ 0 \end{Bmatrix}_k = [T] \begin{Bmatrix} \alpha_x \Delta T \\ \alpha_y \Delta T \\ \dfrac{1}{2} \alpha_s \Delta T \end{Bmatrix}_k \qquad (9.27)$$

Similarly,

$$\begin{Bmatrix} \beta_1 \Delta C \\ \beta_2 \Delta C \\ 0 \end{Bmatrix}_k = [T] \begin{Bmatrix} \beta_x \Delta C \\ \beta_y \Delta C \\ \dfrac{1}{2} \beta_s \Delta C \end{Bmatrix}_k \qquad (9.28)$$

Equation (9.23) can be written as

$$\begin{Bmatrix} \varepsilon_1 \\ \varepsilon_2 \\ \dfrac{\varepsilon_6}{2} \end{Bmatrix}_k - \begin{Bmatrix} \alpha_1 \Delta T \\ \alpha_2 \Delta T \\ 0 \end{Bmatrix}_k - \begin{Bmatrix} \beta_1 \Delta C \\ \beta_2 \Delta C \\ 0 \end{Bmatrix}_k = \begin{bmatrix} S_{11} & S_{12} & 0 \\ S_{12} & S_{22} & 0 \\ 0 & 0 & \dfrac{S_{66}}{2} \end{bmatrix}_k \begin{Bmatrix} \sigma_1 \\ \sigma_2 \\ \sigma_6 \end{Bmatrix}_k \qquad (9.29)$$

Using (4.49), (4.60), (9.27) and (9.28), (9.29) becomes

$$[T] \begin{Bmatrix} \varepsilon_x \\ \varepsilon_y \\ \dfrac{\varepsilon_s}{2} \end{Bmatrix}_k - [T] \begin{Bmatrix} \alpha_x \Delta T \\ \alpha_y \Delta T \\ \dfrac{\alpha_s}{2} \Delta T \end{Bmatrix}_k - [T] \begin{Bmatrix} \beta_x \Delta C \\ \beta_y \Delta C \\ \dfrac{\beta_s}{2} \Delta C \end{Bmatrix}_k = \begin{bmatrix} S_{11} & S_{12} & 0 \\ S_{12} & S_{22} & 0 \\ 0 & 0 & S_{66}/2 \end{bmatrix}_k [T] \begin{Bmatrix} \sigma_x \\ \sigma_y \\ \sigma_s \end{Bmatrix}_k \qquad (9.30)$$

Premultiplying both sides by $[T]^{-1}$ and rearranging the terms

$$\begin{Bmatrix} \varepsilon_x - \alpha_x \Delta T - \beta_x \Delta C \\ \varepsilon_y - \alpha_y \Delta T - \beta_y \Delta C \\ \varepsilon_s - \alpha_s \Delta T - \beta_s \Delta C \end{Bmatrix}_k = \begin{bmatrix} S_{xx} & S_{xy} & S_{xs} \\ S_{yx} & S_{yy} & S_{ys} \\ S_{sx} & S_{sy} & S_{ss} \end{bmatrix}_k \begin{Bmatrix} \sigma_x \\ \sigma_y \\ \sigma_s \end{Bmatrix}_k \tag{9.31a}$$

or,

$$\begin{Bmatrix} \varepsilon_x \\ \varepsilon_y \\ \varepsilon_s \end{Bmatrix}_k = \begin{bmatrix} S_{xx} & S_{xy} & S_{xs} \\ S_{yx} & S_{yy} & S_{ys} \\ S_{sx} & S_{sy} & S_{ss} \end{bmatrix}_k \begin{Bmatrix} \sigma_x \\ \sigma_y \\ \sigma_s \end{Bmatrix}_k + \begin{Bmatrix} \alpha_x \Delta T + \beta_x \Delta C \\ \alpha_y \Delta T + \beta_y \Delta C \\ \alpha_s \Delta T + \beta_s \Delta C \end{Bmatrix}_k \tag{9.31b}$$

The terms on the left-hand side represent total strain. Thus the last matrix on the right-hand side indicates the hygrothermal strains.

The off-axis mechanical strains thus are

$$\begin{Bmatrix} \varepsilon_{xe} \\ \varepsilon_{ye} \\ \varepsilon_{se} \end{Bmatrix}_k = \begin{Bmatrix} \varepsilon_x - \alpha_x \Delta T - \beta_x \Delta C \\ \varepsilon_y - \alpha_y \Delta T - \beta_y \Delta C \\ \varepsilon_s - \alpha_s \Delta T - \beta_s \Delta C \end{Bmatrix}_k \tag{9.32}$$

In compact form, (9.31b) can be written as

$$\{\varepsilon\}_{x,y}^k = [S]\{\sigma\}_{x,y}^k + \{e\}_{x,y}^k \tag{9.33}$$

where $\{e\}_{x,y}^k$ represents the free hygrothermal strains of the kth layer of the lamina.

After adjustment of different terms, (9.31a) can be written as

$$\begin{Bmatrix} \sigma_x \\ \sigma_y \\ \sigma_s \end{Bmatrix}_k = \begin{bmatrix} Q_{xx} & Q_{xy} & Q_{xs} \\ Q_{yx} & Q_{yy} & Q_{ys} \\ Q_{sx} & Q_{sy} & Q_{ss} \end{bmatrix}_k \begin{Bmatrix} \varepsilon_x - e_x \\ \varepsilon_y - e_y \\ \varepsilon_s - e_s \end{Bmatrix}_k \tag{9.34}$$

Noting the relations of (5.4), (9.34) can be written as

$$\begin{Bmatrix} \sigma_x \\ \sigma_y \\ \sigma_s \end{Bmatrix}_k = \begin{bmatrix} Q_{xx} & Q_{xy} & Q_{xs} \\ Q_{yx} & Q_{yy} & Q_{ys} \\ Q_{sx} & Q_{sy} & Q_{ss} \end{bmatrix}_k \begin{Bmatrix} \varepsilon_x^0 + z\kappa_x - e_x \\ \varepsilon_y^0 + z\kappa_y - e_y \\ \varepsilon_s^0 + z\kappa_s - e_s \end{Bmatrix}_k \tag{9.35}$$

Equation (9.35) in compact form becomes

$$\{\sigma\}_{x,y}^k = [Q]_{x,y}^k \{\varepsilon^0\}_{x,y} + z[Q]_{x,y}^k \{\kappa\}_{x,y} - [Q]_{x,y}^k \{e\}_{x,y}^k \tag{9.36}$$

As the variation of the stress across the laminate is discontinuous, it is preferable to deal with stress-resultants rather than stresses. From (5.11), we get the inplane forces as

$$\{N\}_{x,y} = \sum_{k=1}^{n} \int_{z_{k-1}}^{z_k} \{\sigma\}_{x,y}^k \, dz \tag{9.37}$$

Combining (9.35) and (9.37), yields

$$\begin{Bmatrix} N_x \\ N_y \\ N_s \end{Bmatrix} = \sum_{k=1}^{n} \int_{z_{k-1}}^{z_k} \begin{bmatrix} Q_{xx} & Q_{xy} & Q_{xs} \\ Q_{yx} & Q_{yy} & Q_{ys} \\ Q_{sx} & Q_{sy} & Q_{ss} \end{bmatrix}_k \left(\begin{Bmatrix} \varepsilon_x^0 \\ \varepsilon_y^0 \\ \varepsilon_s^0 \end{Bmatrix} + z \begin{Bmatrix} \kappa_x \\ \kappa_y \\ \kappa_s \end{Bmatrix} - \begin{Bmatrix} e_x \\ e_y \\ e_s \end{Bmatrix}_k \right) dz \qquad (9.38)$$

Combining (5.21) and (9.38), yields

$$\begin{Bmatrix} N_x \\ N_y \\ N_s \end{Bmatrix} = \begin{bmatrix} A_{xx} & A_{xy} & A_{xs} \\ A_{yx} & A_{yy} & A_{ys} \\ A_{sx} & A_{sy} & A_{ss} \end{bmatrix} \begin{Bmatrix} \varepsilon_x^0 \\ \varepsilon_y^0 \\ \varepsilon_s^0 \end{Bmatrix} + \begin{bmatrix} B_{xx} & B_{xy} & B_{xs} \\ B_{yx} & B_{yy} & B_{ys} \\ B_{sx} & B_{sy} & B_{ss} \end{bmatrix} \begin{Bmatrix} \kappa_x \\ \kappa_y \\ \kappa_s \end{Bmatrix} - \begin{Bmatrix} N_x^{HT} \\ N_y^{HT} \\ N_s^{HT} \end{Bmatrix} \qquad (9.39)$$

In compact form, (9.39) can be written as

$$\{N\}_{x,y} = [A]\{\varepsilon^0\}_{x,y} + [B]\{\kappa\}_{x,y} - \{N^{HT}\}_{x,y} \qquad (9.40)$$

$\{N^{HT}\}_{x,y}$ are the hygrothermal force resultants and are given by

$$\begin{Bmatrix} N_x^{HT} \\ N_y^{HT} \\ N_s^{HT} \end{Bmatrix} = \sum_{k=1}^{n} \begin{bmatrix} Q_{xx} & Q_{xy} & Q_{xs} \\ Q_{yx} & Q_{yy} & Q_{ys} \\ Q_{sx} & Q_{sy} & Q_{ss} \end{bmatrix}_k \begin{Bmatrix} e_x \\ e_y \\ e_s \end{Bmatrix}_k d_k \qquad (9.41)$$

where d_k is the thickness of the kth lamina.

Following steps similar steps to that given above, moment stress-resultants are obtained as given below

$$\begin{Bmatrix} M_x \\ M_y \\ M_s \end{Bmatrix} = \begin{bmatrix} B_{xx} & B_{xy} & B_{xs} \\ B_{yx} & B_{yy} & B_{ys} \\ B_{sx} & B_{sy} & B_{ss} \end{bmatrix} \begin{Bmatrix} \varepsilon_x^0 \\ \varepsilon_y^0 \\ \varepsilon_s^0 \end{Bmatrix} + \begin{bmatrix} D_{xx} & D_{xy} & D_{xs} \\ D_{yx} & D_{yy} & D_{ys} \\ D_{sx} & D_{sy} & D_{ss} \end{bmatrix} \begin{Bmatrix} \kappa_x \\ \kappa_y \\ \kappa_s \end{Bmatrix} - \begin{Bmatrix} M_x^{HT} \\ M_y^{HT} \\ M_s^{HT} \end{Bmatrix} \qquad (9.42)$$

where

$$\{M^{HT}\}_{x,y} = \begin{Bmatrix} M_x^{HT} \\ M_y^{HT} \\ M_s^{HT} \end{Bmatrix} = \sum_{k=1}^{n} \begin{bmatrix} Q_{xx} & Q_{xy} & Q_{xs} \\ Q_{yx} & Q_{yy} & Q_{ys} \\ Q_{sx} & Q_{sy} & Q_{ss} \end{bmatrix}_k \begin{Bmatrix} e_x \\ e_y \\ e_s \end{Bmatrix}_k d_k \cdot \bar{z}_k \qquad (9.43)$$

and $\bar{z}_k = \dfrac{z_k + z_{k-1}}{2}$ = z-coordinate of the mid-plane of the kth lamina

Equations (9.39) and (9.42) can be rewritten as follows

$$\begin{Bmatrix} \bar{N}_x \\ \bar{N}_y \\ \bar{N}_s \end{Bmatrix} = \begin{Bmatrix} N_x \\ N_y \\ N_s \end{Bmatrix} + \begin{Bmatrix} N_x^{HT} \\ N_y^{HT} \\ N_s^{HT} \end{Bmatrix} = \begin{bmatrix} A_{xx} & A_{xy} & A_{xs} \\ A_{yx} & A_{yy} & A_{ys} \\ A_{sx} & A_{sy} & A_{ss} \end{bmatrix} \begin{Bmatrix} \varepsilon_x^0 \\ \varepsilon_y^0 \\ \varepsilon_s^0 \end{Bmatrix}$$
$$+ \begin{bmatrix} B_{xx} & B_{xy} & B_{xs} \\ B_{yx} & B_{yy} & B_{ys} \\ B_{sx} & B_{sy} & B_{ss} \end{bmatrix} \begin{Bmatrix} \kappa_x \\ \kappa_y \\ \kappa_s \end{Bmatrix} \qquad (9.44)$$

and

$$\begin{Bmatrix} \overline{M}_x \\ \overline{M}_y \\ \overline{M}_s \end{Bmatrix} = \begin{Bmatrix} M_x \\ M_y \\ M_s \end{Bmatrix} + \begin{Bmatrix} M_x^{HT} \\ M_y^{HT} \\ M_s^{HT} \end{Bmatrix} = \begin{bmatrix} B_{xx} & B_{xy} & B_{xs} \\ B_{yx} & B_{yy} & B_{ys} \\ B_{sx} & B_{sy} & B_{ss} \end{bmatrix} \begin{Bmatrix} \varepsilon_x^0 \\ \varepsilon_y^0 \\ \varepsilon_s^0 \end{Bmatrix}$$
$$+ \begin{bmatrix} D_{xx} & D_{xy} & D_{xs} \\ D_{yx} & D_{yy} & D_{ys} \\ D_{sx} & D_{sy} & D_{ss} \end{bmatrix} \begin{Bmatrix} \kappa_x \\ \kappa_y \\ \kappa_s \end{Bmatrix} \qquad (9.45)$$

In compact form, (9.44) and (9.45) can be written as

$$\{\overline{N}\}_{x,y} = \{N\}_{x,y} + \{N^{HT}\}_{x,y} = [A]_{x,y}\{\varepsilon^0\}_{x,y} + [B]_{x,y}\{\kappa\}_{x,y} \qquad (9.46)$$

and

$$\{\overline{M}\}_{x,y} = \{M\}_{x,y} + \{M^{HT}\}_{x,y} = [B]_{x,y}\{\varepsilon^0\}_{x,y} + [D]_{x,y}\{\kappa\}_{x,y} \qquad (9.47)$$

where

$\{\overline{N}\}_{x,y}$ is the sum of the forces due to mechanical load and hygrothermal forces in the laminate and referred to as the total force and

$\{\overline{M}\}_{x,y}$ is the sum of the moments due to mechanical loading and that due to hygrothermal loading.

Equations (9.46) and (9.47) can be combined into a single equation as

$$\begin{Bmatrix} \{\overline{N}\} \\ \{\overline{M}\} \end{Bmatrix} = \begin{bmatrix} [A] & [B] \\ [B] & [D] \end{bmatrix} \begin{Bmatrix} \{\varepsilon^0\} \\ \{\kappa\} \end{Bmatrix} \qquad (9.48)$$

Note the similarity of (9.48) and (5.21).

Equation (9.48) can be inverted to result in the following equation

$$\begin{Bmatrix} \varepsilon_x^0 \\ \varepsilon_y^0 \\ \varepsilon_s^0 \\ \kappa_x \\ \kappa_y \\ \kappa_s \end{Bmatrix} = \begin{bmatrix} a_{xx} & a_{xy} & a_{xs} & b_{xx} & b_{xy} & b_{xs} \\ a_{yx} & a_{yy} & a_{ys} & b_{yx} & b_{yy} & b_{ys} \\ a_{sx} & a_{sy} & a_{ss} & b_{sx} & b_{sy} & b_{ss} \\ b_{xx} & b_{xy} & b_{xs} & d_{xx} & d_{xy} & d_{xs} \\ b_{yx} & b_{yy} & b_{ys} & d_{yx} & d_{yy} & d_{ys} \\ b_{sx} & b_{sy} & b_{ss} & d_{sx} & d_{sy} & d_{ss} \end{bmatrix} \begin{Bmatrix} \overline{N}_x \\ \overline{N}_y \\ \overline{N}_s \\ \overline{M}_x \\ \overline{M}_y \\ \overline{M}_s \end{Bmatrix} \qquad (9.49)$$

or, in the compact form

$$\begin{Bmatrix} \{\varepsilon^0\} \\ \{\kappa\} \end{Bmatrix} = \begin{bmatrix} [a] & [b] \\ [b] & [d] \end{bmatrix} \begin{Bmatrix} \{\overline{N}\} \\ \{\overline{M}\} \end{Bmatrix} \qquad (9.50)$$

If there is no mechanical loading, $\{\overline{N}\} = \{N^{HT}\}$ and $\{\overline{M}\} = \{M^{HT}\}$ as $\{N\} = \{M\} = \{0\}$ and with this modification, (9.50) can be applied for hygrothermal loading.

9.8 CTE AND CME OF LAMINATES

When a laminate is subjected to pure hygrothermal loading, the strains at the reference plane of the laminate is expressed in terms of the coefficient of thermal expansion and the coefficient of moisture expansion as

$$\begin{Bmatrix} \varepsilon_x^0 \\ \varepsilon_y^0 \\ \varepsilon_s^0 \end{Bmatrix} = \begin{Bmatrix} \overline{\alpha}_x \\ \overline{\alpha}_y \\ \overline{\alpha}_s \end{Bmatrix} \Delta T + \begin{Bmatrix} \overline{\beta}_x \\ \overline{\beta}_y \\ \overline{\beta}_s \end{Bmatrix} \Delta C \tag{9.51}$$

or, in the compact form

$$\{\varepsilon^0\}_{x,y} = \{\overline{\alpha}\}_{x,y} \Delta T + \{\overline{\beta}\}_{x,y} \Delta C \tag{9.52}$$

where

$\{\overline{\alpha}\}_{x,y}$ = coefficients of thermal expansion of the laminate

$\{\overline{\beta}\}_{x,y}$ = coefficients of moisture expansion of the laminate

These coefficients can be determined by comparing (9.49) with (9.51) by considering $\{N\}_{x,y} = \{M\}_{x,y} = \{0\}$ in the latter.

The coefficients of thermal expansion can be obtained by setting $\Delta T = 1$ and $\Delta C = 0$ in the above relations. Thus,

$$\begin{Bmatrix} \overline{\alpha}_x \\ \overline{\alpha}_y \\ \overline{\alpha}_s \end{Bmatrix} = \begin{bmatrix} a_{xx} & a_{xy} & a_{xs} \\ a_{yx} & a_{yy} & a_{ys} \\ a_{sx} & a_{sy} & a_{ss} \end{bmatrix} \begin{Bmatrix} N_x^T \\ N_y^T \\ N_s^T \end{Bmatrix} + \begin{bmatrix} b_{xx} & b_{xy} & b_{xs} \\ b_{ys} & b_{yy} & b_{ys} \\ b_{sx} & b_{sy} & b_{ss} \end{bmatrix} \begin{Bmatrix} M_x^T \\ M_y^T \\ M_s^T \end{Bmatrix} \tag{9.52a}$$

where $\{N^T\}_{xy}$ and $\{M^T\}_{xy}$ are the resultant thermal forces and moments as defined by (9.41) and (9.43) when $\Delta C = 0$ and $\Delta T = 1$.

The coefficients of moisture expansion can similarly be obtained by putting $\Delta T = 0$ and $\Delta C = 1$. Then

$$\begin{Bmatrix} \overline{\beta}_x \\ \overline{\beta}_y \\ \overline{\beta}_s \end{Bmatrix} = \begin{bmatrix} a_{xx} & a_{xy} & a_{xs} \\ a_{yx} & a_{yy} & a_{ys} \\ a_{sx} & a_{sy} & a_{ss} \end{bmatrix} \begin{Bmatrix} N_x^H \\ N_y^H \\ N_s^H \end{Bmatrix} + \begin{bmatrix} b_{xx} & b_{xy} & b_{xs} \\ b_{yx} & b_{yy} & b_{ys} \\ b_{sx} & b_{sy} & b_{ss} \end{bmatrix} \begin{Bmatrix} M_x^H \\ M_y^H \\ M_s^H \end{Bmatrix} \tag{9.53}$$

where $\{N^H\}_{x,y}$ and $\{M^H\}_{x,y}$ are the resultant hygroscopic forces and moments as defined by (9.41) and (9.43) when $\Delta T = 0$.

9.9 DETERMINATION OF STRESSES IN A LAMINATE DUE TO HYGROTHERMO-ELASTIC FORCES

Equation (9.50) relates the hygrothermo-mechanical loading with strains and curvatures at the reference plane.

Total strains at the centre of the kth lamina situated at a distance $\overline{z}_k$ from the reference plane are

$$\begin{Bmatrix} \varepsilon_x \\ \varepsilon_y \\ \varepsilon_s \end{Bmatrix}_k = \begin{Bmatrix} \varepsilon_x^0 \\ \varepsilon_y^0 \\ \varepsilon_s^0 \end{Bmatrix} + \bar{z}_k \begin{Bmatrix} \kappa_x \\ \kappa_y \\ \kappa_s \end{Bmatrix} \tag{9.54}$$

Based on (9.32), mechanical strains at the kth layer are

$$\begin{Bmatrix} \varepsilon_{xe} \\ \varepsilon_{ye} \\ \varepsilon_{se} \end{Bmatrix}_k = \begin{Bmatrix} \varepsilon_x \\ \varepsilon_y \\ \varepsilon_s \end{Bmatrix}_k - \begin{Bmatrix} e_x \\ e_y \\ e_s \end{Bmatrix}_k \tag{9.55}$$

Stresses in the kth layer referred to the loading axis are given by

$$\begin{Bmatrix} \sigma_{xe} \\ \sigma_{ye} \\ \sigma_{se} \end{Bmatrix}_k = \begin{bmatrix} Q_{xx} & Q_{xy} & Q_{xs} \\ Q_{yx} & Q_{yy} & Q_{ys} \\ Q_{sx} & Q_{sy} & Q_{ss} \end{bmatrix}_k \begin{Bmatrix} \varepsilon_{xe} \\ \varepsilon_{ye} \\ \varepsilon_{se} \end{Bmatrix}_k \tag{9.56}$$

These stresses can be transferred to the principal material axis by using the transformation

$$\begin{Bmatrix} \sigma_{1e} \\ \sigma_{2e} \\ \sigma_{6e} \end{Bmatrix}_k = \begin{bmatrix} m^2 & n^2 & 2mn \\ n^2 & m^2 & -2mn \\ -mn & mn & m^2 - n^2 \end{bmatrix} \begin{Bmatrix} \sigma_{xe} \\ \sigma_{ye} \\ \sigma_{se} \end{Bmatrix}_k \tag{9.57}$$

where $m = \cos\theta$ and $n = \sin\theta$ as used.

In compact form, (9.57) can be written as

$$\{\sigma_e\}_{1,2}^k = [T]\{\sigma_e\}_{x,y}^k \tag{9.58}$$

The stresses in the principal material axis direction can be found out from another approach. First, the hygrothermal strains in the principal material axis direction are found out as follows

$$\begin{Bmatrix} \varepsilon_{1e} \\ \varepsilon_{2e} \\ \frac{1}{2}\varepsilon_{6e} \end{Bmatrix}_k = \begin{Bmatrix} \varepsilon_1 - e_1 \\ \varepsilon_2 - e_2 \\ \dfrac{\varepsilon_6}{2} \end{Bmatrix}_k = [T] \begin{Bmatrix} \varepsilon_{xe} \\ \varepsilon_{ye} \\ \frac{1}{2}\varepsilon_{se} \end{Bmatrix}_k \tag{9.59}$$

and

$$\begin{Bmatrix} \sigma_{1e} \\ \sigma_{2e} \\ \sigma_{6e} \end{Bmatrix}_k = \begin{bmatrix} Q_{11} & Q_{12} & 0 \\ Q_{21} & Q_{22} & 0 \\ 0 & 0 & Q_{66} \end{bmatrix}_k \begin{Bmatrix} \varepsilon_{1e} \\ \varepsilon_{2e} \\ \varepsilon_{6e} \end{Bmatrix}_k \tag{9.60}$$

Equations (9.58) and (9.60) give pure hygrothermal stresses in each layer.

9.10 RESIDUAL STRESSES

Residual stresses are introduced due to mismatch of thermal properties of the constituents of the composite. They are introduced in the laminates at the stage of fabrication. Due to the thermal anisotropy of layers, lamination residual stresses may result. Residual stresses are dependent on geometric and material properties such as ply orientation, fibre–volume ratio, stacking sequence and other parameters such as curing processes, other material and processing variables.

For carrying out the elastic analysis of residual stresses, the procedure is as follows

ΔT is the difference between ambient and stress-free temperature. Free thermal strains in each layer are obtained from (9.12). Thermal forces and moments are calculated using (9.41) and (9.43). The hygrothermal stress-strain relations of (9.49) or (9.50) are written in submatrix equations.

$$\begin{Bmatrix} \varepsilon_x^0 \\ \varepsilon_y^0 \\ \varepsilon_s^0 \end{Bmatrix} = \begin{bmatrix} a_{xx} & a_{xy} & a_{xs} \\ a_{yx} & a_{yy} & a_{ys} \\ a_{sx} & a_{sy} & a_{ss} \end{bmatrix} \begin{Bmatrix} N_x^T \\ N_y^T \\ N_s^T \end{Bmatrix} + \begin{bmatrix} b_{xx} & b_{xy} & b_{xs} \\ b_{yx} & b_{yy} & b_{ys} \\ b_{sx} & b_{sy} & b_{ss} \end{bmatrix} \begin{Bmatrix} M_x^T \\ M_y^T \\ M_s^T \end{Bmatrix} \quad (9.61)$$

and

$$\begin{Bmatrix} \kappa_x \\ \kappa_y \\ \kappa_s \end{Bmatrix} = \begin{bmatrix} b_{xx} & b_{xy} & b_{xs} \\ b_{yx} & b_{yy} & b_{ys} \\ b_{sx} & b_{sy} & b_{ss} \end{bmatrix} \begin{Bmatrix} N_x^T \\ N_y^T \\ N_s^T \end{Bmatrix} + \begin{bmatrix} d_{xx} & d_{xy} & d_{xs} \\ d_{yx} & d_{yy} & d_{xs} \\ d_{sx} & d_{sy} & d_{ss} \end{bmatrix} \begin{Bmatrix} M_x^T \\ M_y^T \\ M_s^T \end{Bmatrix} \quad (9.62)$$

$\{\varepsilon^0\}_{x,y}$ and $\{\kappa\}_{x,y}$ are the inplane strains and curvatures of the reference plane respectively. Stresses are given in Section 9.9.

9.11 WARPAGE

Warpage is the out-of-plane deformation of the laminate. When the laminate is asymmetric, this type of out-of-plane deformation may occur due to hygrothermal loading (Fig. 9.7). Theoretical calculations of warpage are possible using classical lamination theory [9.7, 9.8].

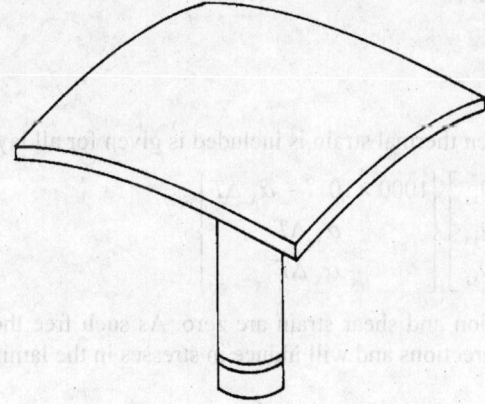

Fig. 9.7 Warpage in a laminate

The curvatures of a laminate are given by

$$\kappa_x = -\frac{\partial^2 w}{\partial x^2}$$

$$\kappa_y = -\frac{\partial^2 w}{\partial y^2} \quad (9.63)$$

$$\kappa_s = -2\frac{\partial^2 w}{\partial x\, \partial y}$$

Integration of curvature terms yield the out-of-plane deflection w as

$$w = \frac{1}{2}\left(\kappa_x \cdot x^2 + \kappa_y \cdot y^2 + \kappa_s \cdot xy\right) + \text{(rigid body motion)} \quad (9.64)$$

Warpage is indicated by the quadratic part of this expression.

Equation (9.62) relates the curvatures to forces and moments. The equation of warpage is obtained by combining (9.62) and (9.64).

9.12 EXAMPLES

Example 9.2

A cross-ply laminate $[0|90]_s$ made from high strength carbon/epoxy unidirectional plies has strains and curvatures at mid-plane as follows

$$\varepsilon_x^0 = 1000 \times 10^{-6},\ \varepsilon_y^0 = 0,\ \varepsilon_s^0 = 0,\ \kappa_x = \kappa_y = \kappa_s = 0$$

Determine the stresses in each layer.

The temperature is changed by $-100°C$ at every point of the laminate.

$$\alpha_1 = -0.9 \times 10^{-6}/°C \quad \text{and} \quad \alpha_2 = 27 \times 10^{-6}/°C$$

$$\begin{aligned}
\varepsilon_x &= \varepsilon_x^0 + z\kappa_x^0 = 1000 \times 10^{-6} \\
\varepsilon_y &= \varepsilon_y^0 + z\kappa_y^0 = 0 \\
\varepsilon_s &= \varepsilon_s^0 + z\kappa_s^0 = 0
\end{aligned} \quad (a)$$

The strain–stress relation when thermal strain is included is given for all layers

$$\begin{Bmatrix} \sigma_x \\ \sigma_y \\ \sigma_s \end{Bmatrix} = \begin{bmatrix} Q_{xx} & Q_{xy} & Q_{xs} \\ Q_{yx} & Q_{yy} & Q_{ys} \\ Q_{sx} & Q_{sy} & Q_{ss} \end{bmatrix} \begin{Bmatrix} 1000 \times 10^{-6} - \alpha_x \Delta T \\ -\alpha_y \Delta T \\ -\alpha_s \Delta T \end{Bmatrix} \quad (b)$$

Inplane strain in the y-direction and shear strain are zero. As such free thermal strains will result in mechanical strains in these directions and will induce in stresses in the laminate.

For 0° layer

$$\begin{aligned}
\alpha_x &= \alpha_1 = -0.9 \times 10^{-6}/°C \\
\alpha_y &= \alpha_2 = 27 \times 10^{-6}/°C \\
\alpha_s &= \alpha_6 = 0
\end{aligned} \quad (c)$$

For 90° layer

$$\alpha_x = 27 \times 10^{-6}/°C$$

$$\alpha_y = -0.9 \times 10^{-6}/°C$$

$$\alpha_s = \alpha_6 = 0 \tag{d}$$

From Example 5A.3, we get
for 0° layer

$$\begin{Bmatrix} \sigma_x \\ \sigma_y \\ \sigma_s \end{Bmatrix} = \begin{bmatrix} 140.9 & 3.0 & 0 \\ 3.0 & 10.1 & 0 \\ 0 & 0 & 5 \end{bmatrix} \begin{Bmatrix} 910 \times 10^{-6} \\ 2700 \times 10^{-6} \\ 0 \end{Bmatrix} = \begin{Bmatrix} 136.31 \\ 30.00 \\ 0 \end{Bmatrix} \text{MPa} \tag{e}$$

and for 90° layer

$$\begin{Bmatrix} \sigma_x \\ \sigma_y \\ \sigma_s \end{Bmatrix} = \begin{bmatrix} 10.1 & 3.0 & 0 \\ 3.0 & 140.9 & 0 \\ 0 & 0 & 5 \end{bmatrix} \begin{Bmatrix} 3700 \times 10^{-6} \\ -90 \times 10^{-6} \\ 0 \end{Bmatrix} = \begin{Bmatrix} 37.10 \\ -1.58 \\ 0 \end{Bmatrix} \text{MPa} \tag{f}$$

There are contributions in two parts for each of the stresses. The first part is the product of the reduced stiffness and the total strain. The second part is the product of the reduced stiffness and the free thermal strains. Therefore, for 0° layer, the stresses can be written in two parts as

$$\sigma_x = 140.9 - 4.58 \text{ MPa}$$
$$\sigma_y = 3.00 - 0.027 \text{ MPa} \tag{g}$$
$$\sigma_s = 0$$

Similarly, stresses for 90° layer can be written as

$$\sigma_x = 10.1 + 27.00 \text{ MPa}$$
$$\sigma_y = 3.00 - 4.581 \text{ MPa} \tag{h}$$
$$\sigma_s = 0$$

Example 9.3

Determine α_1 and α_2 of a unidirectional ply having the following properties.

$E_{1f} = 70$ GPa, $E_m = 4$ GPa, $\quad v_f = 0.2$, $v_m = 0.4$

$\alpha_f = 4.5 \times 10^{-6}$ /°C, $\quad \alpha_m = 90 \times 10^{-6}$ /°C, $V_f = 0.6$

From (9.1), we get

$$\alpha_1 = \frac{70 \times 4.5 \times 10^{-6} \times 0.6 + 4 \times 90 \times 10^{-6} \times (1 - 0.6)}{70 \times 0.6 + 4 \times 0.4}$$

$$= 7.638 \times 10^{-6} /°\text{C}$$

From (9.2), we get

$$\alpha_2 = 4.5 \times 10^{-6} \times 0.6 \times (1 + 0.2) + 90 \times 10^{-6} \times 0.4 (1 + 0.4) - v_{12} \times 7.638 \times 10^{-6}$$

Now, $v_{12} = v_f V_f + v_m V_m = 0.2 \times 0.6 + 0.4 \times 0.4 = 0.28$

Therefore,
$$\alpha_2 = 3.24 \times 10^{-6} + 50.4 \times 10^{-6} - 0.28 \times 7.638 \times 10^{-6} = 51.50 \times 10^{-6} /°C$$

Example 9.4

CTEs of a unidirectional lamina of $V_f = 0.6$ are $\alpha_1 = -1.0 \times 10^{-6} /°C$ and $\alpha_2 = 30 \times 10^{-6} /°C$. Determine the CTEs of the fibre α_{1f} and α_{2f} based on the following properties.

$E_{1f} = 240$ GPa, $E_m = 4$ GPa, $v_f = 0.2$, $v_m = 0.35$ and $\alpha_m = 40 \times 10^{-6} /°C$

From (9.1), we can write

$$-1 \times 10^{-6} = \frac{240 \, \alpha_{1f} \times 0.6 + 4 \times 40 \times 10^{-6} \times 0.4}{240 \times 0.6 + 4 \times 0.6}$$

or, $\alpha_{1f} = -1.45 \times 10^{-6} /°C$

Similarly, from (9.2), we can find out α_{2f}.

$$v_{12} = v_f V_f + v_m V_m = 0.2 \times 0.6 + 0.35 \times 0.4 = 0.26$$

Again,

$$30 \times 10^{-6} = \alpha_{2f} \times 0.6 \times (1+0.2) + 40 \times 10^{-6} \times 0.4 \times (1+0.35) - 0.26(-1.45 \times 10^{-6}) \quad \text{or,}$$

$\alpha_{2f} = 11.31 \times 10^{-6} /°C$

Example 9.5

Determine the thermal forces N_x^T, N_y^T and N_s^T for a $[\pm 45]_s$ laminate whose properties are given below.

$E_1 = 142$ GPa, $E_2 = 10.3$ GPa, $E_6 = 7.2$ GPa, $v_{12} = 0.27$, $v_{21} = 0.02$,

$\alpha_1 = -0.9 \times 10^{-6} /°C$, $\alpha_2 = 27 \times 10^{-6} /°C$.

The temperature difference to be considered is $\Delta T = 56°C$. The result may be indicated in terms of lamina thickness 'd' for each lamina. Also find e_x, e_y and e_s for each lamina.
For a lamina

$$Q_{11} = \frac{E_1}{1 - v_{12} v_{21}} = \frac{142}{1 - 0.27 \times 0.02} = 142.77$$

$$Q_{12} = v_{21} Q_{11} = 0.02 \times 142.77 = 2.86$$

$$Q_{22} = \frac{E_2}{1 - v_{12} v_{21}} = \frac{10.3}{1 - 0.27 \times 0.02} = 10.36$$

$$Q_{66} = E_6 = 7.2$$

For + 45° lamina

$$Q_{xx} = \left(\frac{1}{\sqrt{2}}\right)^4 \times 142.77 + \left(\frac{1}{\sqrt{2}}\right)^4 \times 10.36 + 2\left(\frac{1}{\sqrt{2}}\right)^4 \times 2.56 + 4\left(\frac{1}{\sqrt{2}}\right)^4 \times 7.2$$
$$= 46.91$$

$Q_{yy} = 46.91$

$$Q_{xy} = \left(\frac{1}{\sqrt{2}}\right)^4 \times 142.77 + \left(\frac{1}{\sqrt{2}}\right)^4 \times 10.36 + 2\left(\frac{1}{\sqrt{2}}\right)^4 2.86 - 4\left(\frac{1}{\sqrt{2}}\right)^4 7.2$$
$$= 32.51$$

$$Q_{xs} = \left(\frac{1}{\sqrt{2}}\right)^3 \left(\frac{1}{\sqrt{2}}\right) \times 142 - 77 - \left(\frac{1}{\sqrt{2}}\right)\left(\frac{1}{\sqrt{2}}\right)^3 \times 10.36 + \left(\frac{1}{4} - \frac{1}{4}\right) 2.86$$
$$+ 2\left(\frac{1}{4} - \frac{1}{4}\right) \times 7.2 = 33.10$$

$Q_{ys} = 33.10$

$$Q_{ss} = \frac{1}{4} \times 142.77 + \frac{1}{4} \times 10.36 - 2 \times \left(\frac{1}{4}\right) \times 2.86 + 0$$
$$= 38.28$$

For − 45° lamina, $m = 0.707, n = -0.707$

$Q_{xx} = Q_{yy} = 46.91$

$Q_{xy} = 32.51$

$Q_{xs} = (0.707)^3 (-0.707) \times 142.77 - (0.707)(-0.707)^3 \times 10.36 + 0 + 0$
$\quad = -33.10$

$Q_{ys} = -33.10$

$Q_{ss} = 38.28$

For each lamina, free expansion in the principal material direction is

$e_1 = \alpha_1 \Delta T = -0.9 \times 10^{-6} \times 56 = -50.4 \times 10^{-6}$

$e_2 = \alpha_2 \Delta T = 27 \times 10^{-6} \times 56 = 1512 \times 10^{-6}$

For + 45° lamina, $m = \cos 45° = n = \sin 45° = 0.707$

$e_x = e_1 m^2 + e_2 n^2 = -50.4 \times 10^{-6} \times \frac{1}{2} + 1512 \times 10^{-6} \times \frac{1}{2} = 730.8 \times 10^{-6}$

$e_y = e_1 n^2 + e_2 m^2 = 730.8 \times 10^{-6}$

$e_s = 2(e_1 - e_2) mn$

$\quad = 2(-50.4 - 1512) 10^{-6} \times \frac{1}{2} = -1562.4 \times 10^{-6}$

For $-45°$ lamina, $m = \cos(-45) = 0.707$, $n = \sin(-45) = -0.707$

$$e_x = e_y = 730.8 \times 10^{-6}, \quad e_s = 1562.4 \times 10^{-6}$$

Therefore,

$$\begin{Bmatrix} N_x^T \\ N_y^T \\ N_s^T \end{Bmatrix} = \sum_{k=1}^{4} \begin{bmatrix} Q_{xx} & Q_{xy} & Q_{xs} \\ Q_{yx} & Q_{yy} & Q_{ys} \\ Q_{sx} & Q_{sy} & Q_{ss} \end{bmatrix}_k \begin{Bmatrix} e_x \\ e_y \\ e_s \end{Bmatrix}_k d_k$$

$$= 2 \begin{bmatrix} 46.91 & 32.51 & 33.10 \\ 32.51 & 46.91 & 33.10 \\ 33.10 & 33.10 & 38.28 \end{bmatrix} \begin{Bmatrix} 730.8 \\ 730.8 \\ -1562.4 \end{Bmatrix} 10^{-6} d$$

$$+ 2 \begin{bmatrix} 46.91 & 32.51 & -33.10 \\ 32.51 & 46.91 & -33.10 \\ -33.10 & -33.10 & 38.28 \end{bmatrix} \begin{Bmatrix} 730.8 \\ 730.8 \\ 1562.4 \end{Bmatrix} 10^{-6} d = \begin{Bmatrix} 0.232 \\ 0.232 \\ 0 \end{Bmatrix} d \text{ GPa}$$

Example 9.6

Show that the coefficient of thermal expansion $\bar{\alpha}_x$ of a $[0/90]_s$ laminate is equal to $[E_2(1+\nu_{12})\alpha_2]/E_1$, i.e., it is equal to that of a $[\pm 45]_s$ laminate.

Assume $E_1 \gg E_2$, $\alpha_1 = 0$, $\alpha_2 = \alpha$.

For this material,

$$Q_{11} \cong E_1, \quad Q_{22} \cong E_2, \quad Q_{12} = \nu_{12} E_2 = \nu_{21} E_1, \quad Q_{66} = E_6, \quad \alpha_1 = 0, \quad \alpha_2 = \alpha$$

The transformed coefficients of thermal expansion for the $0°$ layer are

$$\begin{Bmatrix} \alpha_x \\ \alpha_y \\ \frac{1}{2}\alpha_s \end{Bmatrix}^{0°} = \begin{bmatrix} m^2 & n^2 & -2mn \\ n^2 & m^2 & 2mn \\ -2mn & 2mn & (m^2 - n^2) \end{bmatrix} \begin{Bmatrix} \alpha_1 \\ \alpha_2 \\ \frac{\alpha_6}{2} \end{Bmatrix}$$

$$= \begin{bmatrix} 1 & 0 & 0 \\ 0 & 1 & 0 \\ 0 & 0 & 1 \end{bmatrix} \begin{Bmatrix} 0 \\ \alpha \\ 0 \end{Bmatrix}$$

$$(\alpha_x)^0 = 0, \quad (\alpha_y)^0 = \alpha, \quad (\alpha_s)^0 = 0$$

Similarly,

$$(\alpha_x)^{90} = \alpha, \quad (\alpha_y)^{90} = 0, \quad (\alpha_s)^{90} = 0$$

For $0°$ laminate,

$$(Q_{xx})^0 = Q_{11}, \quad (Q_{yy})^0 = Q_{22}, \quad (Q_s)^0 = Q_{12}, \quad (Q_{xs})^0 = (Q_{ys})^0 = 0$$

$(Q_{ss})^0 = Q_{66}$

For 90° lamina

$(Q_{xx})^{90} = Q_{22}, (Q_{yy})^{90} = Q_{11}, (Q_s)^{90} = Q_{12}, (Q_{xs})^{90} = (Q_{ys})^{90} = 0$

$(Q_{ss})^{90} = Q_{66}$

From the definition of thermal forces, we have

$$\begin{Bmatrix} N_x^T \\ N_y^T \\ N_s^T \end{Bmatrix} = \begin{bmatrix} Q_{xx} & Q_{xy} & Q_{xs} \\ Q_{yx} & Q_{yy} & Q_{ys} \\ Q_{sx} & Q_{sy} & Q_{ss} \end{bmatrix}_0 \begin{Bmatrix} \alpha_x \\ \alpha_y \\ \alpha_s \end{Bmatrix}_0 2\Delta T \cdot d$$

$$+ \begin{bmatrix} Q_{xx} & Q_{xy} & Q_{xs} \\ Q_{yx} & Q_{yy} & Q_{ys} \\ Q_{sx} & Q_{sy} & Q_{ss} \end{bmatrix}_{90} \begin{Bmatrix} \alpha_x \\ \alpha_y \\ \alpha_s \end{Bmatrix}_{90} 2\ \Delta T.d$$

where d is the thickness of the lamina

$$\begin{Bmatrix} N_x^T \\ N_y^T \\ N_s^T \end{Bmatrix} = \begin{bmatrix} Q_{11} & Q_{12} & 0 \\ Q_{12} & Q_{22} & 0 \\ 0 & 0 & Q_{66} \end{bmatrix} \begin{Bmatrix} 0 \\ \alpha \\ 0 \end{Bmatrix} 2d \cdot \Delta T + \begin{bmatrix} Q_{22} & Q_{12} & 0 \\ Q_{12} & Q_{11} & 0 \\ 0 & 0 & Q_{66} \end{bmatrix} \begin{Bmatrix} \alpha \\ 0 \\ 0 \end{Bmatrix} 2d \cdot \Delta T$$

$N_x^T = (Q_{12}\alpha + Q_{22}\alpha) 2d \cdot \Delta T$

$N_y^T = (Q_{12}\alpha + Q_{22}\alpha) 2d \cdot \Delta T$

$N_s^T = 0$

The laminate is symmetric, so $[B] = [0]$

We know that

$$\begin{Bmatrix} N_x^T \\ N_y^T \\ N_s^T \end{Bmatrix} = \begin{bmatrix} A_{xx} & A_{xy} & 0 \\ A_{yx} & A_{yy} & 0 \\ 0 & 0 & A_{ss} \end{bmatrix} \begin{Bmatrix} \bar{\alpha}_x \\ \bar{\alpha}_y \\ \bar{\alpha}_s \end{Bmatrix}$$

Note that $\Delta T = 1$

As $N_x^T = N_y^T$, $\bar{\alpha}_x = \bar{\alpha}_y$

Therefore,

$\bar{\alpha}_x = \bar{\alpha}_y = \dfrac{N_x^T}{A_{xx} + A_{xy}}$

$\bar{\alpha}_s = 0$

Noting that,
$$A_{xx} + A_{xy} = (Q_{11} + Q_{12})2d$$
$$= (E_1 + \nu_{12} E_2)2d$$
$$= 2E_1 d \quad \text{as } E_1 \gg E_2$$

Therefore,
$$\bar{\alpha}_x = \bar{\alpha}_y = \alpha \frac{(Q_{12} + Q_{22})2d}{2E_1 d}$$
$$= \frac{(1 + \nu_{12})E_2}{E_1}\alpha$$

The problem can be repeated for $[\pm 45]_s$ laminate.

Example 9.7

Determine the coefficient of thermal expansion $\bar{\alpha}_x$ for a $[\pm 30]_s$ laminate with the following properties.

$E_1 = 142$ GPa, $E_2 = 10.3$ GPa, $E_6 = 7.2$ GPa, $\nu_{12} = 0.27$

$\nu_{21} = 0.02$, $\alpha_1 = -0.9 \times 10^{-6}$ /°C, $\alpha_2 = 27 \times 10^{-6}$ /°C

Let the thickness of the laminate be h.

For each lamina, the free expansion in the principal material axis direction for 30° lamina is

$$e_1 = \alpha_1 \cdot \Delta T = -0.9 \times 10^{-6} \Delta T$$
$$e_2 = \alpha_2 \cdot \Delta T = 27 \times 10^{-6} \cdot \Delta T$$

For + 30° lamina

$$m = \cos 30° = 0.866, \quad n = \sin 30° = 0.5$$

$$e_x = m^2 e_1 + n^2 e_2$$
$$= (0.866)^2 \times (-0.9) \times 10^{-6} \Delta T + (0.5)^2 \times 27 \times 10^{-6} \times \Delta T$$
$$= 6.08 \times 10^{-6} \Delta T$$

$$e_y = n^2 e_1 + m^2 e_2$$
$$= (0.5)^2 \times (-0.9) \times 10^{-6} \times \Delta T + (0.866)^2 \times 27 \times 10^{-6} \times \Delta T$$
$$= 20.03 \times 10^{-6} \Delta T$$

$$e_s = 2(e_1 - e_2) mn$$
$$= 2(-0.9 \times 10^{-6} - 27 \times 10^{-6}) \Delta T (0.9)(0.866)$$
$$= -24.16 \times 10^{-6} \Delta T$$

For $-30°$ lamina,

$$e_x = 6.08 \times 10^{-6} \times \Delta T, \quad e_y = 20.03 \times 10^{-6} \times \Delta T, \quad e_s = 24.16 \times 10^{-6} \times \Delta T$$

For any lamina,

$$Q_{11} = \frac{E_1}{1 - v_{12} v_{21}} = \frac{142}{1 - 0.27 \times 0.02} = 142.77$$

$$Q_{22} = \frac{E_2}{1 - v_{12} v_{21}} = \frac{10.3}{1 - 0.27 \times 0.02} = 10.36$$

$$Q_{12} = v_{21} \cdot Q_{11} = 0.02 \times 142.77 = 2.86$$

$$Q_{66} = E_6 = 7.2$$

For $+30°$ lamina,

$$m = 0.866, \quad n = 0.5$$

$$Q_{xx} = (0.866)^4 \times 142.77 + (0.5)^4 \times 10.36$$
$$+ 2(0.866)^2 (0.5)^2 (2.86) + 4(0.866)^2 (0.5)^2 \times 7.2$$
$$= 87.42$$

$$Q_{yy} = 21.22$$

$$Q_{xy} = 25.10, \quad Q_{xs} = 41.5, \quad Q_{ys} = 16.14, \quad Q_{ss} = 29.44$$

For a $-30°$ lamina,

$$Q_{xx} = 87.42, Q_{yy} = 21.22, Q_{xy} = 25.10, Q_{xs} = -41.5, Q_{ys} = -16.14, Q_{ss} = 29.44$$

From the definition of thermal forces, we have

$$\begin{Bmatrix} N_x^T \\ N_y^T \\ N_s^T \end{Bmatrix} = 2 \begin{bmatrix} 87.42 & 25.10 & 41.50 \\ 25.10 & 21.22 & 16.14 \\ 41.50 & 16.14 & 29.44 \end{bmatrix} \begin{Bmatrix} 6.08 \\ 20.03 \\ -24.16 \end{Bmatrix} 10^{-6} h \, \Delta T$$

$$+ 2 \begin{bmatrix} 87.42 & 25.10 & -41.50 \\ 25.10 & 21.22 & -16.14 \\ -41.50 & -16.14 & 29.44 \end{bmatrix} \begin{Bmatrix} 6.08 \\ 20.03 \\ 24.16 \end{Bmatrix} 10^{-6} \cdot d \cdot \Delta T$$

$$= \begin{Bmatrix} 126.50 \\ 750.81 \\ 0 \end{Bmatrix} 10^{-6} \cdot d \cdot \Delta T$$

For the laminate

$$[A] = 2 \begin{bmatrix} 87.42 & 25.10 & 41.50 \\ 25.10 & 21.22 & 16.14 \\ 41.50 & 16.14 & 29.44 \end{bmatrix} h + 2 \begin{bmatrix} 87.42 & 25.10 & -41.50 \\ 25.10 & 21.22 & -16.44 \\ -41.50 & -16.14 & 29.44 \end{bmatrix} h$$

$$= \begin{bmatrix} 349.68 & 100.48 & 0 \\ 100.48 & 84.88 & 0 \\ 0 & 0 & 117.76 \end{bmatrix} d$$

Therefore

$$100.48 \begin{bmatrix} 349.68 & 100.48 & 0 \\ 10.48 & 84.88 & 0 \\ 0 & 0 & 117.76 \end{bmatrix} h \begin{Bmatrix} \overline{\alpha}_x \\ \overline{\alpha}_y \\ \overline{\alpha}_s \end{Bmatrix} = \begin{Bmatrix} 126.50 \\ 750.81 \\ 0 \end{Bmatrix} h \times 10^{-6}$$

or, $\begin{Bmatrix} \overline{\alpha}_x \\ \overline{\alpha}_y \\ \overline{\alpha}_s \end{Bmatrix} = \begin{Bmatrix} -3.30 \\ 12.74 \\ 0 \end{Bmatrix} \times 10^{-6} \ /° C$

REFERENCES AND SUGGESTED READINGS

9.1 O. Ishai and A. Maza, The effect of environmental loading history on longitudinal strength of glass–fibre reinforced plastics, *Rheological Acta*, V.13, 1974, pp. 381–394.

9.2 O. Ishai and U. Arnon, The effect of hygrothermal history on residual strength of glass-fibre reinforced plastic laminates, *J. Testing Evaluation*, No. 5, No. 4, ASTM Philadelphia, 1977, pp. 320–326.

9.3 R. A. Schapery, Thermal expansion coefficients of composite materials based on energy principles, *Journal of Composite Materials*, V.2, No. 3, 1968, pp. 380–404.

9.4 R. F. Gibson, *Principles of Composite Materials Mechanics*, McGraw Hill, 1994.

9.5 Z. Hashin, Analysis of properties of fibre composites with anisotropic constituents, *Journal of Applied Mechanics*, V.46, 1979, pp. 543–550.

9.6 I. M. Daniel and O. Ishai, *Engineering Mechanics of Composite Materials*, Oxford University Press, 1994.

9.7 I. G. Zerai, I. M. Daniel and J. T. Gotro, Residual stresses and warpage in woven-glass epoxy laminates, *Experimental Mechanics*, V.27, 1987, pp. 44–50.

9.8 T. W. Wang and I. M. Daniel, Thermoviscoelastic analysis of residual stresses and warpage in composite laminates, *Journal of Composite Materials*, V.26, 1992.

EXERCISE 9

9.1 The properties of E-glass-epoxy lamina are
$E_1 = 53.78$ GPa $E_2 = 17.94$ GPa $\nu_{12} = 0.25$ $E_6 = 8.5$ GPa
$\alpha_1 = 6.3 \times 10^{-6}/° C$
$\alpha_2 = 20.52 \times 10^{-6}/° C$

Calculate the thermal forces in terms of ΔT

9.2 A 100 mm × 100 mm × 100 mm graphite-polymer composite is heated 100°C above the reference state. It is restrained in 3–direction. (a) What are the changes in dimensions of the heated element in the 1 and 2 directions? (b) What stress σ_3 is required to restrain the element against deformations in 3-direction? (c) What are the mechanical strains for this case?

The properties of the graphite-polymer composite are:

$E_1 = 155$ GPa $\qquad E_2 = 12$ GPa $\qquad E_3 = 12$ GPa

$\nu_{23} = 0.46 \qquad \nu_{12} = 0.25 \qquad \nu_{31} = 0.25$

$G_{23} = 3.2$ GPa $\qquad G_{13} = 4.4$ GPa $\qquad G_{12} = 4.4$ GPa

$\alpha_1 = -0.018 \times 10^{-6}/°C \quad \alpha_2 = 24.0 \times 10^{-6}/°C \quad \alpha_3 = 24.0 \times 10^{-6}/°C$

9.3 Consider a 100 mm × 100 mm × 100 mm composite material with its fibres oriented at 45° and constrained against deformation in the y-direction. The element is heated to 100°C. What is the stress σ_y required to enforce this constraint and what are the strains $\varepsilon_x, \varepsilon_y$ and ε_z? Take the same properties of Example 9.2.

9.4 Determine the coefficients of thermal expansion α_1 and α_2 of a unidirectional glass/epoxy lamina of the following properties

$E_f = 70$ GPa $\qquad E_m = 4$ GPa $\qquad \nu_f = 0.20 \qquad \nu_m = 0.35$

$\alpha_f = 4.5 \times 10^{-6}/°C \qquad \alpha_m = 90 \times 10^{-6}/°C \qquad V_f = 0.6$

9.5 For a unidirectional graphite-epoxy composite having fibre volume fraction $V_f = 0.60$, the coefficients of thermal expansion have been obtained as

$\alpha_1 = -0.02 \times 10^{-6}/°C$ and $\alpha_2 = 24 \times 10^{-6}/°C$

Determine α_{1f} and α_{2f} based on the following properties

$E_{1f} = 230$ GPa $\qquad E_m = 4$ GPa $\qquad \nu_f = 0.20 \qquad \nu_m = 0.35$

$\alpha_m = 40 \times 10^{-6}/°C$

9.6 Determine the thermal forces N_x^T, N_y^T and N_s^T for a $[\pm 30]_s$ laminate, having the properties given below for a temperature difference of 50°C. Lamina thickness is 0.1 mm.

$V_f = 0.63 \qquad E_1 = 142$ GPa $\qquad E_2 = 10$ GPa $\qquad E_6 = 7$ GPa

$\nu_{12} = 0.25 \qquad \alpha_1 = 0.9 \times 10^{-6}/°C \qquad \alpha_2 = 27 \times 10^{-6}/°C$

9.7 Calculate $\bar{\alpha}_x$ and $\bar{\alpha}_y$ for the laminate of Example 9.6.

9.8 A unidirectional off-axis lamina is subjected to a temperature rise of ΔT. The fibre orientation is θ. Determine σ_y so that there is no shear deformation of the lamina. Express σ_y in terms of material properties, ΔT and θ.

9.9 A $[\pm 45]$ lamina having properties given in example 9.6 was cooled down from 200°C to 50°C during curing and the deformation measured are as follows

$\varepsilon_x^0 = \varepsilon_y^0 = -7.5 \times 10^{-6}, \quad \varepsilon_s^0 = 0$

$k_x = k_y = 0 \quad k_s = 2/m$

Compute the lamina residual stresses σ_{1e}, σ_{2e} and σ_{6e}. $d = 1$ mm.

CHAPTER 10

FAILURE THEORIES AND STRENGTH OF A UNIDIRECTIONAL LAMINA

10.1 INTRODUCTION

A look into composite materials will reveal the following characteristics. First, both the basic constituents of composites – the fibre and the matrix – have different strengths. Further, the behaviour of the interface of the fibre and the matrix is different from that of the bulk matrix. During the manufacture of the FRP, flaws or defects which get introduced into the system may act as stress raisers or failure initiators. Therefore, these factors are to be accounted for in order to identify important features and mechanisms controlling the micro-failure of the composites. Local failure modes at the micro-level are of the following types [10.1].

1. Fibre dominated failure (breakage, micro-buckling and dewetting)
2. Bulk matrix dominated failures (voids, crazing)
3. Interface/flaw dominated failures (crack propagation, edge delamination)

Though failure in the composite is initiated due to the above causes, they have no role to play when failure is treated globally. Failure is discussed with respect to that of a unidirectional lamina as the datum or otherwise considering the lamina as a part of the composite orthotropic laminates.

Failure of a FRP laminate is gradual. When a particular lamina fails in a laminate, a redistribution of stresses takes place in the remaining laminae. Strength of a lamina is the basic element considered for the strength and failure analysis of a laminate.

A composite laminate has many typicalities. The strength of the lamina is directionally dependent. The strength in tension and compression of a unidirectional composite differs significantly. Further, the direction of shear stress with respect to the fibre orientation has considerable bearing on its strength.

10.2 MICROMECHANICS OF FAILURE OF UNIDIRECTIONAL LAMINA

The fundamental case of a unidirectional lamina, how they fail and how to predict the strength from properties of fibres, resins and interface will be considered first.

10.2.1 Longitudinal Tension

The fibre and the matrix have different ultimate strains (at failure) in tension. As such in case of longitudinal tension, the constituent with lower ultimate strain will fail first. For the unidirectional lamina, the longitudinal stress is given by the rule of mixtures as

$$\sigma_{\text{IT}} = \sigma_{\text{I}fT} V_f + \sigma_{m\text{T}} V_m \tag{10.1}$$

where

$\sigma_{\text{I}fT}$ is the average longitudinal tensile stress in the fibre

σ_{mT} is the average longitudinal tensile stress in the matrix

V_f, V_m are volume fractions of the fibre and matrix respectively

Two cases arise

Case I : When the ultimate strain of the fibre is lower than that of the matrix, that is, $\varepsilon^u_{1fT} < \varepsilon^u_{mT}$, the unidirectional composite lamina will fail when its longitudinal strain reaches the ultimate strain of the fibre. The longitudinal strength of the lamina is thus approximated as [Fig.10.1]

$$F_{1T} = F_{1fT}.V_f + \sigma'_m V_m \qquad (10.2)$$

where

F_{1T} is the longitudinal tensile strength of unidirectional lamina

F_{1fT} is the longitudinal fibre tensile strength

σ'_m is the average longitudinal matrix stress when the ultimate fibre strain is reached (Fig. 10.1).

If the constituents are linearly elastic, (10.2) becomes

$$F_{1T} = F_{1fT} V_f + E_m \varepsilon^u_{1fT} V_m$$

or, $\qquad F_{1T} = F_{1fT} \left(V_f + V_m \cdot \dfrac{E_m}{E_{1f}} \right) \qquad (10.3)$

If the fibres are very stiff, then $E_{1f} \gg E_m$ and (10.3) becomes

$$F_{1T} \cong F_{1fT} \cdot V_f \qquad (10.4)$$

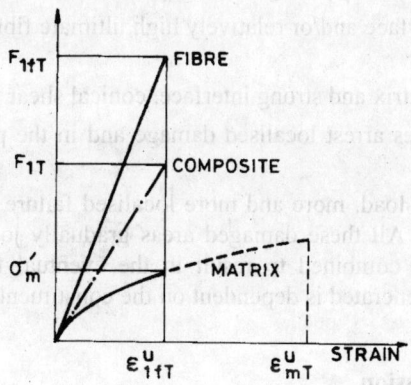

Fig. 10.1 Longitudinal stress–strain curves for the unidirectional composite lamina and its constituents when

$$\varepsilon^u_{fT} < \varepsilon^u_{mT}$$

Case II : In this case, the ultimate strain of the matrix is lower than that of the fibre,

$$\varepsilon^u_{mT} < \varepsilon^u_{1fT} \qquad (10.5)$$

The unidirectional composite lamina fails when the longitudinal tensile strain of the composite attains the ultimate longitudinal strain of the matrix. Then, the longitudinal tensile strength of the composite is approximately written as

$$F_{1T} \cong \sigma'_f V_f + F_{mT} \cdot V_m \tag{10.6}$$

where

F_{mT} is the tensile strength of the matrix

σ'_f is the longitudinal fibre stress when the ultimate matrix strain is reached

This can be further approximated as

$$F_{1T} \cong F_{mT} \left(V_f \cdot \frac{E_{1f}}{E_m} + V_m \right) \tag{10.7}$$

In fact the strengths of the fibre and the matrix are statistically distributed, a fact not considered in the above results [10.2]. The longitudinal tensile strength is basically a case of fibre-dominated strength. Out of several fibres, one fibre is considered as broken and obviously at its weak points [Fig. 10.2]. The reason of this breakage of the fibre is due to a stress high enough to start fracture. This causes a redistribution of stress around the fracture. Stress must be transferred across the broken fibres and this is accomplished by developing a high interfacial shear stress. The longitudinal fibre stress is zero at the break, but gradually increases to a stress level of any other fibre of the composite at a characteristic distance δ from the break. The interfacial shear stress drops to zero after traversing this distance δ. There is a local increase in both fibre stress and interfacial shear stress in adjacent fibres due to the fibre break. Thus, if a fibre breaks, its load carrying length is reduced by 2δ.

Different types of failure will result depending on the properties of the constituents [10.1]. Failure of composites can occur in three ways [Fig. 10.3].

1. For a brittle matrix and relatively strong interface, the matrix will transversely crack.
2. For a relatively weak interface and/or relatively high ultimate fibre strain, fibre-matrix debonding will take place.
3. For a relatively ductile matrix and strong interface, conical shear failures in matrix occur.

In general, adjacent fibres arrest localised damage and in the process the ineffective length of the fibre is increased.

With the increase of the load, more and more localised failure takes place. This also results in the failure of adjacent fibres. All these damaged areas gradually join together as localised failures interact. Finally, they all are combined to result in the eventual failure. The sequence at which different failure patterns are generated is dependent on the constituent properties and the fibre-volume ratio.

10.2.2 Longitudinal Compression

In attempting to model the unidirectional strength of composites, the first problem is how to measure experimentally what is being modelled. Indeed, designers might argue that structures rarely experience true compression failures as instability usually precedes compression failure. In recent years, however, significant advances have been made in an attempt to deal with compressive strength of materials [10.4].

Failure Theories and Strength of a Unidirectional Lamina 257

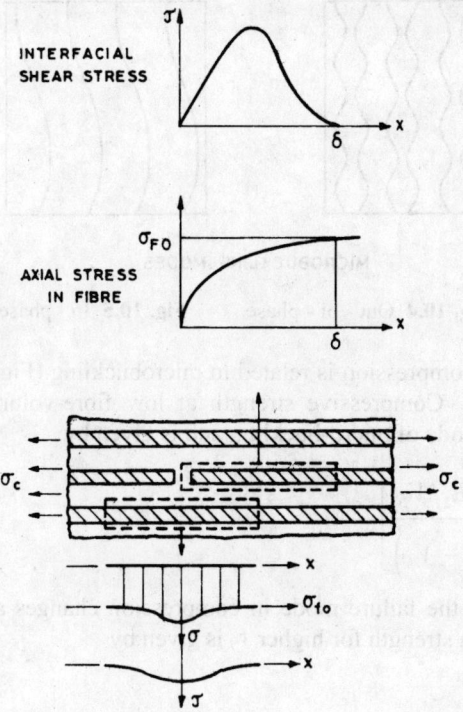

Fig. 10.2 Local stress distribution around a fibre break

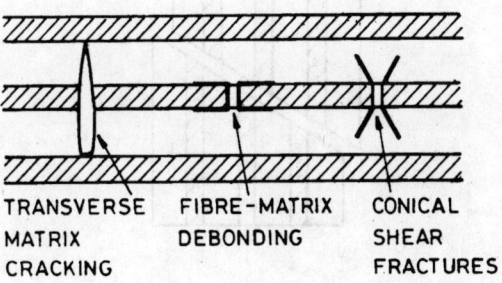

Fig. 10.3 Failure mechanism around a single fibre break

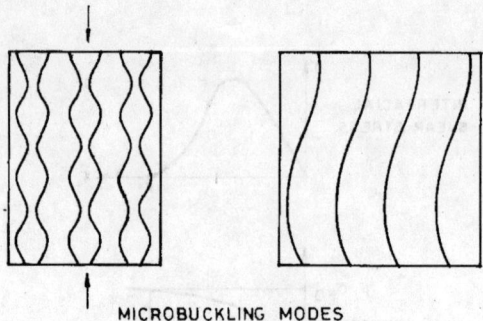

Fig. 10.4 Out - of - phase   **Fig. 10.5** In - phase

Failure in longitudinal compression is related to microbuckling (Figs. 10.4 and 10.5) or kinking of fibres within the matrix. Compressive strength at low fibre-volume ratio is dictated by the extensional or out-of-phase mode of micro-buckling and is given by

$$F_{1C} \cong 2V_f \left[\frac{E_m E_{1f} V_f}{3(1 - V_f)} \right]^{\frac{1}{2}} \tag{10.8}$$

At higher values of V_f, the failure mode in compression changes and it becomes shear or in-phase mode. The compression strength for higher V_f is given by

$$F_{1C} \cong \frac{G_m}{1 - V_f} \tag{10.9}$$

where G_m is the shear modulus of elasticity of the matrix.

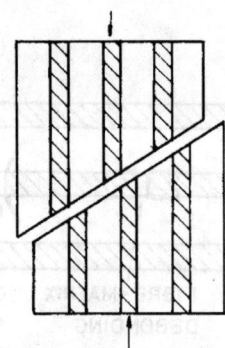

Fig. 10.6 Shear failure mode under longitudinal compression

Another mode of failure may result at higher V_f. It may be dictated by the shear strength of the fibre (Fig. 10.6). The compressive strength for this case is given by

$$F_{1C} = 2F_{sf} \left[V_f + (1 - V_f) \frac{E_m}{E_{1f}} \right] \tag{10.10}$$

when F_{sf} is the shear strength of fibre.

10.2.3 Transverse Tension

Transverse tensile loading is undoubtedly the most critical loading of a unidirectional composite. Due to this loading pattern, high stress or strain concentration in matrix and interface/interphase occurs. Stress distribution around the fibre can be evaluated theoretically or experimentally. Among the theoretical approaches are the analytical method using the complex variables, the finite difference method, the finite element method and the boundary element method. Two or three-dimensional photoelastic method has been used as the experimental technique. The most critical zone is the fibre-matrix interface where critical stresses and strains are likely to occur.

The transverse tensile strength of the unidirectional lamina is determined using a stress concentration factor. It is assumed that fibres are much stiffer than the matrix and are perfectly bonded to it. The stress and strain concentration factors are related as follows [10.5].

$$K_\varepsilon = \frac{\varepsilon_{2\max}}{\varepsilon_2} \cong K_\sigma \left(\frac{E_{2T}}{E_m}\right) \frac{(1+v_m)(1-2v_m)}{1-v_m} \tag{10.11}$$

where

- K_ε is the strain concentration factor
- K_σ is the stress concentration factor
- $\varepsilon_{2\max}$ is the maximum transverse strain
- ε_2 is the average transverse strain
- v_m is the matrix Poisson's ratio

Thermal stresses are induced in the composite lamina due to various reasons. Residual stresses and strains are set up due to the curing of the matrix. These stresses result in the constituents having different values of coefficients of thermal expansion. In order to predict failure of a thermally loaded composite lamina, the stress–strain relationship of the lamina has been assumed to be linear. Using the maximum stress theory criterion, the transverse tensile strength with the inclusion of thermal stress is given by

$$F_{2T} = \frac{1}{K_\sigma}\left(F_{mT} - \sigma_{rm}\right) \tag{10.12}$$

where σ_{rm} is the maximum residual stress

Otherwise, using the maximum strain theory criterion, the transverse tensile strength for a thermally loaded composite is

$$F_{2T} = \frac{1-v_m}{K_\sigma(1+v_m)(1-2v_m)}\left(F_{mT} - \varepsilon_{rm}E_m\right) \tag{10.13}$$

where ε_{rm} is the maximum residual strain.

10.2.4 Transverse Compression

A unidirectional composite may fail in transverse compression under a number of failure mechanisms. The high compressive stress concentration at the interface may cause compression failure of the matrix and/or crushing of the fibre. The strength in transverse compression for a unidirectional lamina is

$$F_{2C} = \frac{F_{mC} + \sigma_{rm}}{K_\sigma} \tag{10.14}$$

where F_{mC} is the compressive strength of the matrix.

High compressive stress may result in an overall shear failure due to the presence of high interfacial shear stresses which cause matrix shear failure and/or debonding.

10.2.5 Inplane Shear

The most susceptible part of the unidirectional composite in inplane shear is the fibre-matrix interface (Fig. 10.7). Assuming that the matrix shear failure takes place in the unidirectional lamina, the inplane shear strength is predicted as

$$F_6 = \frac{F_{mS}}{K_\tau} \qquad (10.15)$$

where
- K_τ is the shear strength concentration factor
- F_{mS} is the matrix shear strength

The variation of K_τ with material properties and V_f has been determined by the finite difference method [10.6].

10.3 ANISOTROPIC STRENGTH AND FAILURE THEORIES

The micromechanical strength predictions are accurate with regard to failure initiation at critical points. It is only approximate as far as the global failure of the lamina is concerned.

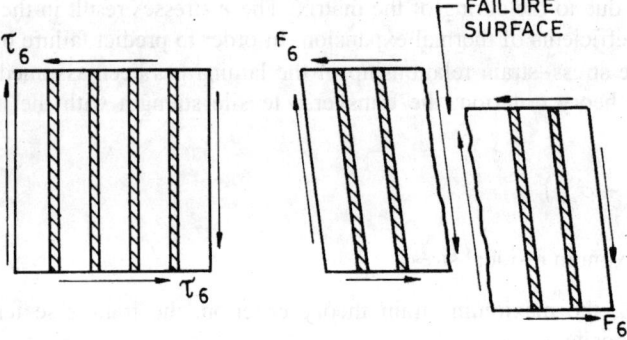

Fig. 10.7 Failure mode in inplane shear

From macromechanical point of view, the strength of a lamina is an anisotropic property, that is, it varies with the orientation of the fibre. Correlation of the strength along an arbitrary direction to some basic strength parameters is highly desirable. A lamina may be characterised by a number of basic strength parameters defined before. For an inplane loading, a lamina may be characterised by five strength parameters – F_{1T}, F_{1C}, F_{2T}, F_{2C} and F_6.

Macromechanical failure theories for composites are generally an extension and adoption of isotropic failure theories to account for anisotropy in strength and stiffness of the composite. A comprehensive review of various theories of anisotropic material is available [10.7].

There are four commonly used failure criteria for fibre reinforced plastics

1. Maximum stress theory
2. Maximum strain theory

3. Deviatoric strain energy theory (Tsai-Hill)
4. Interactive tensor polynomial theory (Tsai-Wu)

10.3.1 Maximum Stress Theory

The maximum stress theory states that failure will occur when at least one stress component along one of the principal material axes reaches the magnitude of corresponding strength in that direction. The complex stress system in the lamina is transformed into stresses along the principal material axes. Failure is defined by

$$\sigma_1 = \begin{cases} F_{1T} & \text{when } \sigma_1 > 0 \\ -F_{1C} & \text{when } \sigma_1 < 0 \end{cases} \quad (10.16a)$$

$$\sigma_2 = \begin{cases} F_{2T} & \text{when } \sigma_2 > 0 \\ -F_{2C} & \text{when } \sigma_2 < 0 \end{cases} \quad (10.16b)$$

$$|\sigma_6| = F_6 \quad (10.16c)$$

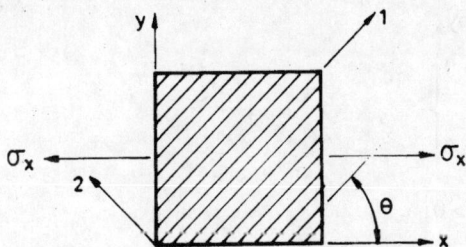

Fig. 10.8 Off-axis loading of a unidirectional lamina

For a general case of loading, referring to Fig. 10.8, for the off-axis loading of the lamina, the stress components along the principal material axes are

$$\sigma_1 = \sigma_x \cos^2 \theta$$
$$\sigma_2 = \sigma_x \sin^2 \theta \quad (10.17)$$
$$\sigma_6 = -\sigma_x \cos\theta \sin\theta$$

Following off-axis strength F_x is obtained (by equating the stress components to the corresponding strengths).

When σ_x is in tension, i.e., $\sigma_x > 0$

$$F_{xT} = \frac{F_{1T}}{\cos^2 \theta}$$

$$F_{xT} = \frac{F_{2T}}{\sin^2 \theta} \quad (10.18)$$

$$F_{xT} = \frac{F_6}{\sin\theta \cos\theta}$$

and when σ_x is in compression, i.e., $\sigma_x < 0$

$$F_{xC} = \frac{F_{1C}}{\cos^2\theta}$$

$$F_{xC} = \frac{F_{2C}}{\sin^2\theta} \qquad (10.19)$$

$$F_{xC} = \frac{F_6}{\sin\theta \cos\theta}$$

Therefore, out of five subcriteria, three are applicable according to the maximum stress theory whether σ_x is in tension or compression. It is a non-interactive theory in that it assesses that the failure in any perpendicular direction is caused independently of the stresses acting in other directions. Thus, the nature of interaction in a biaxial state of stress is not considered.

10.3.2 Maximum Strain Theory

The maximum strain theory states that failure will occur when at least one strain component along one of the principal material axes reaches the value of the corresponding ultimate strain in that direction. Failure is defined by

$$\varepsilon_1 = \begin{cases} \varepsilon_{1T}^u & \text{when } \varepsilon_1 > 0 \\ \varepsilon_{1C}^u & \text{when } \varepsilon_1 < 0 \end{cases} \qquad (10.20a)$$

$$\varepsilon_2 = \begin{cases} \varepsilon_{2T}^u & \text{when } \varepsilon_2 > 0 \\ \varepsilon_{2C}^u & \text{when } \varepsilon_2 < 0 \end{cases} \qquad (10.20b)$$

$$\varepsilon_6 = \varepsilon_6^u \qquad (10.20c)$$

ε_{1T}^u is the ultimate longitudinal tensile strain

ε_{1C}^u is the ultimate longitudinal compressive strain

ε_{2T}^u is the ultimate transverse tensile strain

ε_{2C}^u is the ultimate transverse compressive strain

ε_6^u is the ultimate inplane shear strain

For the case of a general biaxial state of stress, the stresses in the xy-axis system are transformed into stresses in the principal material axis system 1-2, that is, stress components σ_1, σ_2 and σ_6. The corresponding strain components in the 1-2 direction can be obtained from the stress–strain relationship.

$$\varepsilon_1 = \frac{1}{E_1}\left(\sigma_1 - v_{12}\sigma_2\right) \qquad (10.21a)$$

$$\varepsilon_2 = \frac{1}{E_2}(\sigma_2 - \nu_{21}\sigma_1) \tag{10.21b}$$

$$\varepsilon_6 = \frac{\sigma_6}{E_6} \tag{10.21c}$$

Signs of shear stress in the principal material directions are shown in Fig. 10.9.

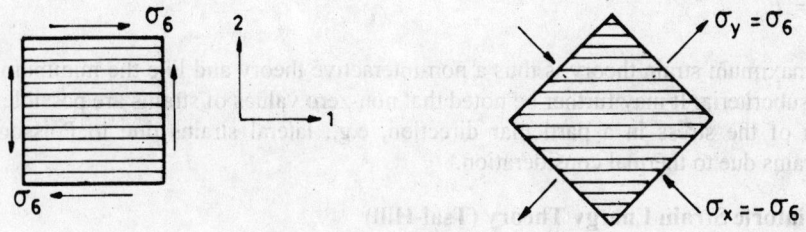

POSITIVE SHEAR STRESS

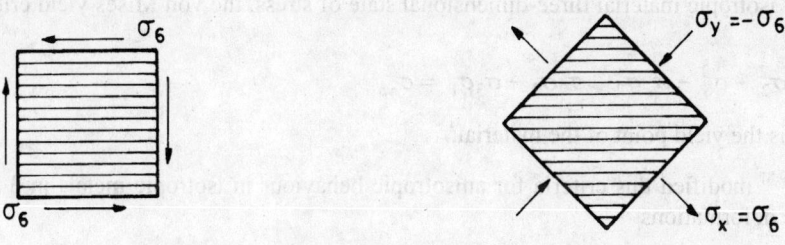

NEGATIVE SHEAR STRESS

Fig. 10.9 Positive and negative shear stress acting along the principal material directions

If the basic strength parameters are known, the ultimate strains of the unidirectional lamina can be obtained from that information

$$\varepsilon_{1T}^u = \frac{F_{1T}}{E_1}$$

$$\varepsilon_{1C}^u = -\frac{F_{1C}}{E_1} \tag{10.22}$$

$$\varepsilon_{2T}^u = \frac{F_{2T}}{E_2}$$

$$\varepsilon_{2C}^u = -\frac{F_{2C}}{E_2}$$

$$\varepsilon_6^u = \frac{F_6}{E_6}$$

For the general biaxial state of stress the failure criteria can then be expressed as

$$\sigma_1 - \nu_{12}\sigma_2 = \left\{\begin{array}{ll} F_{1T} & \text{when } \varepsilon_1 > 0 \\ -F_{1C} & \text{when } \varepsilon_1 < 0 \end{array}\right\} \quad (10.23a)$$

$$\sigma_2 - \nu_{21}\sigma_1 = \left\{\begin{array}{ll} F_{2T} & \text{when } \varepsilon_2 > 0 \\ -F_{2C} & \text{when } \varepsilon_2 < 0 \end{array}\right\} \quad (10.23.b)$$

$$|\sigma_6| = F_6 \quad (10.23c)$$

The maximum strain theory is thus a non-interactive theory and like the minimum stress theory it has five subcriteria. It may further be noted that non-zero values of strains are possible even without application of the stress in a particular direction, e.g., lateral strains due to Poisson's effect and residual strains due to thermal consideration.

10.3.3 Deviatoric Strain Energy Theory (Tsai-Hill)

For isotropic materials, there exists the distortional or deviatoric energy theory proposed by various investigators such as von Mises, Huber, Hencky and others in different forms for the prediction of failure. This is indeed a very popular approach for isotropic material.

For an isotropic material three-dimensional state of stress, the von Mises yield criterion is given by

$$\sigma_1^2 + \sigma_2^2 + \sigma_3^2 - \sigma_1\sigma_2 - \sigma_2\sigma_3 - \sigma_3\sigma_1 = \sigma_{yp}^2 \quad (10.24)$$

where σ_{yp} is the yield point of the material.

Hill [10.8] modified this criteria for anisotropic behaviour in isotropic metals in 3-dimension for large plastic deformations

$$A\sigma_1^2 + B\sigma_2^2 + C\sigma_1\sigma_2 + D\sigma_6^2 + E\sigma_3^2 + F\sigma_2\sigma_3 + G\sigma_3\sigma_1 + H\sigma_4^2 + I\sigma_5^2 = 1 \quad (10.25)$$

For a 2-dimensional problem, (10.25) reduces to

$$A\sigma_1^2 + B\sigma_2^2 + C\sigma_1\sigma_2 + D\sigma_6^2 = 1 \quad (10.26)$$

In (10.25) and (10.26), A, B, C, D etc. are constants and are dependent on the material and its nature of anisotropy. In von Mises' criterion, the amount of energy is used to distort the body rather than to change the volume, whereas for anisotropic materials distortion cannot be separated from dilatation. So strictly speaking, this theory is not a distortional energy failure theory.

Two-dimensional stress results of Hill were simplified by Azzi and Tsai [10.9] for the case of fibre reinforced composites – considering the composite to be transversely isotropic. The 1,2,3-directions are assumed to be the principal material axes of transversely isotropic lamina with 1-direction being along the fibre direction. The parameters A, B, C and D are related to failure strengths F_{1T}, F_{1C} etc.

Thus, for inplane shear loading to failure,

$$\sigma_6^u = F_6, \quad \sigma_1 = \sigma_2 = 0, (10.26) \text{ yields}$$

$$D = \frac{1}{F_6^2} \quad (10.27)$$

For uniaxial transverse loading to failure, $\sigma_2^u = F_2$, $\sigma_1 = \sigma_6 = 0$, (10.26) yields

$$B = \frac{1}{F_2^2} \tag{10.28}$$

For uniaxial longitudinal loading to failure, $\sigma_1^u = F_1, \sigma_2 = \sigma_6 = 0$, (10.26) yields

$$A = \frac{1}{F_1^2} \tag{10.29}$$

Thus, all the parameters except C are determined. The value C can be obtained from the interaction of normal stresses σ_1 and σ_2. This can only be done from a biaxial test. Otherwise, if equal biaxial loading is applied, that is, $\sigma_1 = \sigma_2 \neq 0$ and $\sigma_6 = 0$, the maximum stress criteria is assumed to define the failure of the material; that is, failure will take place, when transverse normal stress σ_2 reaches the transverse strength of the composite, F_2 is much lower than the longitudinal strength F_1. Equation (10.26) then becomes.

$$C = -\frac{1}{F_1^2} \tag{10.30}$$

On substitution of the above values of A, B, C and D into eqn. (10.26), the Tsai-Hill criteria assumes the following form

$$\frac{\sigma_1^2}{F_1^2} + \frac{\sigma_2^2}{F_2^2} + \frac{\sigma_6^2}{F_6^2} - \frac{\sigma_1 \sigma_2}{F_1^2} = 1 \tag{10.31}$$

It may be mentioned that (10.31) written as follows will appear as more rational

$$\frac{\sigma_1^2}{F_1^2} + \frac{\sigma_2^2}{F_2^2} + \frac{\sigma_6^2}{F_6^2} - \left(\frac{\sigma_1}{F_1}\right)\left(\frac{\sigma_2}{F_2}\right) = 1 \tag{10.32}$$

where $F_1 = F_{1T}$ or F_{1C}

$F_2 = F_{2T}$ or F_{2C}

The values of F_1 and F_2 will depend on the nature of σ_1 and σ_2. This implies that if applied stress is tension, then the tensile strength is to be used and if the applied stress is compression, then the compressive strength is to be used.

When the uniaxial load applied is off-axis (Fig. 10.8), the stresses are to be transformed in the direction of the principal material axis (10.17) which when substituted in (10.32) results in

$$\frac{1}{F_x^2} = \frac{m^4}{F_1^2} + \frac{n^4}{F_2^2} + \left[\frac{1}{F_6^2} - \frac{1}{F_1 F_2}\right] m^2 n^2 \tag{10.33}$$

where $m = \cos\theta$, $n = \sin\theta$ and $F_x = \sigma_x^u$

For advanced composites $F_1 \gg F_6$ and (10.33) becomes

$$\frac{1}{F_x^2} = \frac{m^4}{F_1^2} + \frac{n^4}{F_2^2} + \frac{m^2 n^2}{F_6^2} \tag{10.34}$$

The main advantage of the Tsai-Hill criterion is that it is expressed as a single criterion instead of a number of subcriteria required in the maximum stress or maximum strain theories. As such there is a considerable interaction between F_1, F_2 and F_6. In the Tsai-Hill criterion, however, distinction has not been made between tensile and compressive strengths.

10.3.4 Interactive Tensor Polynomial Theory [Tsai-Wu]

Tsai and Wu [10.10] have proposed a modified tensor polynomial theory by postulating that a failure surface in stress space exists in the following form in contracted notation

$$f_i \sigma_i + f_{ij} \sigma_{ij} = 1 \qquad (10.35)$$

where f_i and f_{ij} are strength tensors of second and fourth rank.

Equation (10.35) is indeed very complicated. We restrict ourselves to the case of an orthotropic lamina in plane stress and then (10.35) in expanded form becomes

$$f_1 \sigma_1 + f_2 \sigma_2 + f_6 \sigma_6 + f_{11} \sigma_1^2 + f_{22} \sigma_2^2 + f_{66} \sigma_6^2$$
$$+ 2 f_{12} \sigma_1 \sigma_2 + 2 f_{16} \sigma_1 \sigma_6 + 2 f_{26} \sigma_1 \sigma_6 = 1 \qquad (10.36)$$

The linear terms of stresses in (10.36) enables one to make a distinction between tensile and compressive strengths.

The term f_{12} treats the interaction between normal stresses σ_1 and σ_2.

All linear terms associated with σ_6 must vanish, as the shear strength in principal material directions is independent of the sign of the shear stress. Thus,

$$f_6 = f_{16} = f_{26} = 0 \qquad (10.37)$$

The coefficients which are not zero are to be evaluated. The uniaxial loading applied in the longitudinal direction of tensile stress at failure, gives $\sigma_1^u = F_{1T}, \sigma_2 = \sigma_6 = 0$ and which on substitution in (10.36) yields

$$f_1 . F_{1T} + f_{11} F_{1T}^2 = 1 \qquad (10.38)$$

For the case of the compressive force in the longitudinal direction, to failure

$$-f_1 . F_{1C} + f_{11} F_{1C}^2 = 1$$

Solving (10.38) and (10.39) yields

$$f_1 = \frac{1}{F_{1T}} - \frac{1}{F_{1C}} \qquad (10.40)$$

$$f_{11} = \frac{1}{F_{1T} . F_{1C}} \qquad (10.41)$$

Similarly, considering uniaxial tensile and compressive loading in the transverse direction as above yields

$$f_2 = \frac{1}{F_{2T}} - \frac{1}{F_{2C}} \qquad (10.41)$$

$$f_{22} = \frac{1}{F_{2T} . F_{2C}} \qquad (10.43)$$

For pure shear loading at failure, $\sigma_6^u = F_6, \sigma_1 = \sigma_2 = 0$ yields

$$f_{66} = \frac{1}{F_6^2} \tag{10.44}$$

The coefficient f_{12} is still to be determined. The biaxial test is used for this purpose. Equal normal stresses in the two principal material directions are applied to failure, that is, $\sigma_1^u = \sigma_2^u = F_{12}, \sigma_6 = 0$. Accordingly, we obtain

$$(f_1+f_2)F_{12} + (f_{11}+f_{22}+2f_{12})F_{12}^2 = 1 \tag{10.45}$$

where F_{12} is the biaxial tensile failure stress and is to be determined experimentally for equal biaxial tensile loading. Solving (10.45), yields

$$f_{12} = \frac{1}{2F_{12}^2}\left[1 - F_{12}\left(\frac{1}{F_{1T}} - \frac{1}{F_{1C}} + \frac{1}{F_{2T}} - \frac{1}{F_{2C}}\right)\right.$$
$$\left. - F_{12}^2 \left(\frac{1}{F_{1T}F_{1C}} + \frac{1}{F_{2T}F_{2C}}\right)\right] \tag{10.46}$$

Therefore, f_{12} is a function of biaxial tensile strength F_{12} in addition to the basic strength parameters.

To experimentally obtain F_{12}, the biaxial tensile strength is a difficult task. A simpler option of determining f_{12} is to perform an off-axis tensile strength, that is, the fibres are inclined at an angle θ to the loading axis. For $\theta = 45°$ (10.17) becomes

$$\sigma_1 = \sigma_2 = |\sigma_6| = \frac{1}{2}F_{45T} \tag{10.47}$$

where F_{45T} is the off-axis strength in tension for 45° lamina.

Combining (10.36) and (10.47) yields

$$\frac{F_{45T}}{2}(f_1+f_2) + \frac{F_{45T}^2}{4}(f_{11} + f_{22} + f_{66} + 2f_{12}) = 1$$

or, $$f_{12} = \frac{2}{F_{45T}^2}\left[1 - \frac{F_{45T}}{2}\left(\frac{1}{F_{1T}} - \frac{1}{F_{1C}} + \frac{1}{F_{2T}} - \frac{1}{F_{2C}}\right)\right.$$
$$\left. - \frac{F_{45T}^2}{4}\left(\frac{1}{F_{1T}F_{1C}} + \frac{1}{F_{2T}F_{2C}} + \frac{1}{F_6^2}\right)\right] \tag{10.48}$$

The coefficient f_{12} may not have a very significant effect on the assessment of overall strength. Its expression given by (10.48) is not very elegant. As such it is approximated as follows [10.11]

$$f_{12} \cong -\frac{1}{2}(f_{11}f_{22})^{\frac{1}{2}} \tag{10.49}$$

Out of all the failure theories discussed so far, the Tsai-Wu failure theory is most general in character. It improves upon the Tsai-Hill criterion in that it clearly distinguishes between the tensile and compressive strengths. It has been shown that the Tsai-Hill and Tsai-Wu criteria predicts almost identical values for the tensile strength. For the compressive strength these two theories do not differ significantly (Fig. 10.10).

After eliminating all zero terms, the Tsai-Wu criterion assumes the expression

$$f_1\sigma_1 + f_2\sigma_2 + f_{11}\sigma_1^2 + f_{22}\sigma_2^2 + f_{66}\sigma_6^2 + 2f_{12}\sigma_1\sigma_2 = 1 \tag{10.50}$$

10.4 EXAMPLES

Example 10.1

An E-glass/epoxy composite has the following properties of its constituents

$V_f = 0.65$ $E_{1f} = 70$ GPa $E_m = 3.5$ GPa

$F_{1fT} = 3500$ MPa   and $F_{mT} = 100$ MPa

Determine the longitudinal modulus of the composite E_1 and longitudinal tensile strength F_{1T} of the composite.

It is assumed that both the fibre and the matrix behave in a linear elastic manner. The ultimate failure strains for the fibre and matrix are

$$\varepsilon_{fT}^u = \frac{F_{1fT}^u}{E_{1f}} = \frac{3500 \times 10^6}{70 \times 10^9} = 0.05$$

$$\varepsilon_{mT}^u = \frac{F_{mfT}^u}{E_m} = \frac{100 \times 10^6}{35 \times 10^9} = 0.0286 < \varepsilon_{fT}^u$$

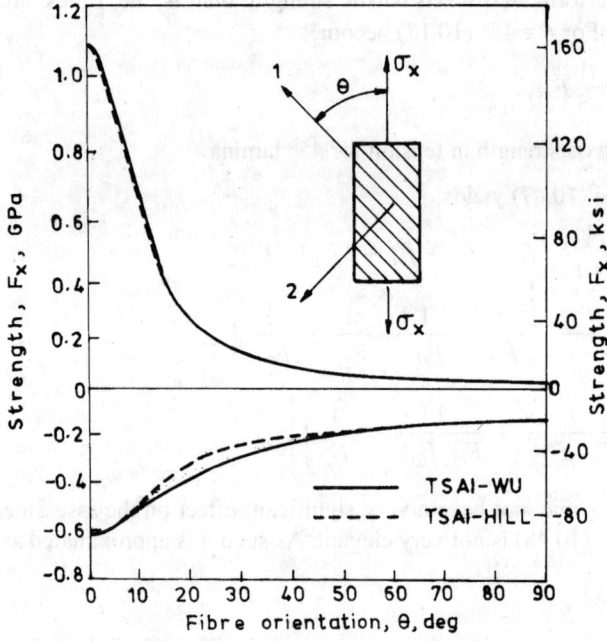

Fig. 10.10 Comparison of the Tsai-Hill and Tsai-Wu Criteria

Therefore, the matrix will fail first and the maximum fibre stress at failure will not be reached. The maximum stress developed in the fibre at the instant of matrix cracking is

$$F_{1T} = F_{mT}^u \left(V_f \cdot \frac{E_{1f}}{E_m} + V_m \right)$$

$$= 100 \left(0.65 \cdot \frac{70}{3.5} + 0.35 \right)$$

$$= 1335 \quad \text{MPa}$$

Similarly,

$$E_1 = V_f \cdot E_{1f} + V_m \cdot E_m$$

$$= 0.65 \times 70 + 3.5 \times 0.35$$

$$= 46.725 \quad \text{GPa}$$

Example 10.2

An angle–ply lamina made of S-glass/epoxy has the following properties in the principal fibre direction

$F_{1T} = 1280$ MPa $\quad F_{1C} = 622$ MPa $\quad F_{2T} = 49$ MPa

$F_{2C} = 245$ MPa $\quad F_6 = 69$ MPa

$E_1 = 35$ GPa $\quad E_2 = 7$ GPa $\quad E_6 = 3$ GPa $\quad \nu_{12} = 0.3$

A tensile load of $\sigma_x = 2$ MPa is applied at an angle $60°$ to the principal fibre direction. Check the safety of the laminate as per failure theories.

Maximum stress theory

Stresses are resolved in the principal material directions

$$\sigma_1 = \sigma_x \cos^2 \theta = 2 \times \frac{1}{4} = 0.5 \quad < 1280$$

$$\sigma_2 = \sigma_x \sin^2 \theta = 2 \times \frac{3}{4} = 1.5 \quad < 49$$

$$\sigma_6 = \sigma_x \cos \theta \sin \theta = 0.866 \quad < 69$$

The laminate is safe according to the maximum stress theory.

Maximum strain theory

Strains in the principal material directions are

$$\varepsilon_1 = \frac{1}{E_1} \left[\cos^2 \theta - \nu_{12} \sin^2 \theta \right] 2 \times 10^6$$

$$= \frac{1}{35 \times 10^9} \left[\frac{1}{4} - \frac{0.3 \times 3}{4} \right] 2 \times 10^6$$

$$= 1.43 \times 10^{-6} \quad < \frac{1280 \times 10^6}{35 \times 10^9} \quad < 0.036$$

$$\nu_{21} = 0.3 \times \frac{7}{35} = 0.06$$

$$= 1.386 \times 10^{-4} < \frac{49 \times 10^6}{7 \times 10^9} < 0.007$$

$$\varepsilon_6 = \frac{1}{3 \times 10^9} \left[\frac{\sqrt{3}}{4} \right] 2 \times 10^6$$

$$= 2.886 \times 10^{-4} < \frac{69 \times 10^6}{3 \times 10^9} < 0.023$$

The laminate is safe according to the maximum strain theory.

Tsai-Hill theory

The Tsai-Hill failure criteria is given by the following equations.

$$\frac{\sigma_1^2}{F_1^2} + \frac{\sigma_2^2}{F_2^2} + \frac{\sigma_6^2}{F_6^2} - \frac{\sigma_1 \sigma_2}{F_1 F_2}$$

$$= \left[\left(\frac{\cos\theta}{F_1} \right)^2 + \left(\frac{\sin\theta}{F_2} \right)^2 + \left(\frac{\sin\theta \cos\theta}{F_6} \right)^2 - \frac{\cos\theta \sin\theta}{F_1 F_2} \right] \sigma_x^2$$

$$= \frac{1}{10^{12}} \left[\left(\frac{0.5}{1280} \right)^2 + \left(\frac{0.866}{49} \right)^2 + \left(\frac{0.5 \times 0.866}{69} \right)^2 - \left(\frac{0.5 \times 0.866}{1280 \times 49} \right) \right] 2^2 \times 10^{12}$$

$$= 3.489 \times 10^{-4} < 1$$

The Tsai-Hill theory indicates the safety of the laminate.

The maximum stress theory and maximum strain theory give higher strength values than the Tsai-Hill theory.

Example 10.3

The unidirectional lamina is same as Example 10.2. The loading is

$$\sigma_x = -2 \text{ MPa}, \quad \sigma_y = 4 \text{ MPa}, \quad \sigma_s = -1 \text{ MPa} \quad \theta = 60°.$$

Check the safety based on different failure theories.
We know

$$\begin{Bmatrix} \sigma_1 \\ \sigma_2 \\ \sigma_6 \end{Bmatrix} = \begin{bmatrix} m^2 & n^2 & 2mn \\ n^2 & m^2 & -2mn \\ -mn & mn & m^2 - n^2 \end{bmatrix} \begin{Bmatrix} \sigma_x \\ \sigma_y \\ \sigma_s \end{Bmatrix}$$

$$= \begin{bmatrix} \dfrac{1}{4} & \dfrac{3}{4} & \dfrac{\sqrt{3}}{2} \\ \dfrac{3}{4} & \dfrac{1}{4} & -\dfrac{\sqrt{3}}{2} \\ -\dfrac{\sqrt{3}}{4} & \dfrac{\sqrt{3}}{4} & -\dfrac{1}{2} \end{bmatrix} \begin{Bmatrix} -2 \\ 4 \\ -1 \end{Bmatrix} = \begin{Bmatrix} 1.634 \\ 0.366 \\ 3.098 \end{Bmatrix} \text{ MPa}$$

Maximum stress theory

Comparing the stresses with allowable values, the lamina is safe for all cases.

Maximum strain theory

Strains in the principal material directions are

$$\varepsilon_1 = \dfrac{\sigma_1}{E_1} - \dfrac{\nu_{12}}{E_2}\sigma_2$$

$$= \dfrac{1.634 \times 10^6}{35 \times 10^9} - \dfrac{0.06 \times 0.366 \times 10^6}{7 \times 10^9}$$

$$= 0.4355 \times 10^{-4} < 0.036$$

$$\varepsilon_2 = \dfrac{\sigma_2}{E_2} - \dfrac{\nu_{12}}{E_1}\sigma_1$$

$$= \dfrac{0.366 \times 10^6}{7 \times 10^9} - \dfrac{0.3 \times 1.634 \times 10^6}{35 \times 10^9}$$

$$= 3.828 \times 10^{-5} < 0.007$$

$$\varepsilon_6 = \dfrac{3.098 \times 10^6}{3 \times 10^9}$$

$$= 1.633 \times 10^{-3} < 0.023$$

Tsai-Hill theory

$$\left(\dfrac{1.634}{1280}\right)^2 + \left(\dfrac{0.366}{49}\right)^2 + \left(\dfrac{3.098}{69}\right)^2 - \dfrac{(1.634)(0.366)}{1280 \times 49}$$

$$= 1.083 \times 10^{-3} < 1$$

Hence, based on the above three failure criteria, the unidirectional lamina is safe.

Example 10.4

A carbon/epoxy unidirectional lamina is subjected to a stress system of Fig. 10.11. The ply properties are

$E_1 = 150$ kN/mm^2 $\quad E_2 = 12$ kN/mm^2 $\quad E_6 = 6$ kN/mm^2

$\nu_{12} = 0.3$ $\quad F_{1T} = 1550$ N/mm^2 $\quad F_{1C} = 1150$ N/mm^2

272 *Mechanics of Composite Materials and Structures*

$$F_{2T} = 60 \text{ N/mm}^2 \quad F_{2C} = 240 \text{ N/mm}^2 \quad F_6 = 75 \text{ N/mm}^2$$

Based on different failure theories check the safety

$$\nu_{21} = 0.3 \times \frac{12}{150} = 0.024$$

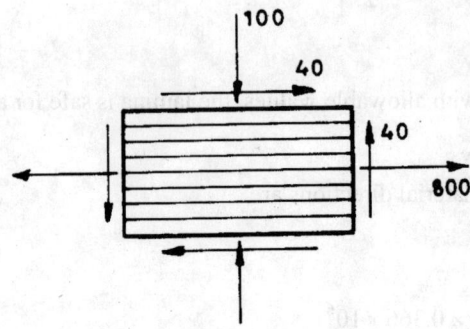

Fig. 10.11 Example 10.4

The reduced stiffness and compliances matrices of the ply are

$$[Q]_{12} = \begin{bmatrix} 151.08 & 3.63 & 0 \\ 3.63 & 12.09 & 0 \\ 0 & 0 & 6 \end{bmatrix} \text{ kN/mm}^2$$

$$[S]_{12} = \begin{bmatrix} 6.67 & -2 & 0 \\ -2 & 83.31 & 0 \\ 0 & 0 & 166.7 \end{bmatrix} \times 10^{-3} \quad 1/\left(\text{kN/mm}^2\right)$$

Stresses in the principal material directions are

$$\sigma_1 = 800 \text{ N/mm}^2, \quad \sigma_2 = -100 \text{ N/mm}^2, \quad \sigma_6 = 40 \text{ N/mm}^2$$

Strains in the principal material axis are given by

$$\begin{Bmatrix} \varepsilon_1 \\ \varepsilon_2 \\ \varepsilon_6 \end{Bmatrix} = \begin{bmatrix} 6.67 & -2 & 0 \\ -2 & 83.31 & 0 \\ 0 & 0 & 166.7 \end{bmatrix} \begin{Bmatrix} 800 \\ -100 \\ 40 \end{Bmatrix} \times 10^{-6}$$

$$= \begin{Bmatrix} 5536 \\ -9931 \\ 6668 \end{Bmatrix} \times 10^{-6}$$

Maximum stress theory

$$\sigma_1 (= 800) < 1550$$
$$\sigma_2 (= -100) < 240$$
$$\sigma_6 (= 40) < 70$$

Hence the material is safe as per maximum stress theory.

Maximum strain theory

$$\varepsilon_1 \left(= 5536 \times 10^{-6}\right) < \frac{1550}{150 \times 10^3} < 0.0103$$

$$\varepsilon_2 \left(= -9931 \times 10^{-6}\right) < \frac{240}{12 \times 10^3} < 0.02$$

$$\varepsilon_6 \left(= 6668 \times 10^{-6}\right) < \frac{75}{6 \times 10^3} < 0.0125$$

Therefore, the material is safe as per maximum strain theory.

Tsai-Hill theory

$$\left(\frac{\sigma_1}{F_1}\right)^2 + \left(\frac{\sigma_2}{F_2}\right)^2 + \left(\frac{\sigma_6}{F_6}\right)^2 - \frac{\sigma_1 \sigma_2}{F_1 F_2}$$

$$= \left(\frac{800}{1550}\right)^2 + \left(\frac{-100}{240}\right)^2 + \left(\frac{40}{75}\right)^2 - \left(\frac{800}{1550}\right)\left(\frac{-100}{240}\right)$$

$$= 0.939 < 1$$

Therefore, safe.

Tsai-Wu theory

$$f_1 = \frac{1}{F_{1T}} - \frac{1}{F_{1C}} = \frac{1}{1550} - \frac{1}{1150} = -2.244 \times 10^{-4}$$

$$f_2 = \frac{1}{F_{2T}} - \frac{1}{F_{2C}} = \frac{1}{60} - \frac{1}{240} = 0.0125$$

$$f_{11} = \frac{1}{F_{1T} F_{1C}} = \frac{1}{1550 \times 1150} = 5.61 \times 10^{-7}$$

$$f_{22} = \frac{1}{F_{2T} F_{2C}} = \frac{1}{60 \times 240} = 6.94 \times 10^{-5}$$

$$f_{66} = \frac{1}{F_6^2} = \frac{1}{75^2} = 1.78 \times 10^{-4}$$

$$f_{12} = -\frac{1}{2}\left(f_{11} f_{22}\right)^{1/2} = -\frac{1}{2}\left(5.61 \times 10^{-7} \times 6.94 \times 10^{-5}\right)^{1/2}$$

$$= -3.12 \times 10^{-6}$$

Substituting the above values in (10.50), yields

$$-2.244 \times 10^{-4}(800) + 0.0125(-100) + 5.61 \times 10^{-7}(800)^2$$
$$+ 6.95 \times 10^{-5}(-100)^2 + 1.78 \times 10^{-4}(40)^2 + 2(-3.12 \times 10^{-6})(800)(-100)$$
$$= 0.41 < 1$$

Hence safe.

Example 10.5

The carbon/epoxy unidirectional ply of Example 10.4 is oriented at 45° to the reference xy- axis. The stress system in the reference axis is shown in Fig. 10.12. Determine if the ply failure has occurred using Tsai-Hill failure criteria.

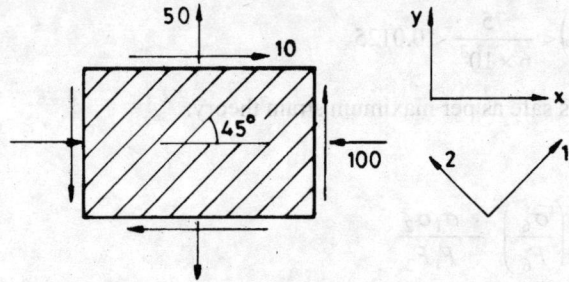

Fig. 10.12 Example 10.5

Stresses in the principal material axis direction is given by

$$\begin{Bmatrix} \sigma_1 \\ \sigma_2 \\ \sigma_6 \end{Bmatrix} = \begin{bmatrix} 0.5 & 0.5 & 1 \\ 0.5 & 0.5 & -1 \\ -0.5 & 0.5 & 0 \end{bmatrix} \begin{Bmatrix} -100 \\ 50 \\ 10 \end{Bmatrix}$$

or, $\begin{Bmatrix} \sigma_1 \\ \sigma_2 \\ \sigma_6 \end{Bmatrix} = \begin{Bmatrix} -15 \\ -35 \\ 75 \end{Bmatrix}$ N/mm^2

According to the Tsai-Hill failure criteria,

$$\left(\frac{-15}{1150}\right)^2 + \left(\frac{-35}{240}\right)^2 + \left(\frac{75}{75}\right)^2 - \left(\frac{-15}{1150}\right)\left(\frac{-35}{240}\right)$$
$$= 1.02 > 1$$

Therefore, the lamina will fail

Example 10.6

A cylindrical pressure vessel is subjected to an internal pressure p. (Fig. 10.13) The mean diameter of the cylinder is $d = 1.5$ m and wall thickness = 25 mm. The vessel is filament wound, the filament winding angle is 49.46° with the longitudinal axis of the pressure vessel. The glass/epoxy material has the following properties.

E_1 = 38 GPa E_2 = 8 GPa E_6 = 4 GPa   and ν_{12} = 0.2

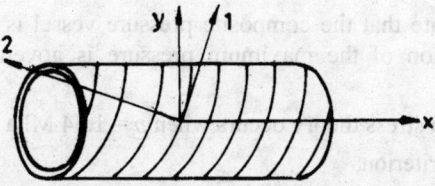

Fig. 10.13 Example 10.6

For this material, permissible stresses are as follows.

F_{1T} = 1100 MPa F_{1C} = 600 MPa F_{2T} = 30 MPa F_{2C} = 145 MPa

F_6 = 85 MPa

Determine the internal pressure that would cause the failure of the vessel according to the (a) maximum stress theory and (b) Tsai-Hill failure theory.

In a thin-walled cylindrical pressure vessel, stresses are given by

$$\sigma_x = \frac{pr}{2t} = \frac{0.5p \times 1.5}{2(0.025)} = 15p, \; \sigma_s = 0$$

$$\sigma_y = \frac{pr}{t} = \frac{p \times 1.5}{0.025 \times 2} = 30p$$

$\cos 49.46° = 0.65, \quad \sin 49.46° = 0.76$

$\sigma_s = 0.0$

Stress among the principal material axes are

$\sigma_1 = \sigma_x \cos^2\theta + \sigma_y \sin^2\theta + 2\sigma_s \sin\theta \cos\theta$

$ = (15p)(0.65)^2 + 30p(0.76)^2 + 2 \times 0 \times (0.65)(0.76)$

$ = 23.67p \quad$ MPa

$\sigma_2 = \sigma_x \sin^2\theta + \sigma_y \cos^2\theta - 2\sigma_s \sin\theta \cos\theta$

$ = (15p)(0.76)^2 + (30p)(0.65)^2 - 0$

$ = 21.34p$ MPa

$\sigma_6 = -\sigma_x \sin\theta \cos\theta + \sigma_y \sin\theta \cos\theta + \sigma_s(\cos^2\theta - \sin^2\theta)$

$ = -(15p)(0.65)(0.76) + (30p)(0.65)(0.76)$

$ = 7.41p \quad$ MPa

Let the failure theories be applied now.

(a) According to the maximum stress theory, the value of p will be the least of the following.

$$\sigma_1 = 23.67p = F_{1T} = 1100, \quad \text{therefore}, \quad p = 46.47 \text{ MPa}$$
$$\sigma_2 = 21.34p = F_{2T} = 30, \quad \text{therefore}, \quad p = 1.41 \text{ MPa}$$
$$\sigma_6 = 7.41p = F_6 = 85, \quad \text{therefore}, \quad p = 11.47 \text{ MPa}$$

The above results indicate that the composite pressure vessel is the weakest in the transverse direction and the determination of the maximum pressure is governed by the transverse tensile strength.

Failure according to maximum stress theory occurs when $p = 1.54$ MPa

(b) According to Tsai-Hill criterion,

$$\left(\frac{23.67p}{1100}\right)^2 + \left(\frac{21.34p}{30}\right)^2 + \left(\frac{7.41p}{85}\right)^2 - \left(\frac{23.67p}{1100}\right)\left(\frac{21.34p}{30}\right) = 1$$

or, $p = 1.416$ MPa

It may be interesting to note that the value of p obtained by the two failure criteria are pretty close.

10.5 IMPORTANCE OF SHEAR STRESS

Shear stresses induced in FRP materials have some special features. The sign of shear stress in homogeneous isotropic material plays no role in the determination of the strength of the material. The same argument is not valid for the case of orthotropic lamina and composites.

In Fig. 10.14 are shown two cases of unidirectional lamina subjected to pure shear condition. Shear stresses act on the opposite direction in each case. In both cases, the stresses are transformed in the direction of principal material axes. In the case of the positive shear, fibres are subjected to compression and tensile stresses are developed in the transverse direction. The reverse is the case for negative shear.

The transverse strength of the composite governs the strength of the lamina. As such the stress system in the positive shear will yield lower apparent strength than the negative shear. Thus off-axis loading of the lamina needs careful investigation.

10.6 EXAMPLE ON SHEAR STRENGTH

Example 10.7

A unidirectional lamina has fibres inclined at 45° to the x-axis. The properties of the lamina are as follows (in GPa)

$$E_1 = 145, \quad E_2 = 12, \quad E_6 = 6, \quad v_{12} = 0.3$$

and in MPa

$$F_{1T} = 1500, \quad F_{1C} = 1200, \quad F_{2T} = 50, \quad F_{2C} = 250, \quad F_6 = 70$$

Determine the off-axis positive and negative shear strengths using the Tsai-Hill failure criterion.

$$m = n = \cos 45° = \sin 45° = \frac{1}{\sqrt{2}}; \quad m^2 = n^2 = \frac{1}{2}; \quad 2mn = 1$$

σ_s is the applied shear stress and $\sigma_x = \sigma_y = 0$

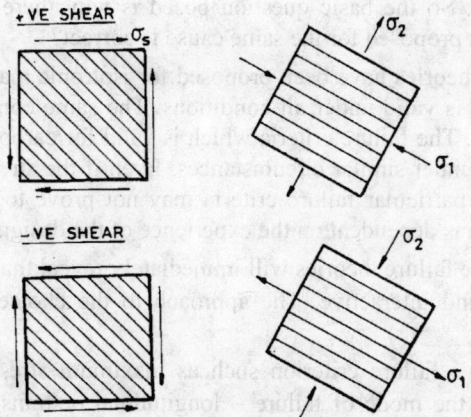

Fig. 10.14 Effect of sign of shear stress on composites

For positive shear
$$\sigma_1 = 2mn \quad \sigma_s = \sigma_s$$
$$\sigma_2 = -2mn \quad \sigma_s = -\sigma_s$$
$$\sigma_6 = 0$$

Tsai-Hill criterion yields

$$\left(\frac{\sigma_1}{F_1}\right)^2 + \left(\frac{\sigma_2}{F_2}\right)^2 + \left(\frac{\sigma_6}{F_6}\right)^2 - \left(\frac{\sigma_1}{F_1}\right)\left(\frac{\sigma_2}{F_2}\right) = 1$$

or, $\left(\dfrac{\sigma_s}{1500}\right)^2 + \left(\dfrac{-\sigma_s}{250}\right)^2 + 0 - \left(\dfrac{\sigma_s}{1500}\right)\left(\dfrac{-\sigma_s}{250}\right) = 1$

or, $\sigma_s = 228.75$ N/mm^2

For negative shear
$$\sigma_1 = -\sigma_s$$
$$\sigma_2 = \sigma_s$$
$$\sigma_6 = 0$$

Tsai-Hill criterion yields

$\left(\dfrac{-\sigma_s}{1200}\right)^2 + \left(\dfrac{\sigma_s}{50}\right)^2 + 0 - \left(\dfrac{-\sigma_s}{1200}\right)\left(\dfrac{\sigma_s}{50}\right) = 1$

or, $\sigma_s = 48.95$ N/mm^2

The example clearly demonstrates that the strengths in positive and negative shear for a unidirectional lamina are significantly different.

10.7 CHOICE OF FAILURE CRITERIA

In the preceding section, four failure criteria are discussed – the maximum stress theory, the maximum strain theory, the Tsai-Hill failure criterion and the Tsai-Wu failure criterion. But a few more failure criteria exists. So the basic question posed is why there are so many failure criteria. Which of the many theories proposed for the same cause is correct?

So far, many failure theories have been proposed for isotropic materials. It has been found that no particular failure theory is valid under all conditions. The same conclusion is arrived at for fibre reinforced plastic materials. The failure criteria which is valid for carbon/epoxy ply may not be valid for glass/epoxy composite under similar circumstances. Even if the stress conditions are changed for the same FRP lamina, the particular failure criteria may not prove to be satisfactory. As such, the choice of the failure criteria is dependent on the experience of the designer.

A careful look into the failure theories will immediately reveal that they can be divided into two groups – non-interactive and interactive. The approach to the choice of any failure criteria is as follows.

First, a non-interactive failure criterion such as maximum stress theory or maximum strain theory is applied to check the mode of failure – longitudinal or transverse, tensile or compressive mode or shear mode. The lamina may indicate failure using a non-interactive theory. If not, the lamina should be checked using an interactive failure criterion. It may be that independent stresses may not initiate a failure, but their interaction may do.

10.8 TYPICAL STRENGTH PROPERTIES

Table 10.1 indicates the strength of some typical unidirection lamina. For the unidirectional composites given in the table, the fibre-volume fraction has been considered as 60 percent and the values indicated are at ambient temperature.

TABLE 10.1

Typical strength properties of unidirectional composites (MPa)

	F_{1T}	F_{1C}	F_{2T}	F_{2C}	F_6
High strength carbon /epoxy	1500	1200	50	250	70
High modulus carbon/epoxy	1000	850	40	200	60
E-glass/epoxy	1000	600	30	150	40
Kevlar/epoxy	1300	280	30	140	60

F_{1T} = Longitudinal tensile strength $\quad F_{1C}$ = Longitudinal compressive strength
F_{2T} = Transverse tensile strength $\quad F_{2C}$ = Transverse compressive strength
F_6 = Inplane shear strength

REFERENCES AND SUGGESTED READINGS

10.1 J.R. Vinson and R.L. Sierakowski, *The Behaviour of Structures Composed of Composite Materials*, Martinus Nijhoff Publishers, Dordrecht, The Netherlands, 1987.

10.2 B.W. Rosen, Tensile failure of fibrous composites, *AIAA, J.*, April, 1965.

10.3 N.F. Dow and B.W. Rosen, Evaluation of filament-reinforced composites for aerospace structural applications, *NASA CR*-207 April, 1965.

10.4 B.W. Rosen, Mechanics of compressive strengthening, Chapter 3, in *Fibre Composite Materials*, ASM, Metals Park, Ohio, USA, 1965.

10.5 I.M. Daniel, *Photoelastic Investigation of Composites in Mechanics of Composite Materials*, Vol. 2, G.P. Sendeckyj (Editor), Academic Press, 1974, pp 433–489.

10.6 D.F. Adams and D.R. Doner, Longitudinal shear loading of unidirectional composite, *J. Composite Materials*, V.1, No.1, 1967, pp. 4–17.

10.7 G.C. Sih and A.M. Skudra, *Failure Mechanics of Composites*, Elsevier, London, 1985.

10.8 R. Hill, A theory of yielding and plastic flow of anisotropic materials, *Proceedings of Royal Society*, Series A, Vol. 193, 1948.

10.9 V.D. Azzi and S.W. Tsai, Anisotropic strength of composites, Experimental Mechanics, V.5, 1965, pp. 286-288.

10.10 S.W. Tsai and E.M.Wu, A general theory of strength of anisotropic materials, *J. Composite Materials*, V.5, 1971, pp. 58–80.

10.11 O. Hoffman, The brittle strength of orthotropic materials, *J. Composite Materials*, V.1, 1967, pp. 200–206.

10.12 I. M. Daniel and O. Ishai, *Engineering Mechanics of Composite Materials*, Oxford University Press, 1994

EXERCISE 10

10.1 Determine E_1 and F_{1T} for the composite with the following properties.

(a) $V_f = 0.70$, $E_f = 70$ GPa, $E_m = 4$ GPa,

$F_{1fT} = 3500$ MPa, $F_{mT} = 100$ MPa

(b) $V_f = 0.70$, $E_f = 240$ GPa, $E_m = 4$ GPa,

$F_{1fT} = 3500$ MPa, $F_{mT} = 100$ MPa

10.2 Stress concentration factor is 2 for a E-glass/epoxy composite lamina which is subjected to transverse tension. Determine the transverse tensile strength given the following properties.

$V_f = 0.70$, $E_f = 70$ GPa, $E_m = 4$ GPa,

$\nu_m = 0.35$, $F_{mT} = 100$ MPa

10.3 Find the Tsai-Hill failure criteria for pure shear at various angles θ to the principal material directions.

10.4 A unidirectional lamina is subjected to biaxial stresses: $\sigma_x = 3\sigma_0$, $\sigma_y = 2\sigma_0$ - x-axis is inclined at 60° to the fibre direction. The material properties are as follows.

$F_{1T} = F_{1C} = 2F_{2C} = 4F_6 = 8F_{2T} = 500$ MPa.

Determine σ_0 at failure of the lamina according to

(a) Maximum stress theory

(b) Deviatoric strain energy theory.

10.5 The loading axis is inclined at an angle θ to the fibre axis. Tension along the y-axis is the only load applied. Using maximum strain criteria, determine F_{mT} and θ so that prediction of inplane shear and tensile failure load coincide.

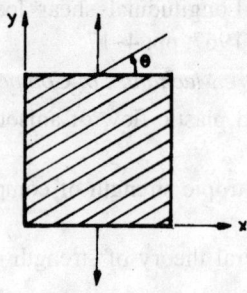

Prob. 10.5

10.6 Using the Tsai-Hill failure criterion for a unidirectional lamina subjected to pure shear, find an expression for the shear stress at failure in terms of F_1, F_2 and F_6. The loading axis is inclined at θ to the principal material axis.

10.7 A unidirectional lamina is subjected to σ_x and σ_s. The tensile load and the fibre are inclined at 45° degrees. Magnitude of both the stresses are equal to F_0. Determine in terms of F_1, F_2 and F_0 by using the Tsai-Hill failure criterion.

10.8 A lamina is subjected to pure shear loading which is inclined at 45° to the pure shear direction. Express shear at failure in terms of Tsai-Hill coefficients.

10.9 A lamina is loaded as shown. Using Tsai-Wu failure criterion, determine F_0.

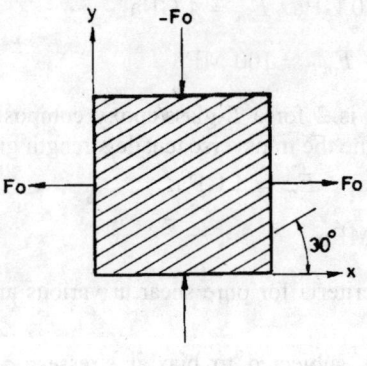

Prob. 10.9

10.10 For the lamina of Prob. 10.9, determine F_0 using maximum stress failure criterion and Tsai-Hill failure criteria for the material with the following properties

$F_{1T} = 2500$ MPa, $F_{2T} = 60$ MPa, $F_{1C} = 1500$ MPa,
$F_{2C} = 250$ MPa, $F_6 = 70$ MPa

CHAPTER 11

ANALYSIS OF LAMINATE STRENGTH

11.1 INTRODUCTION

We move on to the analysis of laminate strength. The failure of a lamina in a laminate does not signify failure, as the failure of the laminate is progressive. Each lamina in a laminate is assumed to be in a state of plane stress. Stresses in each lamina are evaluated separately by transforming the stresses in the principal material axis direction and applying a suitable or selected failure criterion. The analysis of failure of a laminate is therefore more complicated than a single lamina. Various parameters such as orientation of fibres in a lamina, stacking sequence, fabrication process and others dictate the strength of the laminate.

11.2 POSSIBLE MODES OF FAILURE

A laminate may fail in one of the following types of failure modes: (1) first-ply failure (FPF), (2) ultimate laminate failure (ULF) and (3) inter-laminar failure. FPF is defined as the failure of the laminate with the failure of the first layer. ULF is defined as the failure of the laminate when ultimate load capacity is reached following failure of all the plies. Inter-laminar failure is defined as the failure resulting from the separation of adjacent layers though the individual lamina remains intact.

The first-ply failure is similar to treating the laminate as having reached the 'yield stress' and the ultimate laminate failure is analogous to the ultimate stress criterion. In FPF, higher safety factors may be considered in the analysis, as the whole approach is rather conservative. In ULF analysis, however, more precision is needed about loading and stress distribution. In primary structures, the working loads are generally kept below that producing FPF.

11.3 STRESS ANALYSIS AT FIRST-PLY FAILURE

In order to carry out laminate stress analysis, the first step is to evaluate laminate stiffnesses. All the applied loads such as N_x, N_y, N_s, M_x, M_y and M_s should be known. Based on the known geometry and loading conditions, the stresses and strains in the individual lamina can be determined as follows

(1) Obtain all laminate stiffnesses: A_{ij}, B_{ij}, D_{ij}.

(2) Determine the laminate compliance a_{ij}, b_{ij}, d_{ij}.

(3) The reference plane strains $\varepsilon_x^0, \varepsilon_y^0$ and ε_s^0 and curvatures κ_x, κ_y and κ_s of the laminate are to be calculated from (11.1).

$$\begin{Bmatrix} \varepsilon_x^0 \\ \varepsilon_y^0 \\ \varepsilon_s^0 \\ \hline \kappa_x \\ \kappa_y \\ \kappa_s \end{Bmatrix} = \begin{bmatrix} a_{xx} & a_{xy} & a_{xs} & b_{xx} & b_{xy} & b_{xs} \\ a_{yx} & a_{yy} & a_{ys} & b_{yx} & b_{yy} & b_{ys} \\ a_{sx} & a_{sy} & a_{ss} & b_{sx} & b_{sy} & b_{ss} \\ \hline b_{xx} & b_{xy} & b_{xs} & d_{xx} & d_{xy} & d_{xs} \\ b_{yx} & b_{yy} & b_{ys} & d_{yx} & d_{yy} & d_{ys} \\ b_{sx} & b_{sy} & b_{ss} & d_{sx} & d_{sy} & d_{ss} \end{bmatrix} \begin{Bmatrix} N_x \\ N_y \\ N_s \\ \hline M_x \\ M_y \\ M_s \end{Bmatrix} \tag{11.1}$$

When the laminate is symmetric, the coupling matrix is zero. The inplane and bending compliance can be evaluated separately.

(4) Strains at the centroid of the kth lamina situated at a distance $\bar{z}_k$ from the reference plane are obtained as

$$\begin{Bmatrix} \varepsilon_x \\ \varepsilon_y \\ \varepsilon_s \end{Bmatrix}_k = \begin{Bmatrix} \varepsilon_x^0 - \bar{z}_k \cdot \kappa_x \\ \varepsilon_y^0 - \bar{z}_k \cdot \kappa_y \\ \varepsilon_s^0 - \bar{z}_k \cdot \kappa_s \end{Bmatrix}_k \tag{11.2}$$

(5) From the strains in the xy- axis system, strains in the material axis system are obtained for the kth lamina

$$\begin{Bmatrix} \varepsilon_1 \\ \varepsilon_2 \\ \dfrac{\varepsilon_6}{2} \end{Bmatrix}_k = \begin{bmatrix} m^2 & n^2 & 2mn \\ n^2 & m^2 & -2mn \\ -mn & mn & (m^2 - n^2) \end{bmatrix}_k \begin{Bmatrix} \varepsilon_x \\ \varepsilon_y \\ \dfrac{\varepsilon_s}{2} \end{Bmatrix}_k \tag{11.3}$$

(6) The corresponding stresses in the kth lamina are then determined as follows

$$\begin{Bmatrix} \sigma_1 \\ \sigma_2 \\ \sigma_6 \end{Bmatrix}_k = \begin{bmatrix} Q_{11} & Q_{12} & 0 \\ Q_{21} & Q_{22} & 0 \\ 0 & 0 & Q_{66} \end{bmatrix}_k \begin{Bmatrix} \varepsilon_1 \\ \varepsilon_2 \\ \varepsilon_6 \end{Bmatrix}_k \tag{11.4}$$

(7) A failure criterion is chosen and using it, ply strength analysis is carried out. From the calculation it will be obvious if any ply of the laminate has failed.

In the FPF approach, each layer of the laminate is checked with the selected failure criteria. We multiply the state of stress in each layer by a safety factor S_k.

Applying Tsai–Hill criterion for the failure state of stress for layer k yields

$$\frac{S_k^2 \sigma_{1k}^2}{F_1^2} + \frac{S_k^2 \sigma_{2k}^2}{F_2^2} + \frac{S_k^2 \sigma_{6k}^2}{F_6^2} - \frac{S_k^2 \sigma_{1k} \sigma_{2k}}{F_1 F_2} = 1 \tag{11.5}$$

or, $\quad aS_k^2 = 1$ \hfill (11.6)

where

$$a = \frac{\sigma_{1k}^2}{F_1^2} + \frac{\sigma_{2k}^2}{F_2^2} + \frac{\sigma_{6k}^2}{F_6^2} - \frac{\sigma_{1k} \sigma_{2k}}{F_1 F_2} \tag{11.7}$$

Solution of (11.6) is

$$S_k = \pm \frac{1}{\sqrt{a}} \qquad (11.8)$$

The positive sign refers to the actual state of stress.

The procedure is repeated for each ply of the laminate. The minimum value thus obtained from the different plies is chosen.

11.4 ULTIMATE LAMINATE FAILURE OR ANALYSIS OF LAST–PLY FAILURE

In many cases, the first–ply failure does not mean that the ultimate capacity of the laminate has been reached. When the first ply fails, other plies may remain intact. But with the failure of the first ply, a redistribution of stresses takes place in the remaining plies. The failure mode of the first ply may be any of the three types – fibre failure, matrix failure and shear failure.

The stiffness of the laminate is reduced after the first-ply failure. After FPF, the load may be increased till the second ply fails. A further reduction of stiffness results and a further redistribution of stresses in the intact ply take place. With further increase of load, a third-ply failure results and the process is repeated till the last ply fails. The problem now is how to simulate the characteristics of the laminate so as to predict the progressive failure in such a way that it be close to the real situation.

The basic question arises regarding the reduction of stiffness of the failed ply and its estimated value. A number of investigators have studied the problem and come up with different suggestions [11.1 - 11.12]. The position of the failed ply is assumed to remain in the same position. Some or all of the elastic constants E_1, E_2 or E_6 are made zero and accordingly two methods exist for ultimate laminate failure analysis. They are total-ply failure method and the partial-ply failure method.

11.4.1 Total-Ply Failure Method

In this approach, the stiffness of this individual ply after failure is assumed to be zero. Thus all the elastic constants E_1, E_2 and E_6 of the failed ply is assumed to have zero values. This suggests that if ply has shear failure or matrix failure with the fibre unaffected, it is presumed that the ply is unable to carry any load in the fibre direction. The geometry of the laminate however remains unchanged with the failed ply remaining incapacitated.

The stiffness of the laminate will now have reduced values and for inplane loading, the load–strain curve will indicate a kink at FPF. With the increase in load, the curve will continue, but the shape will be much flatter.

It is to be checked to see whether the load at which the first ply fails causes the second ply also to fail. If not, the load is to be increased.

If with the increase of load, a second ply fails, then the stiffness of the second ply is also reduced to zero, by assuming the elastic constants of the ply to be zero. Calculation is to be made to check if any other ply fails in the laminate at this load. If not, the load is increased and the steps given above are to be repeated till LPF is obtained.

11.4.2 Partial-Ply Failure Method

The assumptions made in this approach are as follows. If the ply fails due to the failure of the fibre, then the ply is assumed to become totally ineffective and is incapable of carrying any further load by any mechanism. Then, for this case all the moduli E_1, E_2 and E_6 are zero. However, if the matrix fails or a shear failure occurs, it is then assumed that the load carrying capacity of the ply in the fibre direction is not affected. This suggests that for this case, E_1 remains unchanged, though E_2 and E_6 are set to zero. The procedure of calculation is similar to that mentioned earlier.

11.5 EXAMPLES

Example 11.1

A carbon/epoxy cross-ply laminate (0/90)s consists of unidirectional plies and is subjected to a tensile force $N_x = 200$ N/mm. The ply is 0.1 mm thick and elastic properties in kN/mm^2 are

$$E_1 = 140, \quad E_2 = 10, \quad E_6 = 5, \quad v_{12} = 0.3$$

The ply strengths for the material in N/mm^2 are

$$F_{1T} = 1500, \quad F_{1C} = 1200, \quad F_{2T} = 50, \quad F_{2C} = 250, \quad F_6 = 70$$

Check whether failure will occur according to maximum stress theory.

Solution

The reduced stiffness matrix Q_{ij} of 0° ply, membrane stiffness matrix A_{ij} of the laminate and the compliance matrix a_{ij} of the laminate are as follows

$$[Q]^0_{ij} = \begin{bmatrix} 140.9 & 3 & 0 \\ 3 & 10.1 & 0 \\ 0 & 0 & 5 \end{bmatrix} \text{ kN/mm}^2$$

$$[A]_{x,y} = \begin{bmatrix} 30.2 & 1.2 & 0 \\ 1.2 & 30.2 & 0 \\ 0 & 0 & 2 \end{bmatrix} \text{ kN/mm}$$

$$[a]_{x,y} = \begin{bmatrix} 0.0332 & -0.0013 & 0 \\ -0.0013 & 0.0332 & 0 \\ 0 & 0 & 0.5 \end{bmatrix} \text{ 1/kN/mm}$$

Membrane strains are given by

$$\begin{Bmatrix} \varepsilon^0_x \\ \varepsilon^0_y \\ \varepsilon^0_s \end{Bmatrix} = \begin{bmatrix} 0.0332 & -0.0013 & 0 \\ -0.0013 & 0.0332 & 0 \\ 0 & 0 & 0.5 \end{bmatrix} \begin{Bmatrix} 200 \\ 0 \\ 0 \end{Bmatrix} = \begin{Bmatrix} 3320 \times 10^{-6} \\ -130 \times 10^{-6} \\ 0 \end{Bmatrix}$$

The strain distribution is constant through the laminate thickness.

Ply 1 at 0°

For Ply 1, the loading axis coincides with the principal material axis, strains in the principal material directions are

$$\begin{Bmatrix} \varepsilon_1 \\ \varepsilon_2 \\ \varepsilon_6 \end{Bmatrix} = \begin{Bmatrix} 3320 \times 10^{-6} \\ -130 \times 10^{-6} \\ 0 \end{Bmatrix}$$

The corresponding stresses are

$$\begin{Bmatrix} \sigma_1 \\ \sigma_2 \\ \sigma_6 \end{Bmatrix} = \begin{bmatrix} 140.9 & 3 & 0 \\ 3 & 10.1 & 0 \\ 0 & 0 & 5 \end{bmatrix} \begin{Bmatrix} 3320 \times 10^{-6} \\ -130 \times 10^{-6} \\ 0 \end{Bmatrix} \times 10^3 \quad \text{N/mm}^2$$

$$= \begin{Bmatrix} 467 \\ 8.6 \\ 0 \end{Bmatrix} \quad \text{N/mm}^2$$

Therefore,

$$\sigma_1 = 467 < F_{1T} < 1500$$
$$\sigma_2 = 8.6 < F_{2T} < 50$$

In order to cause failure, σ_1 is to be increased by (1500/467) or 3.21

In order to cause failure, σ_2 is to be increased by (50/8.6) or 5.81

Ply 2 at 90°

Strains in the principal material directions are given by

$$\begin{Bmatrix} \varepsilon_1 \\ \varepsilon_2 \\ \varepsilon_6 \end{Bmatrix} = \begin{bmatrix} 0 & 1 & 0 \\ 1 & 0 & 0 \\ 0 & 0 & -1 \end{bmatrix} \begin{Bmatrix} 3320 \\ -130 \\ 0 \end{Bmatrix} \times 10^{-6} = \begin{Bmatrix} -130 \times 10^{-6} \\ 3320 \times 10^{-6} \\ 0 \end{Bmatrix}$$

The corresponding stresses are

$$\begin{Bmatrix} \sigma_1 \\ \sigma_2 \\ \sigma_6 \end{Bmatrix} = \begin{bmatrix} 140.9 & 3 & 0 \\ 3 & 10.1 & 0 \\ 0 & 0 & 5 \end{bmatrix} \begin{Bmatrix} -130 \times 10^{-6} \\ 3320 \times 10^{-6} \\ 0 \end{Bmatrix} \times 10^3 \quad \text{N/mm}^2$$

$$= \begin{Bmatrix} -8.6 \\ 33.14 \\ 0 \end{Bmatrix} \quad \text{N/mm}^2$$

In order to cause failure, σ_1 for this ply is to be increased by (1200/8.4) or 142.8.

In order to cause failure, σ_2 for this ply is to be increased by (50/33.14) or 1.51.

The minimum of all the above factors by which the load is to be increased is 1.51. Therefore, the load is to be increased by a factor of 1.51 before FPF is predicted by the maximum stress theory. Hence, FPF load is 200 × 1.51 or 302 N/mm² and this will occur in 90° plies in the transverse tensile mode of failure. Note that the analysis predicts that both 90° plies will fail simultaneously.

Example 11.2

If instead of N_x, $M_x = 20$ Nmm/mm is the applied load given in the problem Example 11.1, check whether failure will take place according to the maximum stress theory.

Solution:

$[Q]_{xy}$ matrix is same as given in Example 11.1

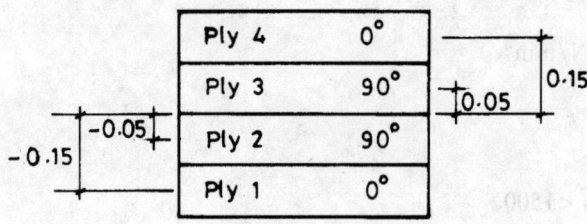

Fig. 11.1 Example 11.2

$$[D]_{xy} = 2 \begin{bmatrix} 140.9 & 3 & 0 \\ 3 & 10.1 & 0 \\ 0 & 0 & 5 \end{bmatrix} \left(0.15^2 \times 0.1 + \frac{0.1^3}{12} \right)$$

$$+ 2 \begin{bmatrix} 10.1 & 3 & 0 \\ 3 & 140.9 & 0 \\ 0 & 0 & 5 \end{bmatrix} \left(0.1 \times .05^2 + \frac{0.1^3}{12} \right)$$

$$= \begin{bmatrix} 0.664 & 0.016 & 0 \\ 0.015 & 0.141 & 0 \\ 0 & 0 & 0.0267 \end{bmatrix} \times 10^3 \text{ N/mm}$$

The compliance matrix $[d]$ is given by

$$[d]_{x,y} = \begin{bmatrix} 1.51 & -0.171 & 0 \\ -0.171 & 7.111 & 0 \\ 0 & 0 & 37.453 \end{bmatrix} \times 10^{-3} \text{ 1/N/mm}$$

Due to the symmetry of the laminate, the inplane and bending compliance have no coupling effect. Only a moment is acting on the laminate. As there is no inplane force externally applied, the axial strains in the mid-plane are zero. The curvatures can be determined as follows

$$\begin{Bmatrix} \kappa_x \\ \kappa_y \\ \kappa_s \end{Bmatrix} = \begin{bmatrix} 1.51 & -0.171 & 0 \\ -0.171 & 7.111 & 0 \\ 0 & 0 & 37.553 \end{bmatrix} \times 10^{-3} \begin{Bmatrix} 20 \\ 0 \\ 0 \end{Bmatrix}$$

which yields

$$\kappa_x = 30200 \times 10^{-6} \quad 1/\text{mm}$$
$$\kappa_y = -3420 \times 10^{-6} \quad 1/\text{mm}$$
$$\kappa_s = 0$$

Ply 1 at $0°$

Referring to Fig. 11.1, the distance of the centroid of the ply is $z = 0.15$ mm. In this case, the loading axes and principal material axes are the same.

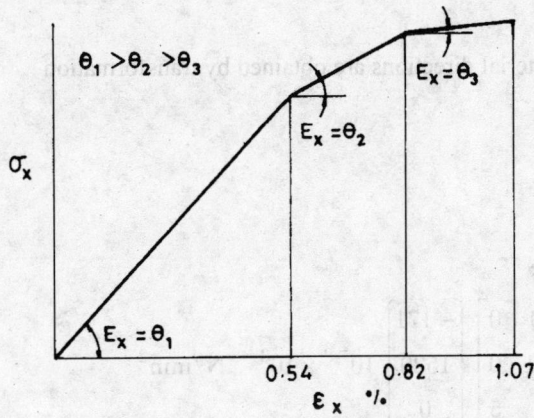

Fig. 11.2 Longitudinal tensile strength by complete ply failure method

$$\varepsilon_1 = \varepsilon_x = -(-0.15) \times 30200 \times 10^{-6} = 4530 \times 10^{-6}$$
$$\varepsilon_2 = \varepsilon_y = -(-0.15) \times (-3420) \times 10^{-6} = -513 \times 10^{-6}$$
$$\varepsilon_6 = \varepsilon_s = 0$$

Corresponding stresses in the principal material directions are

$$\begin{Bmatrix} \sigma_1 \\ \sigma_2 \\ \sigma_6 \end{Bmatrix} = \begin{bmatrix} 140.9 & 3 & 0 \\ 3 & 10.1 & 0 \\ 0 & 0 & 5 \end{bmatrix} \begin{Bmatrix} 4530 \\ -513 \\ 0 \end{Bmatrix} 10^{-6} \times 10^3 \quad \text{N/mm}^2$$

or,

$$\sigma_1 = 636 \quad \text{N/mm}^2$$
$$\sigma_2 = 8.4 \quad \text{N/mm}^2$$

Applying maximum stress criterion, we get

$$\sigma_1 = 636 < F_{1T} < 1500$$
$$\sigma_2 = 8.4 < F_{2T} < 50$$

Therefore, both σ_1 and σ_2 are to be increased to cause failure.

The factor by which σ_1 is to be increased is (1500/636) or 2.36.

The factor by which σ_2 is to be increased is (50/8.4) or 5.95.

Ply 2 at 90°

For ply 2, the distance of the centre of the lamina from the mid-plane is $z = -0.05$ mm

$$\varepsilon_x = -(-0.05) \times 30200 \times 10^{-6} = 1510 \times 10^{-6}$$
$$\varepsilon_y = -(-0.05) \times -3420 \times 10^{-6} = -171 \times 10^{-6}$$
$$\varepsilon_s = 0$$

Strains in the principal material directions are obtained by transformation

$$\varepsilon_1 = -171 \times 10^{-6}$$
$$\varepsilon_2 = 1520 \times 10^{-6}$$
$$\varepsilon_6 = 0$$

Corresponding stresses are

$$\begin{Bmatrix} \sigma_1 \\ \sigma_2 \\ \sigma_6 \end{Bmatrix} = \begin{bmatrix} 140.9 & 3.0 & 0 \\ 3.0 & 10.1 & 0 \\ 0 & 0 & 5 \end{bmatrix} \begin{Bmatrix} -171 \\ 1520 \\ 0 \end{Bmatrix} 10^{-6} \times 10^3 \quad \text{N/mm}^2$$

$$= \begin{Bmatrix} -19.53 \\ 14.84 \\ 0 \end{Bmatrix} \quad \text{N/mm}^2$$

Applying maximum stress criterion,

$$\sigma_1 = -19.53 < F_{1C} < 1200$$
$$\sigma_2 = 14.84 < F_{2T} < 50$$

For failure to occur, σ_1 should be increased by (1200/19.53) or 61.44.

For failure to occur, σ_2 should be increased by (50/14.84) or 3.37.

Ply 3 at 90°

Calculations are similar to ply 2, however, the stresses in this ply will be of opposite sign to that of ply 2. It is not further investigated as the stresses are rather small.

Ply 4 at 0°

For this ply

$$\sigma_1 = -636 \quad \text{N/mm}^2$$

$$\sigma_2 = -8.4 \text{ N/mm}^2$$
$$\sigma_6 = 0$$

Applying maximum stress theory

$$\sigma_1 = -636 < F_{1C} < 1200$$
$$\sigma_2 = -8.4 < F_{2C} < 250$$

For failure to occur, σ_1 should be increased by (1200/636) or 1.89.

For failure to occur, σ_2 should be increased by (250/8.4) or 29.76.

If we look into the different factors obtained from various plies, the minimum is 1.89. Hence, ply failure moment is 20×1.89 or 37.8 Nmm/mm. The maximum stressed ply is the top ply and the failure is in longitudinal compression.

Figure 11.2 presents a typical diagram to demonstrate the stages of the complete ply failure.

Example 11.3

A $(0/45/90)_s$ laminate is subjected to a longitudinal tensile load N_x. Determine N_x for the last-ply failure approach. Each ply has a thickness of 0.125 mm. The elastic properties of the ply are as follows

$$E_1 = 140, \quad E_2 = 10, \quad E_6 = 5 \quad (\text{in kN/mm}^2), \quad \nu_{12} = 0.3$$

$$F_{1T} = 1500, \quad F_{1C} = 1200, \quad F_{2T} = 50, \quad F_{2C} = 250, \quad F_6 = 70 \quad (\text{in N/mm}^2)$$

Solution

Reduced stiffness for different plies are

$$[Q]_{x,y}^{0} = \begin{bmatrix} 140.9 & 3 & 0 \\ 3 & 10.1 & 0 \\ 0 & 0 & 5 \end{bmatrix} \text{ kN/mm}^2$$

$$[Q]_{x,y}^{45°} = \begin{bmatrix} 44.3 & 34.3 & 32.7 \\ 34.3 & 44.3 & 32.7 \\ 32.7 & 32.7 & 36.3 \end{bmatrix} \text{ kN/mm}^2$$

$$[Q]_{x,y}^{90} = \begin{bmatrix} 10.1 & 3 & 0 \\ 3 & 140.9 & 0 \\ 0 & 0 & 5 \end{bmatrix} \text{ kN/mm}^2$$

First-ply failure:

The extensional stiffness is

$$[A] = \begin{bmatrix} 48.83 & 10.08 & 8.20 \\ 10.08 & 48.83 & 8.20 \\ 8.20 & 8.20 & 11.58 \end{bmatrix} \text{ kN/mm}$$

$$A_{xx} = 2\{(0.125 \times 140.9) + (0.125 \times 44.3) + (0.125 \times 10.1)\}$$
$$= 48.83$$

The extensional compliance is

$$[a] = \begin{bmatrix} 0.0235 & -0.00232 & -0.015 \\ -0.00232 & 0.0235 & -0.015 \\ -0.015 & -0.015 & 0.1078 \end{bmatrix} 10^{-3} \quad 1/(kN/mm)$$

Let the applied load be N_x, then the strains are given by

$$\begin{Bmatrix} \varepsilon_x^0 \\ \varepsilon_y^0 \\ \varepsilon_s^0 \end{Bmatrix} = \begin{bmatrix} 0.0235 & -0.00232 & -0.015 \\ -0.00232 & 0.0235 & -0.015 \\ -0.015 & -0.015 & 0.1078 \end{bmatrix} \begin{Bmatrix} N_x \\ 0 \\ 0 \end{Bmatrix} 10^{-3}$$

$$= \begin{Bmatrix} 23.5 \times 10^{-3} N_x \\ -2.32 \times 10^{-3} N_x \\ -15 \times 10^{-3} N_x \end{Bmatrix}$$

Ply 1 and 6 at 0°

Strains in the principal material direction of the plies are

$$\begin{Bmatrix} \varepsilon_1 \\ \varepsilon_2 \\ \varepsilon_6 \end{Bmatrix} = \begin{Bmatrix} 23.5 \times 10^{-3} N_x \\ -2.32 \times 10^{-3} N_x \\ -15 \times 10^{-3} N_x \end{Bmatrix}$$

Stresses in the principal material direction for these plies are

$$\begin{Bmatrix} \sigma_1 \\ \sigma_2 \\ \sigma_6 \end{Bmatrix} = \begin{bmatrix} 140.9 & 3.0 & 0 \\ 3.0 & 10.1 & 0 \\ 0 & 0 & 5.0 \end{bmatrix} \begin{Bmatrix} 23.5 \\ -2.32 \\ -15 \end{Bmatrix} N_x \times 10^{-3} \quad N/mm^2$$

or, $\begin{Bmatrix} \sigma_1 \\ \sigma_2 \\ \sigma_6 \end{Bmatrix} = \begin{Bmatrix} 3.304 \, N_x \\ 0.047 \, N_x \\ -0.075 \, N_x \end{Bmatrix} \quad N/mm^2$

In order to cause failure according to maximum stress theory, the possible N_x values are given below.

From consideration of σ_1,

$$N_x = \frac{1500}{3.304} = 454 \quad N/mm$$

From consideration of σ_2,

$$N_x = \frac{50}{0.047} = 1063.83 \quad \text{N/mm}$$

From consideration of σ_6,

$$N_x = \frac{70}{0.075} = 933.3 \quad \text{N/mm}$$

Ply 2 and 5 at 45°

$$\begin{Bmatrix} \varepsilon_1 \\ \varepsilon_2 \\ \varepsilon_6 \end{Bmatrix} = \begin{bmatrix} 0.5 & 0.5 & 0.5 \\ 0.5 & 0.5 & -0.5 \\ -1 & 1 & 0 \end{bmatrix} \times 10^{-3} \times \begin{Bmatrix} 23.5 \\ -2.32 \\ -15 \end{Bmatrix} N_x$$

$$= \begin{Bmatrix} 3.09 \\ 18.09 \\ -25.82 \end{Bmatrix} N_x \times 10^{-3}$$

Stresses corresponding to the above strains in the principal material directions are

$$\begin{Bmatrix} \sigma_1 \\ \sigma_2 \\ \sigma_6 \end{Bmatrix} = \begin{bmatrix} 140.9 & 3 & 0 \\ 3 & 10.1 & 0 \\ 0 & 0 & 5 \end{bmatrix} \begin{Bmatrix} 3.09 \\ 18.09 \\ -25.82 \end{Bmatrix} N_x \times 10^{-3} \quad \text{N/mm}^2$$

$$= \begin{Bmatrix} 0.4897 \, N_x \\ 0.192 \, N_x \\ -0.129 \, N_x \end{Bmatrix} \quad \text{N/mm}^2$$

In order to cause failure by maximum stress theory, the possible values of N_x are

from the consideration of $\sigma_1 = 1500 / 0.4897 = 3063.1 \quad$ N/mm

from the consideration of $\sigma_2 = 50 / 0.192 = 260.42 \quad$ N/mm

from the consideration of $\sigma_6 = 70 / 0.129 = 542.64 \quad$ N/mm

Ply 3 and 4 at 90°

Strains in the principal material directions are

$$\begin{Bmatrix} \varepsilon_1 \\ \varepsilon_2 \\ \varepsilon_6 \end{Bmatrix} = \begin{Bmatrix} -2.32 \, N_x \\ 23.5 \, N_x \\ -15 \, N_x \end{Bmatrix} \times 10^{-3}$$

Corresponding stresses in the principal material axes direction are

$$\begin{Bmatrix} \sigma_1 \\ \sigma_2 \\ \sigma_6 \end{Bmatrix} = \begin{bmatrix} 140.9 & 3 & 0 \\ 3 & 10.1 & 0 \\ 0 & 0 & 5 \end{bmatrix} \begin{Bmatrix} -2.32 \\ 23.5 \\ -15 \end{Bmatrix} 10^{-3} \, N_x \quad \text{N/mm}^2$$

$$= \begin{Bmatrix} -0.2564 \\ 0.2304 \\ -0.075 \end{Bmatrix} N_x \quad \text{N/mm}^2$$

In order to cause failure by maximum stress theory, the possible values of N_x are

N_x from the consideration of σ_1 = $\dfrac{1200}{0.2564}$ = 4680 N/mm

N_x from the consideration of σ_2 = $\dfrac{50}{0.2304}$ = 217 N/mm

N_x from the consideration of σ_6 = $\dfrac{70}{0.075}$ = 933.33 N/mm

The minimum value of N_x from all the six plies is 217 N/mm. This applies to ply 3 and 4 and fibres for these plies are oriented at 90°.

Second-ply failure

Using the complete-ply failure approach, we now assume that E_1, E_2 and E_6 for two failed 90° plies are set to zero. As such, these two plies do not contribute to the stiffness of the laminate.

The extensional stiffness matrix terms are as follows:

$$A_{xx} = 2\,[(0.125 \times 140.9) + (0.125 \times 44.3) + (0.125 \times 0)]$$
$$= 46.3 \quad \text{kN/mm}$$

$$A_{yy} = 2\,[(0.125 \times 10.1) + (0.125 \times 44.3)] \quad = \quad 13.6 \quad \text{kN/mm}$$

$$A_{ss} = 2 \times 0.125 \,\{5 + 36.3\} \quad = \quad 10.325 \quad \text{kN/mm}$$

$$A_{xy} = 2 \times 0.125 \,\{3 + 34.3\} \quad = \quad 9.325 \quad \text{kN/mm}$$

$$A_{xs} = 2 \times 0.125 \,\{0 \times 32.7 + 0\} \quad = \quad 8.175 \quad \text{kN/mm}$$

$$A_{yx} = 8.175 \quad \text{kN/mm}$$

Therefore,

$$[A] = \begin{bmatrix} 46.3 & 9.316 & 8.175 \\ 9.316 & 13.60 & 8.175 \\ 8.175 & 8.175 & 10.325 \end{bmatrix}$$

The compliance extensional matrix is

$$[a] = \begin{bmatrix} 0.0266 & -0.0103 & -0.0123 \\ -0.0103 & 0.1444 & -0.1062 \\ -0.0123 & -0.1062 & 0.1910 \end{bmatrix}$$

Before we proceed further, it is to be checked whether the laminate with reduced stiffness is able to sustain the load that caused the first-ply failure of the intact laminate.

The strains are

$$\begin{Bmatrix} \varepsilon_x^0 \\ \varepsilon_y^0 \\ \varepsilon_s^0 \end{Bmatrix} = \begin{bmatrix} 0.0266 & -0.0103 & -0.0123 \\ -0.0103 & 0.1444 & -0.0162 \\ -0.0123 & 0.1062 & 0.1910 \end{bmatrix} \begin{Bmatrix} 217 \\ 0 \\ 0 \end{Bmatrix} \times 10^{-3}$$

$$= \begin{Bmatrix} 5.7614 \\ -2.2351 \\ -2.6691 \end{Bmatrix} \times 10^{-3}$$

For plies 1 and 6 at 0°

$$\begin{Bmatrix} \varepsilon_1 \\ \varepsilon_2 \\ \varepsilon_6 \end{Bmatrix} = \begin{Bmatrix} 5761.4 \\ -2235.1 \\ -2269.1 \end{Bmatrix} \times 10^{-6}$$

Corresponding stresses in the ply 1 and 6 are

$$\begin{Bmatrix} \sigma_1 \\ \sigma_2 \\ \sigma_6 \end{Bmatrix} = \begin{bmatrix} 140.9 & 3 & 0 \\ 3 & 10.1 & 0 \\ 0 & 0 & 5 \end{bmatrix} \begin{Bmatrix} 5761.4 \\ -2235.1 \\ -2269.1 \end{Bmatrix} \times 10^{-6} \times 10^3 \quad N/mm^2$$

$$= \begin{Bmatrix} 805.08 \\ -5.29 \\ -11.34 \end{Bmatrix} \quad N/mm^2$$

For causing failure, σ_1 is to be increased by 1500/805.08 or 1.86.

σ_2 to be increased by 250/5.29 or 47.26

σ_6 to be increased by 70/11.34 or 6.17

Plies 2 and 5 at 45°

The strains in the principal material direction of the plies are

$$\begin{Bmatrix} \varepsilon_1 \\ \varepsilon_2 \\ \varepsilon_6 \end{Bmatrix} = \begin{bmatrix} 0.5 & 0.5 & 0.5 \\ 0.5 & 0.5 & -0.5 \\ -1 & 1 & 0 \end{bmatrix} \begin{Bmatrix} 5761.4 \\ -2235.1 \\ -2269.1 \end{Bmatrix} \times 10^{-6}$$

$$= \begin{Bmatrix} 628.6 \\ 2897.7 \\ -7997 \end{Bmatrix} \times 10^{-6}$$

Corresponding stresses in the principal material directions are

$$\begin{Bmatrix} \sigma_1 \\ \sigma_2 \\ \sigma_6 \end{Bmatrix} = \begin{bmatrix} 140.9 & 3 & 0 \\ 3 & 10.1 & 0 \\ 0 & 0 & 5 \end{bmatrix} \begin{Bmatrix} 628.6 \\ 2897.7 \\ -7997 \end{Bmatrix} \times 10^{-6} \times 10^3 \quad N/mm^2$$

$$= \begin{Bmatrix} 97.26 \\ 31.15 \\ -40.0 \end{Bmatrix} \quad N/mm^2$$

The factor by which σ_1 is increased is 1500/97.26 or 21.52

The factor by which σ_2 is to be increased is 50/31.15 or 1.535

The factor by which σ_6 is to be increased is 70/40 or 1.75

We have already assumed that the stresses in plies 3 and 4 are equal to zero.

The minimum value of the different factors considered by which stresses are to be increased to cause failure is 1.605. This occurs in plies 2 and 5.

Therefore, the load to cause second-ply failure is

$$N_x = 1.605 \times 217 = 348.29 \quad N/mm$$

Failure occurs in 45° plies in shear mode.

Third-ply failure

Finally, only two plies are left which have not failed. The elastic properties of all plies are zero except these two. The two intact plies are kept in their original position. The extensional stiffness terms for the intact part is now

$$\begin{aligned}
A_{xx} &= A_{11} = 2[0.125 \times 140.9] = 35.23 \quad kN/mm \\
A_{yy} &= A_{22} = 2[0.125 \times 10.1] = 2.53 \quad kN/mm \\
A_{66} &= A_{ss} = 2[0.125 \times 5] = 1.25 \quad kN/mm \\
A_{12} &= A_{xy} = 2[0.125 \times 3] = 0.75 \quad kN/mm \\
A_{13} &= A_{23} = A_{xs} = A_{ys} = 0
\end{aligned}$$

Therefore,

$$[A] = \begin{bmatrix} 35.23 & 0.75 & 0 \\ 0.75 & 2.53 & 0 \\ 0 & 0 & 1.25 \end{bmatrix} \quad kN/mm$$

The extensional compliance matrix is

$$[a] = \begin{bmatrix} 0.0286 & -0.0085 & 0 \\ -0.0085 & 0.3985 & 0 \\ 0 & 0 & 0.8000 \end{bmatrix} \quad 1/kN/mm$$

Analysis of Laminate Strength

The mid-plane strains are

$$\begin{Bmatrix} \varepsilon_x^0 \\ \varepsilon_y^0 \\ \varepsilon_s^0 \end{Bmatrix} = \begin{bmatrix} 0.0286 & -0.0085 & 0 \\ -0.0085 & 0.3985 & 0 \\ 0 & 0 & 0.8 \end{bmatrix} \begin{Bmatrix} 348.29 \\ 0 \\ 0 \end{Bmatrix} \times 10^{-3}$$

$$= \begin{Bmatrix} 9961.1 \\ -2960.5 \\ 0 \end{Bmatrix} \times 10^{-6}$$

Corresponding stresses in the principal material direction for these plies are

$$\begin{Bmatrix} \sigma_1 \\ \sigma_2 \\ \sigma_6 \end{Bmatrix} = \begin{bmatrix} 140.9 & 3 & 0 \\ 3 & 10.1 & 0 \\ 0 & 0 & 5 \end{bmatrix} \begin{Bmatrix} 9961.1 \\ -2960.5 \\ 0 \end{Bmatrix} \times 10^{-6} \times 10^3 \quad \text{N/mm}^2$$

$$= \begin{Bmatrix} 1394.64 \\ -0.01775 \\ 0 \end{Bmatrix} \quad \text{N/mm}^2$$

The factor by which σ_1 is to be increased to initiate failure is 1500/1394.64 or 1.075

Therefore, applying maximum stress theory, the ultimate failure load is 348.29 × 1.075 or 374.41 N/mm.

Interestingly, the final failure of the laminate is the fibre failure.

Example 11.4

Repeat Example 11.3, but with a partial-ply failure approach.

Upto first-ply failure approach, the treatment is same as given in Example 11.3. Therefore, for FPF,

$N_x = 217$ N/mm

The plies 3 and 4 (90° plies) fail in transverse tension (due to σ_2).

In partial-ply failure approach, E_2 and E_6 of the failed 90° plies are made equal to zero.

The value of E_1 remains unaffected, as it is assumed that failure in the transverse tension does not affect the module in the fibre direction.

Let us discuss about the possible value of v_{12}. Actually, the result will not be affected by v_{12}.

Suppose, v_{12} is non-zero, then

$$v_{21} = v_{12} \frac{E_2}{E_1}$$

As E_2 is zero, $v_{21} = 0$ irrespective of the value of v_{12}. Thus, the expression $(1 - v_{12} v_{21})$ required for stiffness calculation is equal to unity. For 90° plies, then

$$Q_{11} = \frac{E_1}{1 - \nu_{12}\nu_{21}} = E_1 = 140 \text{ kN/mm}^2$$

$$Q_{22} = \frac{E_2}{1 - \nu_{12}\nu_{21}} = 0$$

$$Q_{66} = E_6 = 0$$

$$Q_{12} = Q_{13} = Q_{23} = 0$$

$$[Q]_{12} = \begin{bmatrix} 140 & 0 & 0 \\ 0 & 0 & 0 \\ 0 & 0 & 0 \end{bmatrix}$$

Transformed reduced stiffness for the 90° ply is

$$Q_{xy} = \begin{bmatrix} 0 & 0 & 0 \\ 0 & 140 & 0 \\ 0 & 0 & 0 \end{bmatrix}$$

$$A_{xx} = 2\{(0.125 \times 140.9) + (0.125 \times 44.3) + (0.125 \times 0)\}$$
$$= 46.3 \text{ kN/mm}$$

$$A_{yy} = 2\{(0.125 \times 10.1) + (0.125 \times 44.3) + (0.125 \times 140)\}$$
$$= 48.825 \text{ kN/mm}$$

$$A_{ss} = 2\{(0.125 \times 5) + (0.125 \times 34.3) + (0.125 \times 0)\}$$
$$= 9.825 \text{ kN/mm}$$

$$A_{xy} = 2\{(0.125 \times 3) + (0.125 \times 34.3) + (0.125 \times 0)\}$$
$$= 9.325 \text{ kN/mm}$$

$$A_{xs} = 2\{(0.125 \times 0) + (0.125 \times 32.7) + (0.125 \times 0)\}$$
$$= 8.175 \text{ kN/mm} = A_{ys}$$

$$[A]_{x,y} = \begin{bmatrix} 46.3 & 9.325 & 8.175 \\ 9.325 & 48.825 & 8.175 \\ 8.175 & 6.175 & 9.825 \end{bmatrix} \text{ kN/mm}$$

The extension compliance matrix is

$$[a]_{x,y} = \begin{bmatrix} 0.0254 & -0.0015 & -0.02 \\ -0.0015 & 0.0238 & -0.0186 \\ -0.02 & -0.0186 & 0.1336 \end{bmatrix} \text{ 1/(kN/mm)}$$

Strains in the middle plane in the global directions are

$$\begin{Bmatrix} \varepsilon_x^0 \\ \varepsilon_y^0 \\ \varepsilon_s^0 \end{Bmatrix} = \begin{bmatrix} 0.0254 & -0.0015 & -0.200 \\ -0.0015 & 0.0238 & -0.0186 \\ -0.02 & -0.0186 & 0.1336 \end{bmatrix} \begin{Bmatrix} 217 \\ 0 \\ 0 \end{Bmatrix} \times 10^{-3}$$

$$= \begin{Bmatrix} 5512 \\ -326 \\ -4340 \end{Bmatrix} \times 10^{-6}$$

For plies 1 and 6 at 0°

$$\begin{Bmatrix} \varepsilon_1 \\ \varepsilon_2 \\ \varepsilon_6 \end{Bmatrix} = \begin{Bmatrix} 5512 \\ -326 \\ -4340 \end{Bmatrix} \times 10^{-6}$$

Corresponding stresses in the principal material directions are

$$\begin{Bmatrix} \sigma_1 \\ \sigma_2 \\ \sigma_6 \end{Bmatrix} = \begin{bmatrix} 140.9 & 3 & 0 \\ 3 & 10.1 & 0 \\ 0 & 0 & 5 \end{bmatrix} \begin{Bmatrix} 5512 \\ -326 \\ -4340 \end{Bmatrix} \times 10^{-6} \times 10^3 \quad N/mm^2$$

$$= \begin{Bmatrix} 776 \\ 13 \\ -22 \end{Bmatrix} \quad N/mm^2$$

For causing failure

σ_1 is to be increased by 1500/776 or 1.93

σ_2 is to be increased by 50/13 = 3.85

σ_3 is to be increased by 70/22 = 3.18

For plies 2 and 5 at 45°

$$\begin{Bmatrix} \varepsilon_1 \\ \varepsilon_2 \\ \varepsilon_6 \end{Bmatrix} = \begin{bmatrix} 0.5 & 0.5 & 0.5 \\ 0.5 & 0.5 & -0.5 \\ -1 & 1 & 0 \end{bmatrix} \begin{Bmatrix} 5512 \\ -326 \\ -4340 \end{Bmatrix} 10^{-6}$$

$$= \begin{Bmatrix} 423 \\ 4763 \\ -5838 \end{Bmatrix} \times 10^{-6}$$

Corresponding stresses in the principal material directions are

$$\begin{Bmatrix} \sigma_1 \\ \sigma_2 \\ \sigma_6 \end{Bmatrix} = \begin{bmatrix} 140.9 & 3 & 0 \\ 3 & 10.1 & 0 \\ 0 & 0 & 5 \end{bmatrix} \begin{Bmatrix} 423 \\ 4763 \\ -5838 \end{Bmatrix} \times 10^{-6} \times 10^3 \quad N/mm^2$$

$$= \begin{Bmatrix} 73.89 \\ 49.37 \\ -29.19 \end{Bmatrix} \quad N/mm^2$$

As the failure stress in transverse tension is 50 N/mm², the lamina will just escape failure due to the existing load.

As such at $N_x = 246$ N/mm, it is seen that plies 2 and 5 at 45° orientation will partially fail. As the failure is in transverse tension, these plies remain effective in the longitudinal direction. As such a fresh calculation is to be started.

For the 45° plies

$$Q_{xx} = 140 \times \frac{1}{4} = 35$$

$$Q_{yy} = 140 \times \frac{1}{4} = 35$$

$$Q_{xy} = 140 \times \frac{1}{4} = 35$$

$$Q_{xs} = Q_{ys} = Q_{ss} = 35$$

$$[Q]_{x,y}^{45°} = 35 \begin{bmatrix} 1 & 1 & 1 \\ 1 & 1 & 1 \\ 1 & 1 & 1 \end{bmatrix}$$

$A_{xx} = 2\{(0.125 \times 140.9) + (0.125 \times 35) + (0.125 \times 0)\} = 43.98$ kN/mm

$A_{yy} = 2\{(0.125 \times 10.1) + (0.125 \times 35) + (0.125 \times 140.9)\} = 46.5$ kN/mm

$A_{ss} = 2\{(0.125 \times 5) + (0.125 \times 35)\} = 10$ kN/mm

$A_{xy} = 2\{(0.125 \times 3) + (0.125 \times 35)\} = 9.5$ kN/mm

$A_{xs} = A_{ys} = 2 \times 0.125 \times 35 = 8.75$ kN/mm

Therefore,

$$[A]_{x,y} = \begin{bmatrix} 43.98 & 9.5 & 8.75 \\ 9.5 & 46.5 & 8.75 \\ 8.75 & 8.75 & 10 \end{bmatrix}$$

The compliance extensional matrix is

$$[a]_{x,y} = \begin{bmatrix} 0.0276 & -0.0013 & -0.023 \\ -0.0013 & 0.0258 & -0.025 \\ -0.023 & -0.025 & 0.1389 \end{bmatrix}$$

Strains in the middle plane in the global directions are

$$\begin{Bmatrix} \varepsilon_x^0 \\ \varepsilon_y^0 \\ \varepsilon_s^0 \end{Bmatrix} = \begin{bmatrix} 0.0276 & -0.0013 & -0.023 \\ -0.0013 & 0.0258 & -0.025 \\ -0.023 & -0.025 & 0.1389 \end{bmatrix} \begin{Bmatrix} 246 \\ 0 \\ 0 \end{Bmatrix} \times 10^{-3}$$

$$= \begin{Bmatrix} 6790 \\ -320 \\ -5658 \end{Bmatrix} \times 10^{-6}$$

For plies 1 and 6 at 0°, strains in the principal material directions are

$$\begin{Bmatrix} \varepsilon_1 \\ \varepsilon_2 \\ \varepsilon_6 \end{Bmatrix} = \begin{Bmatrix} 6790 \\ -320 \\ -5658 \end{Bmatrix} \times 10^{-6}$$

Corresponding stresses in the principal material direction are

$$\begin{Bmatrix} \sigma_1 \\ \sigma_2 \\ \sigma_6 \end{Bmatrix} = \begin{bmatrix} 140.9 & 3 & 0 \\ 3 & 10.1 & 0 \\ 0 & 0 & 5 \end{bmatrix} \begin{Bmatrix} 6790 \\ -320 \\ -5658 \end{Bmatrix} \times 10^{-6} \times 10^3 \quad N/mm^2$$

$$= \begin{Bmatrix} 955 \\ 17.14 \\ -28.29 \end{Bmatrix} \quad N/mm^2$$

In order to cause failure

σ_1 is to be increased by 1500/955 or 1.579

σ_2 is to be increased by 50/17.14 or 2.917

σ_6 is to be increased by 70/28.29 or 2.47

The lowest factor of those given above is 1.579. Therefore ultimate failure load N_x is given by

at failure $N_x = 246 \times 1.579 = 388.4$ N/mm

11.6 INTERLAMINAR STRESSES

In the category of laminates having plies oriented in different directions, a shear force induces shear stresses in the cross-section perpendicular to the mid-plane. Then equilibrium considerations give rise to the presence of shear stresses in planes parallel to the mid-plane which is at right angles to the cross-section. As there is every likelihood that some of these planes may be interfaces and as the inter-laminar strength is much smaller than the strength of the lamina, this phenomena assumes great importance. In the Classical Laminate Theory (CLT) it is assumed that components of all out-of-plane stresses are zero, that is, $\sigma_z = \tau_{xz} = \tau_{yz} = 0$. Though these assumptions are certainly justified for sections away from the free edges and away from the geometrical discontinuities, they however, fail to satisfy free edge conditions of certain laminates. Therefore, CLT does not appear to provide an answer for such cases.

The CLT is based on an important assumption that a plane section normal to the middle plane of the plate remains normal after deformation. A laminate consisting of different orientations in different lamina will deform as one unit. This requires a transference of forces at the edges. This may result in excessive forces in certain cases which may cause delamination or separation of different laminae. The physical nature of these stresses is first explained qualitatively.

A cross-ply laminate [0/90]s is considered. The laminate consists of four plies having unidirectional fibres. *x*-directional tensile forces act in the laminate. A strip of the laminate is shown in Fig. 11.3. We know that for a unidirectional lamina, longitudinal modulus E_1 is much greater than E_2 and v_{12} is also much higher than v_{21}.

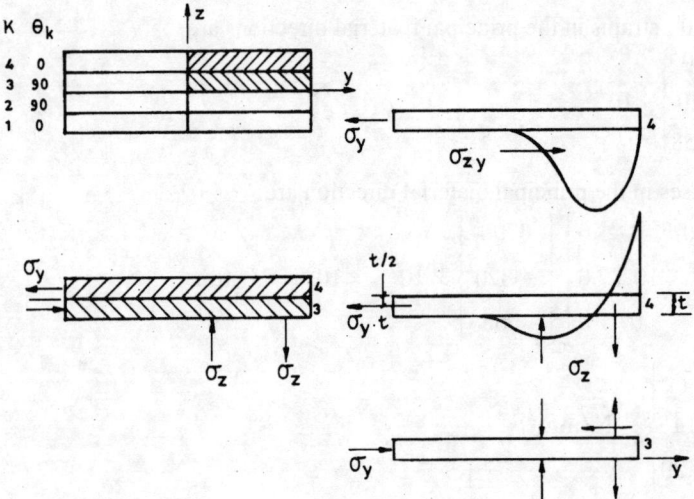

Fig. 11.3 Cross-ply laminate undergoing delamination

But each layer undergoes same axial deformation which suggests that both the 0° and the 90° lamina are equally strained in the *x*-direction. Now, the Poisson's ratios of two layers are different. As such the 0°- lamina will contract more than the 90° lamina in the *y*-direction. This can also be explained physically. The stiff fibres of the 90° lamina will offer much greater resistance in contraction in the *y*-direction than the 0° lamina. But as the two layers are bonded to each other, both the 0° and the 90° layers will undergo the same strain, which means that contraction of both the lamina will be the same. The net effect of it is that the 0° lamina will have tensile stress and the 90° lamina will develop compressive stress. In the 0° lamina, there will thus be the decrease in contraction whereas in the 90° lamina, the contraction will increase. The Classical Laminate Theory enables the calculation of these stresses.

No stresses can act at the free edges. Based on the equilibrium conditions in the *y*-direction of one-half of the fourth lamina, it is evident that there has to be an inter-laminar shear stress σ_{zy}. Let us consider the equilibrium of moments of the yz-plane of that half of the fourth lamina. This necessitates a normal force σ_z to act at the interface between the third and the fourth lamina. It has been found that the maximum value of σ_z occurs at the free edges. However, from the consideration of equilibrium of the z-directional forces, the distribution of σ_z should be such that the resultant will be zero. In order to have that as we move inside the laminate, σ_z decreases, then changes into compression which finally reduces to zero. Inter-laminar tension and compression must balance each other. The normal stress will be very predominant between the second and the third lamina.

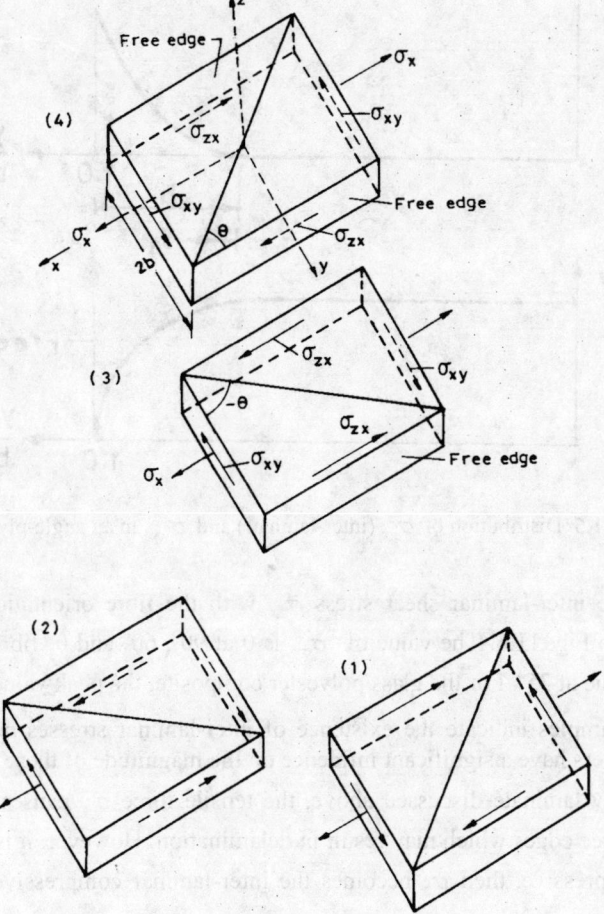

Fig. 11.4 Angle-ply laminate undergoing delamination

Let us look into another situation which gives rise to inter-laminar shear stresses. Consider an angle-ply lamina $[\theta/-\theta]_s$ consisting of four identical unidirectional laminas. It is subjected to a tensile stress σ_x. The tensile stress in each layer is σ_x. Figure 11.4 shows shear deformation in opposite signs of the two layers θ and $-\theta$ when considered independently as they have off-axis orientation. But as the layers are bonded they must have zero shear strain which is deemed possible due to the presence of the inter-laminar shear stress σ_{zx} transmitted from one layer to the other. The variation of σ_{zx} is shown in Fig. 11.5. It may be seen that for most of the central region across the width, σ_{zx} is zero while it increases significantly near the edges. The moment in the lamina due to σ_{xy} is balanced by the moment due to σ_{zx}. It is seen that σ_{xy} remains constant in most parts of the central region and drops to zero at the stress free edges.

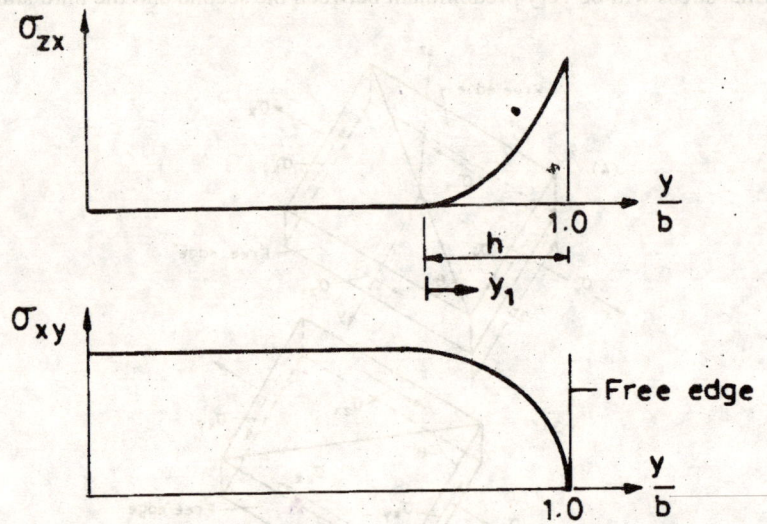

Fig. 11.5 Distribution of σ_{zx} (inter-laminar) and σ_{xy} in an angle-ply laminate

The variation of the inter-laminar shear stress σ_{zx} with the fibre orientation of the carbon/epoxy laminate is shown in Fig. 11.6. The value of σ_{zx} is 0 at 90°, 60° and 0° fibre orientation while σ_{zx} reaches the peak value at 35°. For the glass/polyester composite, the peak value of σ_{zx} is at 15°.

The above examples indicate the existence of inter-laminar stresses near the free edges. The particulars of the layers have a significant influence on the magnitude of these stresses.

In the cross-ply laminate discussed above, the tensile force σ_x causes an interlaminar tensile stress σ_z near the free edges which may result in delamination. However, it is interesting to note that if σ_x becomes compressive, then σ_z becomes the inter-laminar compressive stress which prevents delamination.

Again, keeping σ_x as tensile, if the stacking sequence is changed to [90/0]s, then it will give inter-laminar compressive stress σ_z which will prevent delamination.

On the basis of the finite element analysis carried out, it has been seen that inter-laminar stresses act on a small zone near the free edge of width equal to the laminate thickness. Certain steps may be taken to avoid delamination. One measure is to reinforce the edge. Another option is to make an integral construction.

Analysis of Laminate Strength 303

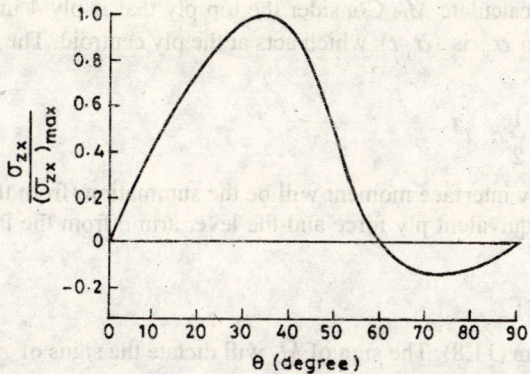

Fig. 11.6 Interlaminar shear stress as a function of fibre orientation for carbon/epoxy laminate

11.7 PREDICTION METHODS

The only way to predict three-dimensional stresses accurately is to perform a three-dimensional analysis. An accurate three-dimensional analysis of a laminated composite is a tedious task. It is performed only at the final stage. As such, some simpler approach is needed which will help the designer. A large number of such methods are available. In this section we are going to discuss those due to Pagano and Pipes [11.13] and Ueng and Zhang [11.14]. The basic assumption in these methods is that the laminate is specially orthotropic in the inplane mode ($A_{xs} = A_{ys} = 0$) and symmetric about its mid-plane.

Pagano and Pipes [11.13] have assumed that the maximum inter-laminar direct stress σ_z caused by the ply interference M_P is given by

$$\sigma_z = \frac{90}{7} \frac{M_P}{h^2} \tag{11.8}$$

where h is the laminate thickness.

The variation of the direct stress within the boundary layers of one laminate thickness (from the free edge) is shown as Fig. 11.7. It may be noted that the area under σ_z sums up to zero.

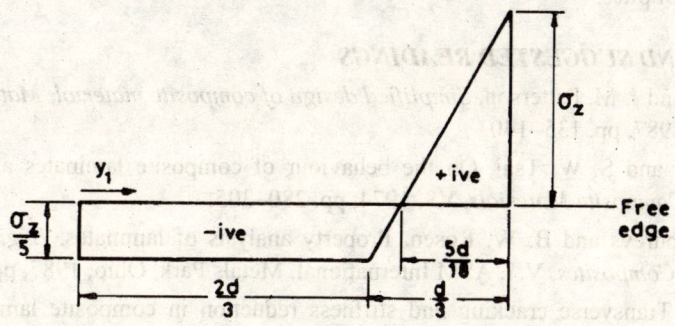

Fig. 11.7 Assumed inter-laminar direct stress σ_z profile

It is now desired to calculate M_P. Consider the top ply that is ply 4 in Fig. 11.3. The force per unit width of the ply due to σ_y is $(\sigma_y t)$ which acts at the ply centroid. The ply interface moment M_P is

$$M_P = \sigma_y t \cdot \frac{t}{2} = \frac{1}{2} \sigma_y t^2 \tag{11.9}$$

Generally speaking, the ply interface moment will be the summation (from the bottom ply to the midplane) of the product of equivalent ply force and the lever arm c from the interface plane considered to the ply centroid.

$$M_P = \sum \sigma_y \cdot t \cdot c \tag{11.10}$$

σ_z can be determined from (11.8). The sign of M_P will dictate the signs of σ_z. The knowledge of the stress value at the free edge is of significant interest to the designer.

For the estimation of inter-laminar shear stress σ_{zx}, a profile for the inplane ply shear stress σ_{xy} is assumed by Ueng and Zhang [11.14] as

$$\sigma'_{xy} = (1 - d_0^{n_1})^{(1-k_1)} \sigma_{xy} \tag{11.11}$$

where

σ'_{xy} = the assumed profile for the ply shear stress in the boundary layer (which is at a distance from the free edge where inter-laminar stresses are effective)

σ_{xy} = the inplane shear stress predicted by CLT

d_0 = y_1/h, where y_1 is the distance of the boundary layer from the free edge ($0 \leq y_1 \leq d$) (Fig. 11.5)

n_1, k_1 = constants

It may be noted that when $y_1 = 0$, $d_0 = 0$ and $\sigma'_{xy} = \sigma_{xy}$ which is exactly the shear stress on the boundary layer. At the free edge, $y_1 = h$, then $d_0 = 1$ and $\sigma_{xy} = 0$ (Fig. 11.5). Between $d_0 = 0$ and $d = 1$, the distribution of shear stress follows a decaying profile (Fig. 11.3), not predicted by CLT, but assumed by (11.11). This however is limited to a symmetric configuration and is limited to a minimum number of plies.

REFERENCES AND SUGGESTED READINGS

11.1 S. W. Tsai and J. M. Patterson, *Simplified design of composite material*, Materials and Design, V8, No. 3, 1987, pp. 135–140.

11.2 H. T. Hahn and S. W. Tsai, On the behaviour of composite laminates after initial failure, *Journal of Composite Materials*, V8, 1974, pp. 280–305.

11.3 E. A. Humphreys and B. W. Rosen, Property analysis of laminates. *Engineering Materials Handbook: Composites*, V.1, ASM International, Metals Park, Ohio, 1987, pp. 218–235.

11.4 R. Talreja, Transverse cracking and stiffness reduction in composite laminates, *Journal of Composite Materials*, V.19, 1985, pp. 355–375.

11.5 Z. Hashin, Analysis of cracked laminates: a variational approach, *Mechanics of Materials*, V.4, 1985, pp. 121–136.

11.6 N. Laws and G. J. Dvorak, Progressive transverse cracking in compositive laminates, *Journal of Composite Materials*, V.22, pp. 900–916.

11.7 S. G. Lim and C. S. Hong, Prediction of transverse cracking and stiffness in cross-ply laminated composites, *Journal of Composite Materials*, V.24, 1990, pp. 695–713.

11.8 J. W. Lee and I. M. Daniel, Progressive transverse cracking of cross-ply composites laminates, *Journal of Composite Materials*, V.24, 1990, pp. 1225–1243.

11.9 G. S. Padhi, R. A. Shenoi S. S. J. Moy and G. L. Hawkins, Progressive collapse and ultimate collapse of laminated plates in bending, *Composite Structures*, V. 40, No. 3–4, 1998, pp. 277–291.

11.10 F. K. Chang and K. Y. Chang, A progressive damage model for laminated composites containing stress concentrations, *Journal of Composite Materials*, V. 21, 1987, pp. 834–857.

11.11 I. Shakid and F. K. Chang, An accumulated damage model for tensile and shear failure of laminated composites, *Journal of Composite Materials*, V.29, 1995, pp. 926–981.

11.12 S. Tolson and N. Zabaras, Finite element analysis of progressive failure in laminated composite plates, *Computers and Structures*, V.38, pp. 361–376.

11.13 N. J. Pagano and B. R. Pipes, Some observation on the inter-laminar strength of composite laminates, *International Journal of Mechanical Sciences*, V.15, 1975, pp. 679–688.

11.14 C. E. Ueng and K. Zhang, A simplified approach for inter-laminar stresses in orthotropic laminated strips, *Journal of Reinforced Plastics and Composites*, V.4, 1985, pp. 273–286.

11.15 K. L. Reifsnidor, Ed., Damage in Composite Materials, STP 775, ASTM, Philadelphia, 1982.

11.16 R. Talreja, Transverse cracking and stiffness reduction in composite laminates, *Journal of Composite Materials*, V.4, 1985, pp. 355–375.

11.17 Z. Hashing, Analysis of cracked laminates: variational approach, *Mechanics of Materials*, V.4, 1985, pp. 121–135.

11.18 J. W. Lee and I. M. Daniel, Progressive transverse cracking of cross-ply composite laminates, *Journal of Composite Materials*, V.24, 1990, pp. 1225–1243.

11.19 J. N. Reddy and A. K. Pandey, A first-ply failure analysis of composite laminates, *Computers and Structures*, V.25, No. 3, 1987, pp. 371–393.

11.20 S. C. Tan, A progressive failure model of composite laminates containing openings, *Journal of Composite Materials*, V.25, 1991, pp. 556–577.

EXERCISE 11

11.1 Determine the axial tensile load N_x of a $[\pm 45]_s$ laminate using the Tsai–Hill criteria for the first-ply failure for high strength composites with $F_1 \gg F_2$ and $F_1 \gg F_6$ in terms of material and strength properties $E_1, E_2, E_6, \nu_{12}, F_2$ and F_6, subjected to uniaxial stress N_x.

11.2 Determine the FPF strength of a $[0/90]_s$ laminate subjected to uniaxial tension or compression based on (a) maximum stress theory, (b) Tsai–Hill theory and (c) Tsai–Wu theory. Give the following properties of each lamina

$E_1 = 39$ GPa, $E_2 = 8.6$ GPa, $E_6 = 3.8$ GPa, $\nu_{12} = 0.28$, $t = 0.1$ mm

$F_{1T} = 1080$ MPa, $F_{1C} = 620$ MPa, $F_{2T} = 39$ MPa, $F_{2C} = 128$ MPa,

$F_6 = 89$ MPa

11.3 Determine the strength at ULF of the laminate of Problem 11.2, using maximum the stress criteria and the total-ply failure method.

11.4 A $(0/45/-45)_s$ is subjected to a longitudinal tensile load N_x. Determine N_x for the last-ply failure approach. Each ply has a thickness of 0.125 mm. The elastic properties of the ply are as follows

$E_1 = 140, E_2 = 10, E_6 = 5$ (in kN/mm^2)

$F_{1T} = 1500, F_{1C} = 1200, F_{2T} = 50, F_{2C} = 250, F_6 = 70$ (N/mm^2)

11.5 A $[\pm 60]_s$ laminate is loaded in uniaxial compression as shown. Determine the compressive strength N_y at FPF according to the maximum strain theory for the following given properties

$\varepsilon_{1T}^u = 0.02 \quad \varepsilon_{2T}^u = 0.005 \quad \varepsilon_{1C}^u = 0.02$

$\varepsilon_{2C}^u = -0.022 \quad \varepsilon_6^u = 0.015 \quad E = 60\ GPa; \bar{v}_{xy} = 1.0$

11.6 A cantilever beam of glass/epoxy $[0/\pm 45]_s$ having the thickness of each lamina as 0.1 mm is subjected to a concentrated force P at the free end. Determine the magnitude of force P when the delamination starts. The properties of each lamina are the same as that given in Problem 11.2.

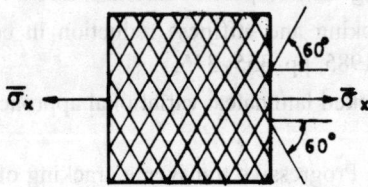

Problem 11.6

11.7 A $[\pm 45]_s$ laminate is biaxially loaded as shown. The properties are the same as those of Prob. 11.4.

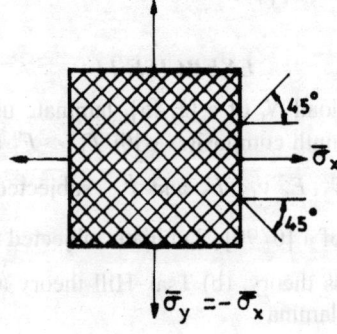

Problem 11.7

Determine the biaxial stress at FPF using the maximum stress criterion.

CHAPTER 12

DESIGN OF FIBRE REINFORCED COMPOSITE STRUCTURES

12.1 INTRODUCTION

The application of FRP composites is increasing rapidly due to the vast strides in technology made in many disciplines. They are looked upon as potential replacements of the metal structures in naval, aerospace, civil and mechanical engineering industries. The micro-mechanics, macro-mechanics and lamination theory have been discussed in the earlier chapters. This chapter deals with the design process for composite structures with a few examples to demonstrate the methods applied to calculate the design of composite structures.

The design of FRP composite structures plays a vital role in ensuring the integrity of the structure without compromising set targets. By the term optimum design, the cost and the size become essential factors for the judicious application of the design methodology. To achieve this optimum, the need and the importance for a proper design goes without saying. Keeping this in view, the present chapter focuses on the general design philosophy, configuration selection, material selection and structural analysis that are pertinent to the composite structural design process.

In case of composite structural design, there are many aspects, which are distinguishable from that of conventional materials. The composite designer encounters a high degree of freedom vis-à-vis the selection of constituent materials, configuration, ply lay-up, fabrication techniques, and the method of analysis. The designer needs to be judicious in selecting the various parameters that govern the composite structural design. No doubt, the design process for the composite structure is quite tricky as compared to a conventional metal structure. Composite design requires greater understanding of the constituent material fundamentals, which the constitute the composite structure. The designer should be able to comprehend material properties like anisotropy in moduli, variation in tensile and compressive strengths, ease of fabrication and the possibility of tailoring the laminate to suit a particular shape. It is important for the designer to know the behaviour of the structure under adverse environmental effects and different failure modes particularly the inter-laminar stress and delamination failures which are relevant to the composite structure alone. In a nutshell, a composite laminate design problem can be diagrammatically represented as shown in Fig 12.1.

The functional requirements are the starting point for any structural design process. These requirements should be clearly understood by the structural designers. The main functional inputs are presented in Fig 12.2.

The input variables may be interpreted as functional variables for the composite structural design process. To this functional variable, performance and economic considerations are introduced to start the design process which is iterative in nature. The structural designer is required to identify the parameters correctly and alter them in each cycle to improve the performance. The criteria for the structural design are required to be clearly defined for the entire iterative design process. The design criteria and the other governing parameters of the design process form the basis for the design methodology.

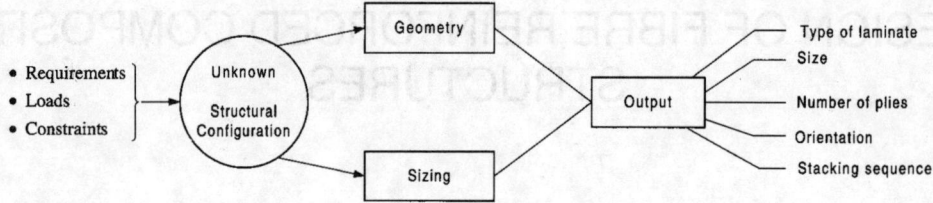

Fig. 12.1 Initiation of the structural design process

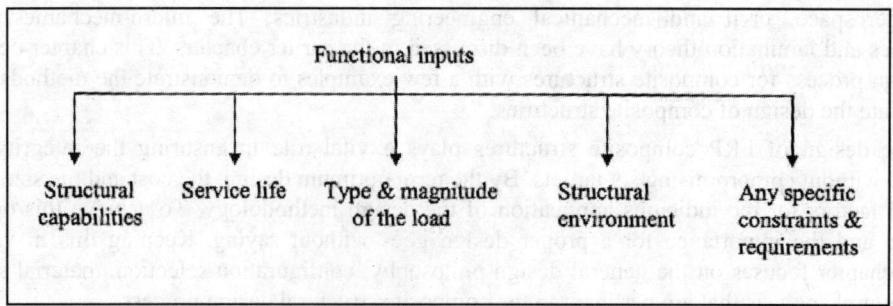

Fig 12.2 Functional inputs

12.2 COMPOSITE STRUCTURAL DESIGN

A composite structural design can be defined as the process of finding an optimum configuration with structural dimensions and materials to support the given load and to economically perform the assigned task.

Any design process including the composite structural design starts with an objective. Objectives are nothing but set targets which are required to be achieved at the end of the design process. The objective may be limited to a single target or multiple targets with certain conflicting requirements. The designer should be able to comprehend the design requirements, the goal and the failure criteria in terms of design objectives. The next step in the composite structural design process is to identify the critical constraints, which affect the performance of the structure directly or indirectly. Most constraints in the structural design process appear to be in the form of size, load, weight, maintenance requirements and the cost depending on the area of application.

With the objectives and constraints as inputs, the process of structural configuration and sizing is commenced. The chosen configuration is then analysed for the behaviour of the structure under the applied loads. Depending on the response analysis, the configuration and the size of the structure are altered and the process is repeated till the satisfactory performance of the structure is obtained.

At this stage, it is important for the structural designer to understand the difference between the analysis and the design. The design is a process of evolving particular structural configurations by varying required parameters to meet a specific purpose. To the designer, options of innumerable combination of sizes and configurations are available to meet design requirements. As options are wide open to the designer, it can be concluded that the structural design is an indeterministic problem. An analysis on the other hand is an overall assessment of the response and capabilities of the particular chosen structure. Here, the option is limited to one chosen structure and hence it can be concluded that the analysis is a deterministic problem.

12.2.1 The Design Spiral

The most important aspect of the design process is to converge to the best optimum design. For composites, this convergent process can be represented by a design spiral. The convergent point, that is the acceptable optimum design is represented as the centre of the spiral as shown in Fig. 12.3. The design spiral indicates the iterative and interactive nature of the whole composite structural design. By satisfying the design criteria, we attempt to produce the desired design. However, if the design criteria are not satisfied, alteration and modification of the required parameters are made and we move cyclically to reach the desired design. There are many variables shown in Fig.12.3, which can alter the performance of the structure. The decision to admit or omit any of the variables, is a matter of judgment based on requirements and experience.

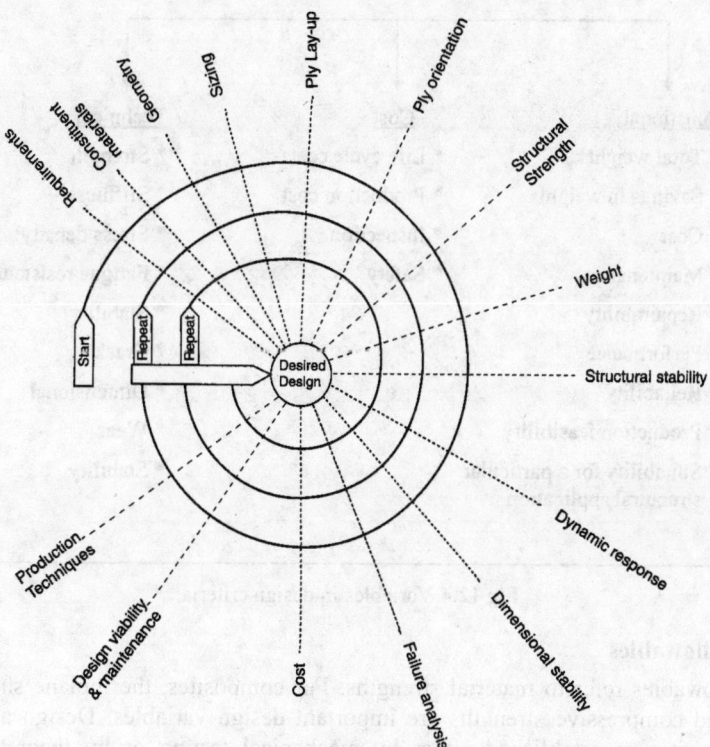

Fig.12.3 The composite design spiral

12.2.2 Design Criteria

The composite structural design is based on certain criteria. The design criteria are important to integrate various factors like loading, responses and durability with functional and operational requirements. While integrating these factors to derive the design criteria, the cost, the weight and the material availability are also taken into consideration along with safety and reliability as counterchecks to obtain a satisfactory design. Generally, the operational design criteria govern the basic structural configuration. Presently composite structural designers employ the same design criteria as those used for metals. In general, design criteria can be classified as operational criteria, environmental criteria and other miscellaneous criteria. The composite structural design is either based on the allowable deflection or on the ultimate strength. To this, sustainability under the ultimate design load and the fatigue behaviour are imposed to further ensure satisfactory performance.

The ultimate design load is arrived at by considering the maximum load that the structure is likely to experience during its service life. To this design load, a factor of safety (FOS) is added to ensure safety under any unforeseen loads. The value of the factor of safety for composite structures depends on the area of application. For composites, FOS is still being evaluated owing to the lack of design experience, though a factor of safety of 2 or more is generally used depending on the criticality of the structural component. The usual design criteria for composite structure is summarised in Fig.12.4.

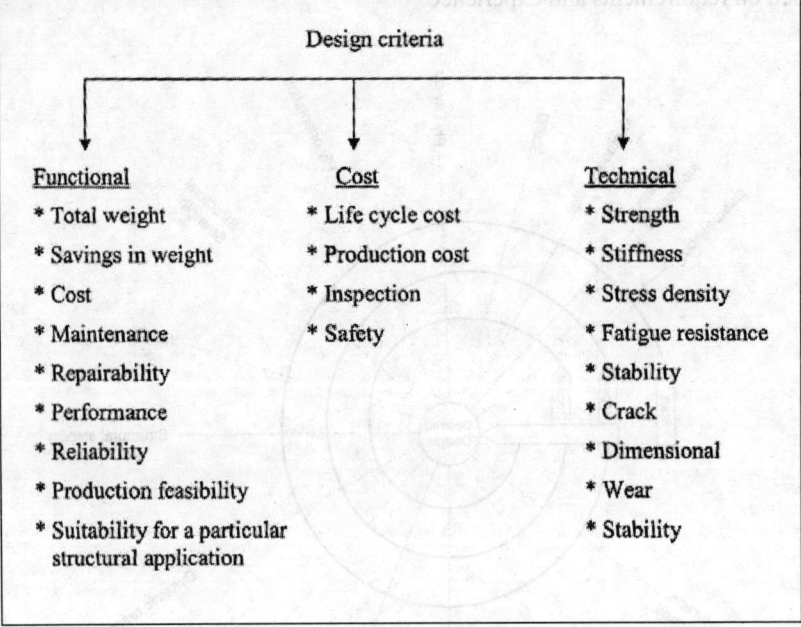

Fig 12.4 Variables in design criteria

12.2.3 Design Allowables

Design allowables refer to material strengths. For composites, the inplane shear, inter-laminar shear, tensile and compressive strengths are important design variables. Design allowables for the laminated composite are established either by mechanical testing or by theoretical calculations. Theoretical estimation of the design allowable requires ply level database of the constituent material properties such as elastic modulus, Poisson's ratio, strength and strain which are generated using mechanical testing procedures at the lamina level. Theoretical procedures using the rule of mixture or empirical formulae can also be employed for approximate evaluation of the above. It has to be verified later by mechanical tests.

The database thus generated can be effectively used for preparing design charts, which are sometimes referred to as carpet plots. These design charts can be used in the design decision making process. The usual ply level design database generated is given in Fig. 12.5.

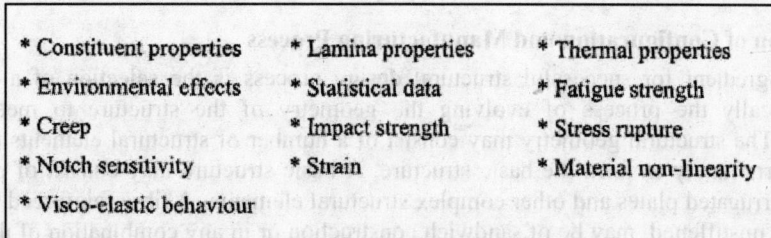

Fig. 12.5. Design data base

12.2.4 Material Selection in Composite Design

The material selection in composite structures is very involved and is a difficult process. The difficulty arises mainly due to the possibilities of combining different types of fibre and resin to produce different types of composite laminae. Each type may possess different properties and be of varying cost. Composite material selection is an optimal decision making process. A material with some specific properties may be too costly from the economic point of view but may have very low specific weight. Here, the decision has to be made whether the cost or the weight is the overriding factor. Therefore, a material selection process requires a rational approach. The aim of the selection process is to choose the right material for the right application. Some of the factors, which need to be considered, are listed in Fig. 12.6.

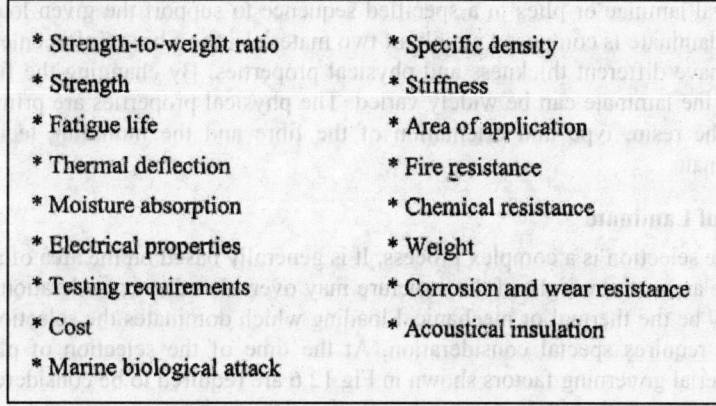

Fig 12.6 Factors governing material selection

These factors are based on those shown in Fig.12.5. Among the various factors, the priority of one over the other depends on the area of structural application. In some structures, the strength-to-weight ratio may be most important while for others it is the thermal or electrical properties of the material. What is required in a material selection process is a rational approach coupled with cost and functional evaluation. A wise approach in any structural application is to categorise these factors as those which are most critical and those which are not so critical in order of importance. This gives a fair idea of what can be expected of the structure. An appropriate consideration of the design requirements aids in categorising the factors governing the material selection as per their importance.

In addition, design requirements help in narrowing down the choice of materials. In composite structural design, fibre and matrix selection is very vital while the selection of all other added products is considered later to further improve the desired properties.

12.2.5 Selection of Configuration and Manufacturing Process

A key ingredient for successful structural design process is the selection of a configuration which is basically the process of evolving the geometry of the structure to meet the design requirements. The structural geometry may consist of a number of structural elements that are interconnected appropriately to form the basic structure. A basic structure may consist of plates, beams, shells, bars, corrugated plates and other complex structural elements. A fibre reinforced structure may be stiffened or unstiffened, may be of sandwich construction or in any combination of these forms. A configuration selection process is a highly involved process. An important favourable property of the composite material is the possibility of tailoring it to a desired shape. Unlike metal structure, no excess metal removal is required in laminated FRP composite structures. This facilitates in fabricating the complex structural configurations.

Another important aspect of the design process for composite structural design is the selection of the manufacturing technique. The prime factors in the manufacturing process selection are cost effectiveness and reliability. In addition, other factors such as the speed of the process (fast or slow), moulding size, structural shape and size, labour intensivity and the possibility of automation are required to be considered for a successful design.

12.3 LAMINATE DESIGN

A panel or a plate, and stiffeners of FRP composite structure, basically defines a laminate. Stacking of several laminae or plies in a specified sequence to support the given load forms a basic laminate. A FRP laminate is composed mainly of two materials: the fibre reinforcement and the resin. Each layer may have different thickness and physical properties. By changing the fibre content, the characteristics of the laminate can be widely varied. The physical properties are primarily dependent on the type of the resin, type and orientation of the fibre and the moulding techniques used to fabricate the laminate.

12.3.1 Selection of Laminate

The laminate selection is a complex process. It is generally based on the area of application. For example, in some areas, the weight of the structure may override other considerations while in some other areas it may be the thermal or mechanical loading which dominates the selection of a laminate. Each application requires special consideration. At the time of the selection of plies to form the laminate, the material governing factors shown in Fig 12.6 are required to be considered depending on the area of application. A thorough structural analysis of the laminate is required for both the plate and the stiffener elements to ensure that the structure is safe and reliable. The resultant stress and deflection thus obtained are compared with the acceptable values for the particular application.

The selection of the type of laminate is one of the most important design drivers. A symmetric laminate type can be analysed with much simpler mathematical formulation while for the unsymmetrical laminate it becomes much more complex. A symmetric laminate is preferred over an unsymmetrical laminate from the point of view of mathematical analysis. In symmetric laminates, the coupling stiffness matrix [B] between extension and bending does not exist. This coupling is spurious as it reduces the effective stiffness of the laminate. The reduction of the effective stiffness means the designer can expect an increased deflection, decreased buckling load and decreased natural frequency of vibration of the structure analysed. However, in symmetric laminates, the coupling between bending and twisting does exist. This coupling has similar effects but to a lesser magnitude. The bending – twisting coupling can be eliminated by choosing cross-ply lay-ups, i.e. [0/90°] sequence. In some special design cases like a laminated shell subjected to aerodynamic loading, the bending-extension coupling can be put into use to take advantage of it. For a symmetric layer with cross ply lay-ups, both shear and bending coupling can be eliminated. On the other hand, for an angle-ply laminates both shear and twisting coupling stiffnesses are present, but the effect can be minimised by increasing the number of layers. A CSM layer is generally treated as an isotropic layer. But the woven roving layer with equal distribution of fibres in both warp and fill directions is treated as an orthotropic layer. Therefore, it is recommended that a judicious approach in the selection of the laminate be made with fine ply inter-dispersion in order to eliminate bending and shear coupling and to minimise torsional coupling. This approach aids in reducing warpage, unexpected distortions, and inter-laminar stresses. In addition, it also leads to simplistic mathematical formulations.

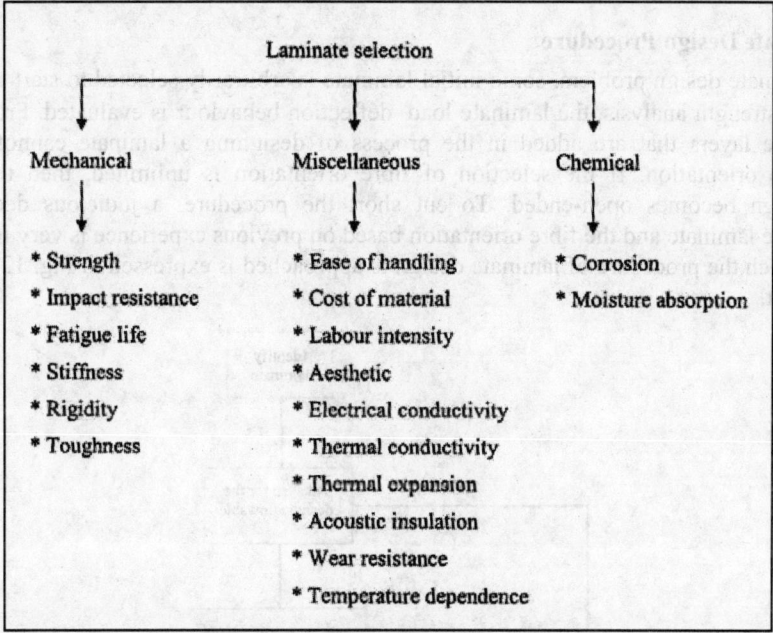

12.7 Factors affecting laminate selection

The stacking sequence is another critical factor in multilayer and multi-dimensional composite structures. The stacking sequence is important in order to avoid delamination. In angle-ply laminates, alternative stacking sequence of plies having fibre orientation of $+\theta$ and $-\theta$ is recommended rather than stacking all $+\theta$ or $-\theta$ together. Arranging in this alternative manner helps in reducing the interlaminar shear stress at the interface compared to any clustered arrangement. Additionally, the fibre orientation also influences the selection of a laminate. Preferably, fibre orientation should be decided on the basis of the direction of the loading. Factors affecting the selection of a laminate are shown in Fig. 12.7

12.3.2 Laminate Design Problem

A laminate design problem involves identifying a laminate that satisfies all the design requirements. The laminate design problem can be presented in the following steps as shown in Fig.12.8.

> | Given: | Design requirements (includes loading, constraints etc.) |
> |---|---|
> | Aim: | Design a laminate to support the load without failure during its service life |
> | Solution: | Select laminate |
> | | Decide on configuration and size |
> | | Carry out a design analysis for strength, stiffness, fatigue and Other structural responses |
> | Result: | Check for satisfactory performance; if not satisfactory, alter stacking sequence, change thickness, increase number of plies to improve the structural response measures |

Fig. 12.8 A laminate design problem

12.3.3 Laminate Design Procedure

In a laminate design problem, some initial laminate is arbitrarily selected to start the procedure. Based on the strength analysis, the laminate load–deflection behaviour is evaluated. From a practical standpoint, the layers that are added in the process of designing a laminate cannot have purely arbitrary fibre orientation. If the selection of fibre orientation is unlimited, then the process of laminate design becomes open-ended. To cut short the procedure, a judicious decision on the selection of the laminate and the fibre orientation based on previous experience is very important. The manner in which the procedure of laminate design is approached is expressed in Fig. 12.9 in the form of a flow-chart.

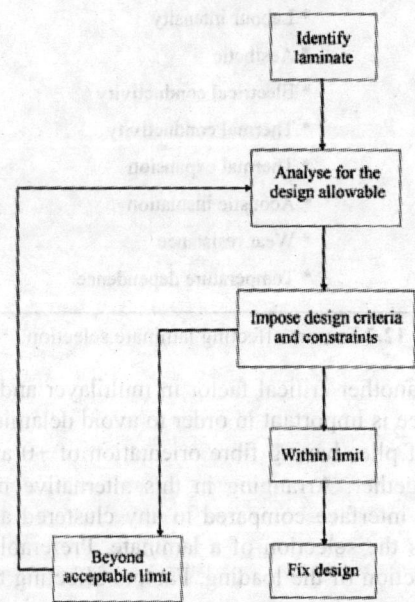

Fig. 12.9 Laminate design procedure

12.4 MATHEMATICAL ANALYSIS OF THE LAMINATE

A structure is built to support a given load without failure. These loads may be static or dynamic in nature. It is important that the structure is analysed mathematically to ensure structural integrity. The mathematical formulation for laminated composites is different from that of the homogeneous metal structures. The mathematical treatment of a laminated composite is explained in this section with a view to help in direct application. In addition, the mathematical approach is presented in such a way that it can be programmed easily for computer application. The mathematical relationships are listed below for direct application to analyse laminate.

The material property of the laminate system i.e. the thickness (d_i) of an individual ply can be estimated from the expression

$$d_i = \frac{m_{fi}\left[\dfrac{\rho_{fi}}{w_{fi}} - (\rho_{fi} - \rho_{mi})\right]}{1000\,\rho_{fi}\,\rho_{mi}} \text{ mm} \qquad (12.1)$$

where

d_i is the thickness of a ply i in mm

m_{fi} is the mass content of the fibre of the individual ply in g/mm^2

ρ_{fi} is the specific gravity of the fibre in individual ply i

ρ_{mi} is the specific gravity of the resin in individual ply i

w_{fi} is the fibre content by weight of individual ply i in Newton

In each term hereafter, in all the mathematical formulations, the suffix 'i' represents the individual ply. As the mechanical properties of the plies of a laminated composite material are different from one another, it is necessary to determine the centroid of the laminate as a whole, for further mathematical evaluation. The neutral axis (z_{NA}) of the multi-layered laminate with regard to any edge of reference is given by the expression (Fig. 12.10).

$$z_{NA} = \frac{\sum E_i d_i z_i}{\sum E_i d_i} \qquad (12.2)$$

where z_i is the distance from the reference edge to the neutral axis of the individual plies and can be calculated from the expression

$$z_i = z_{i-1} + \frac{d_i + d_{i-1}}{2} \qquad (12.3)$$

where $i>1$ and when $i = 1$, i.e., for the first layer from the edge of reference, $z_1 = d_1/2$

The neutral axis $\bar{z}_{NA}$ of the laminate from the other edge reference can be calculated from the expression

$$\bar{z}_{NA} = \left(\sum d_i\right) - z_{NA} \qquad (12.4)$$

The distance between neutral fibre of each layer to the neutral axis of the laminate is calculated from the expression

$$\bar{d}_i = |z_i - z_{NA}| \qquad (12.5)$$

The composite modulus of elasticity $E_c = \dfrac{\sum E_i d_i}{\sum d_i}$ $\qquad (12.6)$

per unit width of the laminate

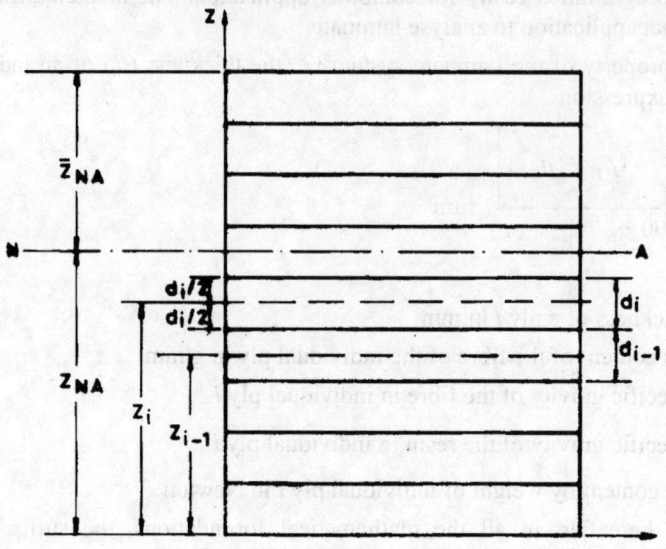

Fig. 12.10 A layered section

Equation (12.6) gives an average elastic modulus of the composite material.

For a composite laminate consisting of different layers, the material being heterogeneous in nature, the expression for the bending stiffness is written as

$$EI = \iint E(y,z) z^2 \, dy \, dz \tag{12.7}$$

In this equation, $E(y,z)$ cannot be taken out of the integral as the modulus varies between the material directions and from one ply to another. In effect, it implies that the geometry and material characters are not independent of each other. Further, it implies that an improvement in E or I in isolation may not improve the bending stiffness of the composite laminate to a desired value.

Therefore, the bending stiffness of the composite laminated plate per unit width about any free edge of the laminate is calculated from the summation of the product of the modulus of elasticity and the second moment of area of the individual plies, which constitute the laminate. Further, the bending stiffness of the composite laminated plate per unit width about its neutral axis can be evaluated by applying the parallel axis theorem.

Stiffness of the overall composite material about any free edge can be calculated.

EI per unit width can be written as

$$EI = \sum E_i \left(I_i + A_i z_i^2 \right) \tag{12.8}$$

Bending stiffness per unit width of the overall laminated composite material i.e., a laminate consisting of a number of plies about the neutral axis is obtained from the expression

$$(EI)_{NA} = \sum E_i \left[I_i + d_i \bar{d}_i^2 \right] \tag{12.9}$$

where $\bar{d}_i$ can be calculated from (12.5).

Equation (12.9) for *EI* can be directly applied to the laminated stiffness evaluation. The second moment of area for a laminated composite plate can be calculated about any free edge i.e., either the top edge or the bottom edge. The bending stiffness *EI* per unit width is then calculated about the neutral axis and is determined by applying the parallel axis theorem as described in (12.9).

The resultant bending stress due to the applied load at the extreme outer fibre of an individual ply is determined from the expression

$$\sigma_i = \frac{E_i z_i^* M}{\sum E_i I_i} \qquad (12.10)$$

where z_i^* is defined by the expression

$$z_i^* = \bar{d}_i + \frac{d_i}{2} \qquad (12.11)$$

Section modulus of the individual ply is determined from the expression

$$Z_i = \frac{\sum (E_i I_i)}{E_i z_i^*} \qquad (12.12)$$

The ultimate moment at which the stress in each ply will reach its maximum acceptable value can be determined from the expression

$$M_i = (\sigma_{max}) Z_i \qquad (12.13)$$

where σ_{max} is the ultimate strength of the individual ply.

Comparing M_i of the individual plies, the ultimate moment of resistance can be evaluated. This would enable to ascertain the ultimate load carrying capacity of the structure. For example, maximum uniformly distributed load q_{max} that can be carried by a simply supported beam is given by the expression

$$q_{max} = \frac{M_{max} \times 8}{L^2} \qquad (12.14)$$

12.4.1 Estimation of Shear Force

In addition to the determination of the load carrying capacity of the laminate in bending, the interface shear stress between the plies of a composite laminate needs to be estimated. Delamination is another important aspect of the laminated composite. Therefore, the evaluation of the inter-laminar shear stress is very important to ensure that the laminate does not fail by delamination. Due to the nature of distribution of the shear force across the cross-section of the laminate, it is necessary to consider the shear force at neutral axis and at all the interface planes between the plies that constitute the laminate. For example, in a WR-CSM laminate, the interface between WR and CSM is vulnerable to delamination failure. When a vertical shear force *Q* acts on a composite beam section, the shear stress τ at any position '*i*' on the section can be found from the expression

$$\tau = \frac{Q}{b_i D} \int_{z_i^*}^{edge} bzE dz \qquad (12.15)$$

where

D is the flexural rigidity of the section and *b* is the total width of the material resisting vertical shear at a distance *z* from the neutral axis,

Q is the applied shear force,

E is the Young's modulus and b_i is the value of b at position 'i' and the limits of integration are taken from the section of interest to the edge.

Shear displacement w_s is estimated by integrating shear strains $Q/(GA_s)$ along the length of the beam.

i) For a clamped or simply supported beam of length L under a uniform load q per unit length, w_s is given by

$$w_s = \frac{qL^2}{(8GA_s)} \qquad (12.16)$$

ii) For the beam under a centrally concentrated load P, w_s is given by

$$w_s = \frac{PL}{(4GA_s)} \qquad (12.17)$$

In case of a composite laminate, the interface shear stress between different composite materials (laminae) at any section can be determined from (12.15).

The integral in (12.15) may be rewritten as a summation over the number of lamina

$$\tau = \frac{Q \sum E_i A_i \overline{d_i}}{b_i D} \qquad (12.18)$$

Maximum shear stress that occurs in the neutral axis of the laminate due to the applied load is given by the expression.

$$\tau_{max} = \frac{Q \sum_{NA}^{edge} E_i A_i \overline{d_i}}{b_i D} \qquad (12.19)$$

$\sum E_i A_i \overline{d_i}$ is the first moment of area of the laminate above the neutral axis for the section being considered multiplied by the elastic modulus. As the elastic modulus varies from one ply to the other, E_i cannot be taken out of the summation. For any other interplanes say at the interface between ith ply and $(i+1)$th ply or for any other sections in the laminate, shear stress can be calculated directly from the expression given in (12.18).

$$\tau = \frac{Q \sum_{i\text{th ply}}^{edge} E_i A_i \overline{d_i}}{b_i D} \qquad (12.20)$$

The distance $\overline{d_i}$ of the NA in (12.20) is obtained from (12.4) and D in (12.20) is calculated from (12.9).

The maximum vertical shear force that the composite laminate is able to support is given by

$$Q_{max} = \frac{\tau_{max} D b_i}{\sum E_i A_i \overline{d_i}} \qquad (12.21)$$

where τ_{max} is the ultimate shear strength of the CSM or WR or UD lamina as the case may be. Using Q_{max} obtained from (12.21), the maximum load (q_{max}) that a composite laminate can support is calculated. For example for a simply supported beam, q_{max} is given by

$$q_{max} = \frac{2 \times Q_{max}}{L} \tag{12.22}$$

12.4.2 Estimation of Deflection

Other than the laminates composed of advanced fibres, the elastic modulus of the laminates composed of conventional fibres are very low compared to steel or other usual structural materials. This requires careful consideration while designing composite laminates to ensure that the deflection under the applied load is kept within limits. The FRP when used in conventional structural form produces large deflections, which may seriously limit the carrying capacity of the structure. To offset this disadvantage, it is essential to use those structural forms which give added stiffness by virtue of their shape thus minimising the deflection of the composite laminates. The deflection of the composite laminates is established either by theoretical calculation or by mechanical testing. For example, for a beam with both edges clamped and uniformly loaded, the deflection along the length is given by the following expression.

$$w_b(x) = \frac{qL^4}{24D}\left[\left(\frac{x}{L}\right)^4 - 2\left(\frac{x}{L}\right)^3 + \left(\frac{x}{L}\right)^2\right] \tag{12.23}$$

and

$$w_{max}(x) = \frac{q}{2S}\left(Lx - x^2\right) \tag{12.24}$$

where S is the shear stiffness and suffices 'b' and 's' indicate deflection due to bending and shear and the maximum deflection is given by

$$w_{max} = w_b + w_s = \frac{1}{384}\frac{qL^4}{EI} + \left(\frac{6}{5}\right)\frac{qL^2}{8dG} \tag{12.25}$$

Both ends clamped beams under hydrostatic loads

Hydrostatic loads are often encountered in marine structures, the deflections are given below

$$w_b(x) = \frac{qL^4}{120D}\left[\left(\frac{x}{L}\right)^5 - 5\left(\frac{x}{L}\right)^3 + 4\right] - \frac{R_l L^3}{6D}\left[\left(\frac{x}{L}\right)^3 - 3\left(\frac{x}{L}\right) + 2\right]$$
$$- \frac{M_l L^2}{2D}\left[\left(\frac{x}{L}\right)^2 - 2\left(\frac{x}{L}\right) + 1\right] \tag{12.26}$$

$$w_s(x) = \frac{1}{S}\left(R_l(x-L) - \frac{q}{6L}^2\left(x^3 - L^3\right)\right) \tag{12.27}$$

θ is the shear factor, $\quad \theta = \dfrac{D}{L^2 S} \tag{12.28}$

D is the bending stiffness.

$$R_L = \frac{qL}{20}\left(\frac{3+40\theta}{1+12\theta}\right) \quad \text{and} \quad M_L = \frac{qL^2}{30}\left(\frac{1+15\theta}{1+12\theta}\right) \tag{12.29}$$

12.4.3 Mathematical Algorithm

i) Obtain the specific gravity of the fibre and the resin from the manufacturer's pamphlet.

ii) Obtain other mechanical and physical properties of the constituent materials from the manufacturer.

iii) Estimate the weight fraction of the fibre and the resin in percentages from sample experiments depending on the manufacturing techniques.

iv) Estimate the void content.

v) Record the mass content in g/m^2.

vi) Evaluate the thickness in mm of each ply from (12.1).

vii) Estimate the total thickness of the laminate Σd_i.

viii) Calculate the distance of the neutral axis of each ply from any one of the edges from (12.3).

ix) Evaluate the theoretical Young's modulus in N/mm^2 of the laminate (12.6).

x) From the reference edge, calculate the NA (neutral axis) of the laminate (12.2).

xi) With reference to the other edge, calculate NA (12.4).

xii) Estimate the distance between the NA of the laminate and the NA of the individual ply constituting the laminate (12.5).

xiii) Calculate the rigidity of the laminate (12.9).

xiv) Calculate the stresses in the individual ply (12.10). Check for $\sigma_i \leq \dfrac{\sigma_{\text{ultimate}}}{F.S}$ for each ply, F.S is the factor of safety depending on the area of the application of the structure.

xv) Calculate the effective modulus of elasticity from (12.6).

xvi) Compare the ultimate load that can be supported by the given structure from both bending and shear considerations. The least of the two would limit the load carrying capacity.

xvii) Calculate the section modulus from (12.12).

xviii) Calculate the moment of resistance of individual ply from (12.13).

xix) Estimate the maximum load that can be supported by the given structure from the bending consideration.

xx) Estimate the shear force from (12.20).

xxi) Estimate the maximum load that can be supported by the given structure from the shear consideration.

12.5 DESIGN EXAMPLES

This section is mainly devoted to demonstrate, how to go about designing a laminated composite structure. Due to the anisotropic nature of the laminated composite materials, it is not always possible to carry out structural analysis in a straightforward manner. However, examples presented in this section are those which can be tackled manually, using hand calculators. As already described, the factors such as the choice of the fibre and the resin, fibre volume fraction, fibre orientation, ply stacking sequence and the number of plies are very important design drivers to the composite structural designer. In addition, the presence of bending, shear and torsional coupling stiffness further complicate the analysis and design of the multilayer and multi-dimensional composite structures. Fibre wet out is an important factor which limits the volume fraction of the fibre in a ply.

12.5.1 Design of a Tension Member

In case of the design of a slender member and in order to simplify calculations, it is assumed that the fibre carries the entire load. Any participation by the resin matrix to support the load is ignored for the analysis of the slender member. However, the resin can influence the manufacturing and environmental considerations of the member. The design of joints or connections for the tension member is very critical, as these may form a very weak point.

Example 12.1

A slender member of 500 mm × 25 mm × 12 mm is required to carry a load of 150 kN. The permissible axial extension of the bar for the given load should not exceed 1.25 mm. Assume that the load is carried entirely by the fibre, determine

  a) laminate type;
  b) fibre orientation;
  c) fibre volume fraction;
  d) number of plies;

Solution

In the design of a slender tension member subjected to axial tension P, the axial elongation (Δ) is

$$\Delta = \frac{PL_0}{EA_0}$$

where EA_0 is the axial stiffness
L_0 is the length of the member

or, $\quad E = \frac{PL_0}{\Delta A_0}$

$$= \frac{150 \times 500}{1.25 \times (25 \times 12)} = 200 \quad \text{kN/mm}^2 = 200 \text{ GPa}$$

$$\sigma_1 = \frac{150000}{25 \times 12} = 500 \text{ N/mm}^2$$

For the present problem, the next step is to choose a ply with an elastic modulus of 200 GPa. This can be done with the help of standard design charts representing the elastic modulus vs. the fibre volume fraction. These charts have been developed for different combinations of fibre and resin such as glass–polyester, glass–epoxy, boron–epoxy etc. [12.1]

Since the modulus required is high, a boron–epoxy composite lamina is considered whose elastic modulus is 201 GPa at a fibre volume fraction of 0.5. As the direction of the loading is uniaxial, unidirectional lamina is selected. Therefore, the fibre orientation is unidirectional.

The thickness of a ply depends on the fibre volume fraction. From the design charts, the thickness of a boron–epoxy lamina with $V_f = 0.5$, is found to be 0.55 mm.

No. of plies required $= \frac{12}{0.55} = 21.82$

22 plies are required to support the given load.

In addition, for a boron-epoxy ply with $V_f = 0.5$, the longitudinal strength is 1380 GPa. i.e., $\sigma_{ult} > \sigma_1$. Hence, the member is adequate to support the load.

Composite type	= boron epoxy laminate
Fibre volume fraction V_f	= 0.5
Fibre orientation	= unidirectional
Number of plies	= 22

12.5.2 Laminate Design for Strength

These are certain structural members, which require high load carrying capacity. The loading may be static and/or dynamic in nature. Here, the strength of the laminate becomes the design driver. Fibres of carbon, aramid, S-glass and boron generally give high strength laminates. In composite laminates inter-laminar strength is also an important factor.

For a preliminary design, a carpet plot may be used for design purposes [Fig12.11–12.12]. A carpet plot is nothing but a design chart showing the uniaxial stiffness or strength as a function of arbitrary ratios of 0, 90 and ± 45 degree plies. The idea here is to get the basic lamina properties, mainly the inplane properties, for the general $[0/90/\pm 45]_s$ laminate in terms of ply ratios which is a quasi-isotropic laminate. A ply ratio is defined as the ratio of the number of $0°/90°/\pm 45°$ plies to the total number of plies. The carpet plots are useful in obtaining first-hand information on the proportion of various ply orientations required to meet the particular design requirements. An example using carpet plots is presented next to understand the utility of carpet plots.

Example 12.2

Using the design charts of Fig.12.11 and 12.12, determine the number of plies of $0°$, $90°$ and $\pm 45°$ orientations in a quasi-isotropic $[0/90/\pm 45]_s$ laminate that meets the following criteria

(a) Minimum modulus in the axial direction = 41.4 MPa
(b) Minimum strength in the axial 0^0 direction = 448.5 MPa

Total laminate thickness not to exceed 2.5 mm

Solution:

Refering to Fig. 12.11 the ply ratio is 0:90:±45 = 20%:60%:20% for tensile modulus of 41.4 MPa.

For this ply ratio, strength from Fig. 12.12 is 352 MPa, which is less than the minimum. The new ply ratio for the minimum strength in 0° of 448.5 MPa is
$0:90:\pm 45 = 30:50:20$

For this ratio, the axial modulus is 48.3 MPa which is higher than the minimum required.

Thus, the laminate configuration selected is $[0_3/90_5/\pm 45_2]_s$.

The laminate thickness is 2.5 mm

0°	30% of 2.5 mm	0.75 mm
90°	50% of 2.5 mm	1.25 mm
+45°	10% of 2.5 mm	0.25 mm
−45°	10% of 2.5 mm	0.25 mm

Ply thickness $= \dfrac{0.25}{2} = 0.125$ mm

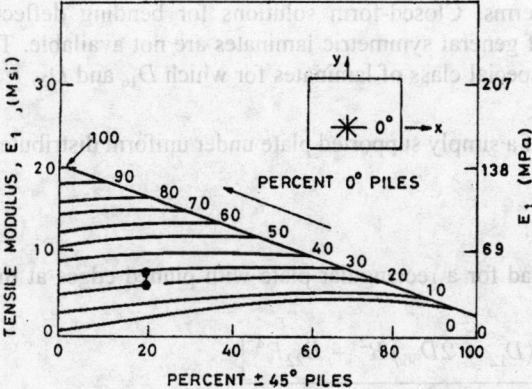

Fig. 12.11 Carpet plot for stiffness

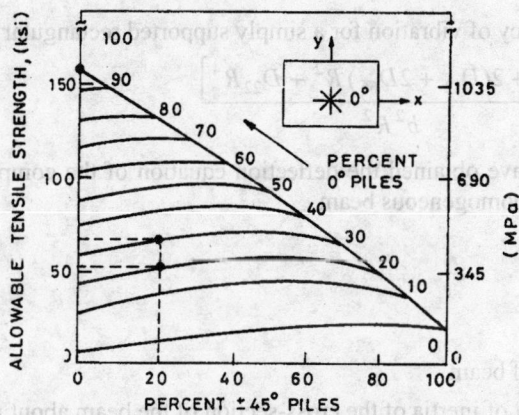

Fig. 12.12 Carpet plot for strength

No of plies in each fibre orientation are

    0°   : 6
    90°   : 10
    45°   : 2
    −45° : 2

12.5.3 Laminate Design for Stiffness

In the design criteria of some structures, the maximum deflection is limited to some specific value. Structures such as aircraft control surfaces, submersibles, sports rackets etc., require small deflections under applied loads. To minimise deflection, the structure with high stiffness is needed. For such cases, fibres with high stiffness such as boron, graphite, carbon, kevlar and aramid are chosen. For the structures made of symmetric laminates, the stiffness is related to the [A] matrix under the inplane loading, while its stiffness against bending, buckling and torsional loads is related to the [D] matrix.

Except for 0°, 90° and 0°/90° combinations, the [D] matrix for all symmetric laminates contains non-zero D_{16} and D_{26} terms. Closed-form solutions for bending deflection, buckling loads and vibrational frequencies of general symmetric laminates are not available. The following closed-form solutions are valid for a special class of laminates for which D_{16} and $D_{26} = 0$ and elements of the [B] matrix are negligible.

1. Central deflection of a simply supported plate under uniform distributed load is

$$w_{max} = \alpha \frac{qa^3 b}{D_{11}} \tag{12.30}$$

2. Critical buckling load for a rectangular plate with pinned edges at the ends of its longitudinal dimensions.

$$N_{cr} = \frac{\pi^2 \left[D_{11} + 2(D_{12} + 2D_{66})R^2 + D_{22}R^4 \right]}{b^2 R^2} \tag{12.31}$$

where $R = \dfrac{a}{b}$

3. Fundamental frequency of vibration for a simply supported rectangular plate

$$f^2 = \frac{\pi^4}{\rho R^4 b^4} \frac{\left[D_{11} + 2(D_{12} + 2D_{66})R^2 + D_{22}R^4 \right]}{b^2 R^2} \tag{12.32}$$

Whitney et al. [12.2] have obtained the deflection equation of the composite symmetric beam in the same form as that of a homogeneous beam

$$\frac{d^2 w}{dx^2} = \frac{bM_x}{E_b I} \tag{12.33}$$

where

- b is the width of beam
- I is the moment of inertia of the cross-section of the beam about midplane
- E_b is the effective bending modulus $= \dfrac{12}{d^3 D_{11}}$
- d is the beam depth
- D_{11} is the first element in the inverse [D] matrix

Transverse shear has been neglected in (12.33).

Example 12.3

It is proposed to replace a steel beam having a bending stiffness of 26.2 kN/m with a hybrid beam made of three laminates consisting of carbon–epoxy plies and 2 laminates consisting of S-glass–epoxy plies. The fibres in all the five laminates are laid unidirectional along the beam axis. The thickness of each carbon ply is 0.15 mm and the S-glass ply is 0.13 mm. The cross-section of the beam is given as (25.4 mm × h). Determine the depth of the beam. Also, determine the number of plies in each laminate. Assume $E_c = 207$ GPa, $E_g = 43$ GPa.

Solution:

Referring to Fig 12.13, two carbon laminates are placed on the two outer surfaces to achieve maximum stiffness. The stacking sequence of these laminates is shown in Fig 12.13. It is assumed that each lamina is of thickness d_0. All the fibres are of 0° orientation.

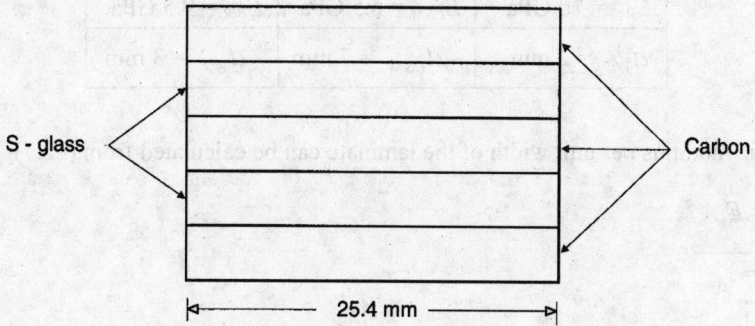

Fig. 12.13 Example 12.4

$$(EI)_{hybrid} = 2E_c I_1 + 2E_g I_2 + E_c I_3$$

$$= 2E_c \left[\frac{bd_0^3}{12} + bd_0 \left(d_0 + \frac{d_0}{2} + \frac{d_0}{2} \right)^2 \right] + 2E_g \left[\frac{bd_0^3}{12} + bd_0 \left(\frac{d_0}{2} + \frac{d_0}{2} \right)^2 \right] + E_c \frac{bd_0^3}{12}$$

$$= bd_0^3 \left(\frac{100 E_c}{12} + \frac{26 E_g}{12} \right)$$

$$10^3 \times 26.2 = 0.0254 d_0^3 \left(\frac{100}{12} \times 207 \times 10^9 + \frac{26}{12} \times 43 \times 10^9 \right)$$

or, $\qquad d_0 = 8.3$ mm

Since, there are five laminates, the total thickness $h = 5 \times 8.3 = 41.5$ mm

In comparison, the thickness of the steel beam of same width is 39 mm

No. of plies in each layer of carbon fibre ply $= \dfrac{8.3}{0.15} = 55.3 = 56$

No. of plies in each layer of glass ply $= \dfrac{8.3}{0.13} = 63.8 = 64$

Example 12.4

An E-glass-polyester composite laminate comprising of woven cloth, CSM and woven roving plies are required to support a tensile load. The particulars of the plies are given below. Assume that the plies are perfectly bonded.

a) Find the elastic modulus per unit width of the composite laminate

b) Determine the maximum load carrying capacity of the laminate in tension per unit width of the laminate.

Given

Woven cloth	CSM	WR
F_{WC} = 125 MPa	F_{CSM} = 50 MPa	F_{WR} = 165 MPa
E_{WC} = 10 GPa	E_{CSM} = 6.5 GPa	E_{WR} = 9.5 GPa
d_{WC} = 2 mm	d_{CSM} = 7 mm	d_{WR} = 3 mm

Solution

a) The elastic modulus per unit width of the laminate can be calculated from (12.7)

$$E_c = \frac{\sum_{i=1}^{n} E_i A_i}{\sum_{i=1}^{n} A_i}$$

$$E_c = \frac{10 \times 2 + 6.5 \times 7 + 9.5 \times 3}{(2 + 7 + 3)} = \frac{94}{12} = 7.833 \text{ GPa per unit mm width}$$

b) Due to perfect bonding of the plies, the axial strain is equal in all the plies

$$\varepsilon_{WC} = \varepsilon_{WR} = \varepsilon_{CSM} \tag{a}$$

The strain in each ply can be expressed as given below

$$\varepsilon_{WC} = \frac{F_{WC}}{E_{WC}}; \quad \varepsilon_{WR} = \frac{F_{WR}}{E_{WR}}; \quad \varepsilon_{CSM} = \frac{F_{CSM}}{E_{CSM}}$$

From the equal strain condition, the expression above can be written as

$$\frac{F_{WC}}{E_{WC}} = \frac{F_{WR}}{E_{WR}} = \frac{F_{CSM}}{E_{CSM}} \tag{b}$$

At this stage, it is required to find out which one of the three plies dominates the design calculations. To predict this, let us assume as a first case, that the ultimate tensile strength of the CSM ply is the critical design driver.

Therefore,

Case I

F_{CSM} = 50 MPa

F_{WR} and F_{WC} are unknowns.

From eqn. (b)

$$F_{WR} = \frac{F_{CSM}}{E_{CSM}} \times E_{WR} = \frac{50}{6.5} \times 9.5 = 73 \text{ MPa}$$

$$F_{WC} = \frac{F_{CSM}}{E_{CSM}} \times E_{WC} = \frac{50}{6.5} \times 10 = 76.9 \text{ MPa}$$

Compare the above-evaluated values of F_{WR}, F_{WC} and F_{CSM} with the ultimate values given in the problem. It is found that the laminate is safe under this assumption.

Case II :

In Case II, let us assume that the ultimate tensile strength of the WR ply is the critical design driver.

F_{WR} = 165 MPa

F_{CSM} and F_{WC} are unknowns.

$$F_{CSM} = \frac{F_{WR}}{E_{WR}} \times E_{CSM} = \frac{165}{9.5} \times 6.5 = 112.89 \text{ MPa}$$

$$F_{WC} = \frac{F_{WR}}{E_{WR}} \times E_{WC} = \frac{165}{9.5} \times 10 = 173.6 \text{ MPa}$$

Compare the above-evaluated values of F_{WR}, F_{WC} and F_{CSM} with the ultimate values given in the problem. It is found that the calculated F_{CSM} and F_{WC} are more than the permissible ultimate tensile strength values. Therefore, this assumption is not valid and the ultimate strength of the WR ply is not the critical design driver.

Case-III :

In Case III, let us assume that the ultimate tensile strength of the woven cloth is the critical stress controlling parameter

F_{WC} = 125 MPa

F_{CSM} and F_{WC} are unknowns.

$$F_{CSM} = \frac{F_{WC}}{E_{WC}} \times E_{CSM} = \frac{125}{10} \times 6.5 = 81.25 \text{ MPa}$$

$$F_{WR} = \frac{F_{WC}}{E_{WC}} \times E_{WR} = \frac{125}{10} \times 9.5 = 98.95 \text{ MPa}$$

Compare the above-evaluated values of F_{WR}, F_{WC} and F_{CSM} with the ultimate values given in the problem. It is found that the calculated F_{WR} is within acceptable limits. However, F_{CSM} is more than the permissible ultimate tensile strength value. Therefore, this assumption is not valid and the ultimate strength of the WC ply is not the critical design driver.

As Case I is acceptable, the ultimate strength of the CSM ply is the critical stress controlling parameter. Therefore, the final stresses in each ply are

F_{CSM} = 50 MPa

F_{WR} = 73 MPa

F_{WC} = 76.9 MPa

The ultimate load carrying capacity of the laminate per unit width is given by:

$$q_{max} = F_{CSM}\ A_{CSM} + F_{WC}\ A_{WC} + F_{WR}\ A_{WR}$$

$$= (50 \times 7) + (76.9 \times 2) + (73 \times 3)$$

$$= 722.8 \text{ N/mm}$$

Example 12.5

A 1m long cross-ply laminated cantilever glass–polyester composite beam having a square cross-section of size 20 mm is required to support a uniformly distributed loading. Assume equal distribution of fibres in both longitudinal and transverse directions. The ultimate inter-laminar shear stress for the beam is equal to 20 MPa. Determine the load carrying capacity of the beam by considering a factor of safety (FOS) of 4. E_{WR} = 10 GPa, σ_{bend} = 195 MPa.

Solution

The beam is subjected to both bending moment and shear force. Therefore, it is to be found out whether the beam is safe from both considerations.

a) Maximum load carrying capacity of the beam based on the maximum bending moment

Maximum bending moment is given by $M_{max} = \dfrac{qL^2}{2}$

$$M = \dfrac{q}{2} \times 10^6 \quad \text{N/mm} \tag{a}$$

Maximum shear stress is given by

$$V = qL = 1000\,q \quad \text{N}$$

Section modulus is given by

$$Z = \dfrac{1}{6}bd^2 = 1333.33 \text{ mm}^3$$

Using standard beam formula

$$M = \sigma Z = 195 \times 1333.33 = 260000 \quad \text{Nmm} \tag{b}$$

Combining (a) and (b) yields

$$q_{max} = \dfrac{260000 \times 2}{10^6} = 0.52 \text{ N/mm}$$

Using a factor of safety of 4, yields

$$q_{max} = 0.13 \text{ N/mm}$$

Based on maximum shear condition

$$\sigma_s = \dfrac{VQ}{Ib}$$

Q is the moment of area of the cross-section above the section considered

$$Q = b \times \dfrac{t}{2} \times \dfrac{t}{4} = 20 \times \dfrac{20}{2} \times \dfrac{20}{4} = 1000 \text{ mm}^3$$

$$\sigma_s = \dfrac{1000\,V}{\dfrac{1}{12} \times 20 \times 20^3 \times 20} = 0.00375\,V$$

or $\quad V = \dfrac{20}{0.00375} = 5333.33 \text{ N} \quad$ which gives

$$q_{max} L = 5333.33;$$

or $\quad q_{max} = \dfrac{5333.33}{1000}$

$$q_{max} = 5.333 \text{ N/mm}$$

With a factor of safety of 4, one obtains

$q_{max} = 1.333 \text{N/mm}$

Therefore, the maximum load carrying capacity is governed by bending and the permissible $q_{max} = 0.13$ N/mm.

Example 12.6

An E-glass–polyester composite laminate system consists of isotropic and orthotropic plies. The thickness of the isotropic laminate is 10 mm and that of the orthotropic laminate is 3 mm. Compute the elongation of the composite system in the direction of the applied tensile load if the load is applied at an angle of 20° off-axis. Assume the plies to be perfectly bonded.

Given

Properties	Isotropic	Orthotropic
F_1 (MPa)	65	123
F_2 (MPa)	65	95
$(F_x)_{45}$ (MPa)	65	58
E_1 (GPa)	7	14.0
E_2 (GPa)	7	12.0
E_6 (GPa)	-	3.75
ν_{12} (GPa)	-	0.20

Solution

The shear strength of the orthotropic laminate system in the off-axis direction i.e., $\theta = 45°$ is obtained from Tsai-Hill criterion.

$$\left(\frac{1}{(F_6)_{45}}\right)^2 = \frac{1}{m^2 n^2}\left[\frac{1}{(F_x)_{45}^2} - \frac{m^4}{(F_1)^2} - \frac{n^4}{(F_2)^2}\right] \tag{a}$$

where $m = \cos\theta = \cos 45° = \dfrac{1}{\sqrt{2}}$ and $n = \sin\theta = \sin 45° = \dfrac{1}{\sqrt{2}}$

Substituting the values in (a), yields

$$\left(\frac{1}{(F_6)_{45}}\right)^2 = \frac{1}{0.25}\left[\left(\frac{1}{58}\right)^2 - \frac{0.25}{(123)^2} - \frac{0.25}{(95)^2}\right]$$

$$\left(\frac{1}{(F_6)_{45}}\right)^2 = 1.012 \times 10^{-3}$$

$F_6 = 31.43$ MPa

The tensile strength of the WR laminate in the off-axis direction i.e., $\theta = 20°$ is evaluated from the Tsai-Hill criterion.

$$\frac{1}{(F_x)^2} = \frac{m^4}{F_1^2} + \frac{n^4}{F_2^2} + \frac{m^2 n^2}{F_6^2}$$

$$= \frac{0.78}{(123)^2} + \frac{0.014}{(95)^2} + \frac{0.1033}{(31.43)^2}$$

or, $F_x = 79.64$ MPa

Elastic modulus of the WR laminate at an angle $\theta = 20°$ is obtained from

$$\frac{1}{E_x} = \frac{m^4}{E_1} + \frac{n^4}{E_2} + \left[\frac{1}{E_6} - \frac{2v_{12}}{E_1}\right] m^2 n^2$$

$$= \frac{0.78}{14} + \frac{0.014}{12} + \left[\frac{1}{3.75} - \frac{2 \times 0.2}{14}\right] \times 0.1033$$

$E_x = 12.27$ GPa

Due to the perfect bonding between the isotropic and orthotropic plies, the axial strain in all the plies are equal.

$$(\varepsilon_{20})_{iso} = (\varepsilon_{20})_{ortho}$$

or at $\theta = 20°$

$$\frac{(\sigma_{ult})_{iso}}{E_{iso}} = \frac{(\sigma_{ult})_{ortho}}{E_{ortho}}$$

As mentioned in Example 12.6, let us assume that the $(\sigma_{ult})_{iso}$ is the maximum stress controlling parameter.

$$(\sigma_{ult})_{ortho} = \frac{(\sigma_{ult})_{iso}}{E_{iso}} \times E_{ortho}$$

$$= \frac{65}{7} \times 12.27 = 113.94 \text{ MPa}$$

Since $(\sigma_{ult})_{ortho} > F_x$ calculated at $\theta = 20°$, $(\sigma_{ult})_{iso}$ can not be the maximum stress controlling parameter.

As a second case, let us assume that the ultimate tensile strength $(\sigma_{ult})_{ortho}$ of the othotropic laminate is the critical stress controlling parameter

$$(\sigma_{ult})_{CSM} = \frac{(\sigma_{ult})_{WR}}{E_{WR}} \times E_{CSM}$$

$$= \frac{79.64}{12.27} \times 7 = 45.43$$

Since $(\sigma_{ult})_{iso} < (F_x)_{iso}$; and hence this assumption is valid for safe stress operations.

$\therefore \sigma_{ortho} = 79.64$ MPa

$\sigma_{iso} = 45.43$ MPa

Maximum tensile load that can be carried in the direction of $\theta = 20°$ can now be calculated.

$$q = \sigma_{\text{ortho}} A_{\text{ortho}} + \sigma_{\text{iso}} A_{\text{iso}}$$
$$= 79.64 \times 3 + 45.43 \times 10$$
$$= 693.22 \text{ N/mm}$$

The elastic modulus of the composite laminate at $\theta = 20°$

$$E_c = \frac{\sum_{i=1}^{n} E_i A_i}{\sum_{i=1}^{n} A_i} = \frac{7 \times 10 + 12.27 \times 3}{13} = 8.22 \text{ GPa per unit width}$$

The elongation of the composite laminate system at $\theta = 20°$

$$\delta l = \frac{PL}{AE} = \frac{693.22 \times L}{13 \times 8220} = 6.49 L \times 10^{-3} \text{ mm}$$

Example 12.7

A beam in Fig. 12.14 is made from the WR and CSM laminate. Find the maximum load carrying capacity of the beam if its ends are fixed. Each ply is 2 mm thick

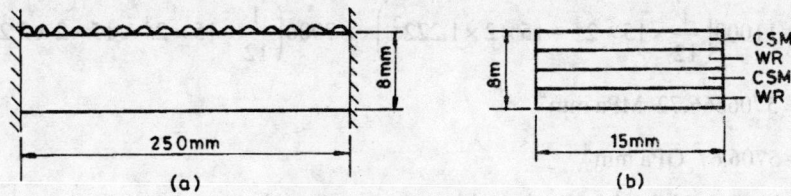

Fig. 12.14 Example 12.7

Given

	CSM	WR
σ_{bend}	= 150 MPa	σ_{bend} = 200 MPa
E_{bend}	= 7 GPa	E_{bend} = 11 GPa
σ_s	= 22 MPa	σ_s = 20 MPa

Solution

The beam is subjected to bending moment and the shear force. Firstly, as the mechanical properties of the WR and CSM are different, the position of the neutral axis is to be first estimated. The distance of NA from the top of the free edge is estimated as follows.

Step 1: The position of the neutral axis from the top reference plane is obtained from (12.2)

$$z_{NA} = \frac{\sum E_i A_i z_i}{\sum E_i A_i}$$

$$= \frac{(11000 \times 2 \times 15 \times 1) + (7000 \times 2 \times 15 \times 3) + (11000 \times 2 \times 15 \times 5) + (7000 \times 2 \times 15 \times 7)}{(11000 \times 2 \times 15) + (7000 \times 2 \times 15) + (11000 \times 2 \times 15) + (7000 \times 2 \times 15)}$$

$$= 3.778 \text{ mm}$$

From (12.5)

$$\overline{d}_i = \overline{z}_{NA} - z_i \quad \text{when} \quad \overline{z}_{NA} \geq z_i$$

$$\overline{d}_i = z_i - \overline{z}_{NA} \quad \text{when} \quad \overline{z}_{NA} < z_i$$

$$d_1 = 3.778 - 1 = 2.778 \text{ mm}$$

$$d_2 = 3.778 - 3 = 0.778 \text{ mm}$$

$$d_3 = 5 - 3.778 = 1.222 \text{ mm}$$

$$d_4 = 7 - 3.778 = 3.222 \text{ mm}$$

The elastic bending stiffness of the composite beam is obtained from (12.7).

$$D_c = [EI]_c = \sum E_i \left(I_i + A_i z_i^2 \right)$$

$$(EI)_c = 11000 \left(\frac{1}{12} \times 15 \times 2^3 + 15 \times 2 \times 2.778^2 \right) + 7000 \left(\frac{1}{12} \times 15 \times 2^3 + 15 \times 2 \times 0.778^2 \right)$$

$$+ 11000 \left(\frac{1}{12} \times 15 \times 2^3 + 15 \times 2 \times 1.222^2 \right) + 7000 \left(\frac{1}{12} \times 15 \times 2^3 + 15 \times 2 \times 3.222^2 \right)$$

$$(EI)_c = 5706666.72 \text{ MPa mm}^4$$

$$= 5706.67 \text{ GPa mm}^4$$

As the elastic modulus in a laminated beam varies from layer to layer, the section modulus (Z_i) is obtained as follows

$$\frac{M}{(EI)_c} = \frac{\sigma_i}{E_i z_i}$$

$$M = \sigma_i \times Z_i$$

where

$$Z_i = \frac{(EI)_c}{E_i z_i}$$

Now applying the above expression to obtain the section modulus of each ply

$$Z_{WR} = \frac{(EI)_c}{E_i z_i} = \frac{5706.67}{11 \times 3.778} = 137.32 \text{ mm}^3$$

$$Z_{CSM} = \frac{5706.67}{7 \times 1.778} = 458.514 \text{ mm}^3$$

$$Z_{WR} = \frac{5706.67}{11 \times 2.222} = 233.48 \text{ mm}^3$$

$$Z_{CSM} = \frac{5706.67}{7 \times 4.222} = 193.093 \text{ mm}^3$$

Permissible bending moment for each ply is now evaluated

$M_{WR} = 200 \times 137.32 = 27464$ Nmm

$M_{CSM} = 150 \times 458.514 = 68777.1$ Nmm

$M_{WR} = 200 \times 233.48 = 46696$ Nmm

$M_{CSM} = 150 \times 193.093 = 28963.95$ Nmm

The maximum permissible bending moment for the composite laminated beam is 27464 Nmm.

The maximum load carrying capacity of the fixed end beam based on the permissible bending moment is given by

$$q_{max} = \frac{M \times 12}{L^2} = \frac{27464 \times 12}{250^2} = 5.273 \text{ N/mm}$$

The inter-laminar shear stress is very critical in case of laminated beams. The shear stress at any section under consideration is given by

$$\sigma_{12} = \frac{V \sum E_i A_i z_i}{(EI_c) b}$$

and the shear force is $\quad V = \dfrac{\sigma_{12} (EI)_c b}{\sum E_i A_i z_i}$

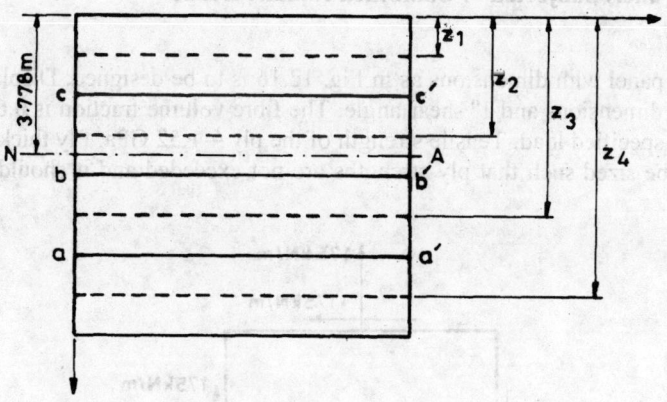

Fig. 12.15 Notations of Example 12.7

At the shear plane aa' as shown in (Fig. 12.15), the shear force is given by

$$V = \frac{20 \times 5706.67 \times 15}{11 \times 2 \times 15 \times 3.222} = 1610.14 \text{ N}$$

The shear force at the neutral axis of the beam, which lies in the CSM ply, is given by

$$V = \frac{22 \times 5706.67 \times 15}{\left(7 \times 0.222 \times 15 \times 0.111\right) + (11 \times 2 \times 15 \times 1.222) + \left(7 \times 15 \times 2 \times 3.222\right)} = 1739 \text{ N}$$

At the shear plane bb', the shear force is given by

$$V = \frac{22 \times 5706.67 \times 15}{7 \times 15 \times 4 \times 1.222 + (11 \times 2 \times 15 \times 3.222)} = 1771.16 \text{ N}$$

At the shear plane cc', the shear force is given by

$$V = \frac{20 \times 5.70667 \times 15}{11 \times 15 \times 2 \times 2.778} = 1867.49 \text{ N}$$

The maximum permissible shear force for the composite laminated beam is 1610.14 N

The maximum load carrying capacity of the fixed end beam based on the permissible shear force is given by

$$q_{max} = \frac{1610.14 \times 2}{250} = 12.88 \text{ N/mm}$$

Based on the comparison of the results obtained for the maximum load carrying capacity from the maximum permissible bending moment condition and the maximum shear force condition, it is concluded that the maximum permissible bending moment condition controls the load carrying capacity of the given beam.

Therefore, the ultimate load carrying capacity q_{max} = 5.273 N/mm

12.5.4 Composite Panels Subjected to Combined Inplane Loads

Example 12.8

A rectangular panel with dimensions as in Fig. 12.16 is to be designed. Displacement limits are 0.5 percent of edge dimensions and 1° shear angle. The fibre volume fraction is 0.6. Assume a factor of safety 2.0 on the specified load. Tensile strength of the ply = 1.52 GPa. Ply thickness is 0.127 mm. The laminate is to be sized such that ply strengths are not exceeded and it should not buckle at the design load.

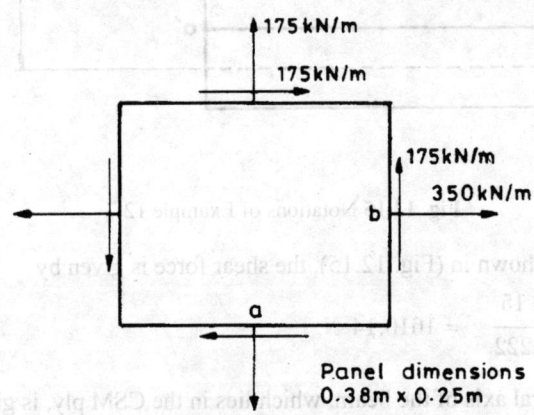

Fig. 12.16 Example 12.8

Solution

The problem has been dealt in Ref. 12.4.

Step 1: Design variables are number of plies, ply orientation and ply stacking sequence.

Step 2: Designed loads are

$$N_x = 2 \times 350 = 700 \text{ kN/m}$$

$$N_y = 2 \times 175 = 350 \text{ kN/m}$$

$$N_s = 2 \times 175 = 350 \text{ kN/m}$$

Step 3: Composite material on-axis properties are obtained from Tables. The ply off-axis properties can be obtained from the coordinate transformation.

Step 4: The laminate configuration is next to be selected

$$N_{l0} = \text{Number of } 0° \text{ plies} = \frac{\text{Designed load } (N_x)}{\text{Longitudinal tensile strength} \times \text{ply thickness}}$$

$$= \frac{700 \times 10^3 \times 10^3}{1.52 \times 10^9 \times 0.127} = 3.63 \cong 4$$

Use $N_{l0} = 8$ (double because of combined loading)

$$N_{l90} = \text{Number of } 90° \text{ plies} = \frac{\text{Designed load } (N_y)}{\text{Longitudinal tensile strength} \times \text{ply thickness}}$$

$$N_{l0} = \frac{350 \times 10^3 \times 10^3}{1.52 \times 10^9 \times 0.127} \cong 1.82 = 2$$

Use $N_{l90} = 4$ (double because of combined loading)

$$\text{No. of } \pm 45° \text{ plies} = \frac{\text{Designed load } (N_{xy}) \times \left[\frac{1}{2}\right] (E_{l1}/G_{\theta 12})}{S_{l1} t_t}$$

E_{l1} is the ply longitudinal modulus = 128 GPa

$G_{\theta 12}$ is the $\pm 45°$ composite shear modulus = 40 GPa

S_{l1} is the long compressive strength = 1.25 GPa

$$N_{l \pm 45} = \frac{350 \times 10^3 \times 128 \times 10^3}{2 \times 40 \times 1.25 \times 10^9 \times 0.127} = 3.53 = 4$$

Use $N_{l \pm 45} = 8$ (double because of combined loading)

The laminate therefore has 20 plies (8 at 0°, 4 at 90°, 8 at $\pm 45°$)

Laminate thickness = 20×0.127 mm = 2.54 mm

Required laminate configuration is $[\pm 45°/0/90/0]_{2s}$

Note that

(1) The scantlings of the laminate have been calculated based on fibre controlled properties. Due to combined loading the number of plies in each orientation have been doubled.

(2) ± 45° plies have been placed on the extreme ends for increasing the shear buckling resistance of the laminate.

(3) The number of ± 45° plies is selected on the basis of longitudinal compression, as it is less than the longitudinal tensile strength.

(4) Once the preliminary scantlings are determined, classical laminate theory is applied for determining force–displacement relationships as also ply stresses.

Checks for Shear Buckling

Thin panels are generally designed on the basis of stability consideration. Instability arises from compressive or inplane shear loads. In the present case, the panel is subjected to tension and inplane shear loading. As such the panel is checked for shear buckling only.

The critical stresses in shear buckling is approximately given by

$$\sigma_{s_{cr}} = \frac{7\pi^2 d^2 E}{12 b^2 (1 - \upsilon_{xy} \upsilon_{yx})} \quad \text{for} \quad \left(1 \le \frac{a}{b} \le 2\right) \tag{a}$$

$$E = (4 E_{xx} E_{yy} G_{xy})^{1/3} \tag{b}$$

The following values are chosen for the laminate

$\upsilon_{xy} = 0.33, \quad \upsilon_{yx} = 0.22, \quad E_{xx} = 66.22$ GPa,

$E_{yy} = 44.83$ GPa, $\quad G_{xy} = 15.86$ GPa

$E = (4 \times 66.22 \times 44.83 \times 15.86)^{1/3} = 57.32$ GPa

$b = 0.25$ m $\quad d = 2.54$ mm $= 0.00254$ m

$$\sigma_{s_{cr}} = \frac{7\pi^2 \times (0.00254)^2 \times 57.32 \times 10^9}{12 \times (0.25)^2 \times (1 - 0.33 \times 0.22)} = 35.58 \text{ MPa}$$

$$\sigma_s \text{ (design)} = \frac{350}{0.00254} \text{ kN/m}^2 = 137.8 \text{ MPa}$$

Therefore, $\quad \sigma_{s_{cr}} < \sigma_s \text{ (design)}$,

Failure is imminent.

Therefore, shear buckling is to be checked in combination with two normal tensile stresses (σ_x and σ_y).

Interaction equation of the stresses is as follows

$$\frac{\sigma_x}{\sigma_{x_{cr}}} + \frac{\sigma_y}{\sigma_{y_{cr}}} - \left[\frac{\sigma_s}{\sigma_{s_{cr}}}\right]^2 + 1.0 \ge 0 \tag{c}$$

σ_s, σ_y, σ_s are the laminate stresses at design load. The approximate estimation of $\sigma_{x_{cr}}$ and $\sigma_{y_{cr}}$ are as follows

$$\sigma_{x_{cr}} = \sigma_{y_{cr}} = \frac{\pi^2 d^2 E}{12 b^2 (1 - \upsilon_{xy} \upsilon_{yx})} \left(\frac{a}{b} + \frac{b}{a} \right)^2 \qquad \text{(d)}$$

$$\sigma_{x_{cr}} = \sigma_{y_{cr}} = \frac{\pi^2 \times (0.00254)^2 \times 57.32 \times 10^9}{12 \times (0.254)^2 \times (1 \times 0.33 \times 0.22)} \left(\frac{0.38}{0.25} + \frac{0.25}{0.38} \right)^2 \text{ N/m}^2$$

$$= 24 \text{ MPa}$$

$$\sigma_x \text{ (design)} = \frac{7000 \times 10^3}{0.00254} \text{ N/m}^2 = 275 \text{ MPa}$$

$$\sigma_y \text{ (design)} = \frac{350 \times 10^3}{0.00254} \text{ N/m}^2 = 137.8 \text{ MPa}$$

Therefore,

$$\frac{275}{24} + \frac{137.8}{24} - \left(\frac{137.8}{35.58} \right)^2 + 1.0 = 11.46 + 5.74 - 15 + 1 = 3.2 > 1$$

If all the loads act simultaneously, the panel will not buckle due to shear loading.

Example 12.9

A simply supported composite laminate of 700 mm length and 100 mm width is required to support a compressive load parallel to the warp direction as shown in Fig.12.17. Compute the critical compressive load. $E_{WR} = 11$ GPa, $E_{CSM} = 7$ GPa.

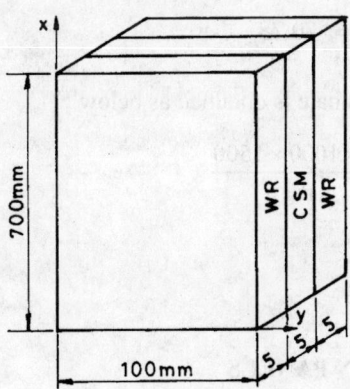

Fig. 12.17 Example 12.9

Solution:

The neutral axis of the whole section from the free edge can be calculated from (12.3)

$$z_{NA} = \frac{\sum A_i E_i z_i}{\sum A_i E_i} = \frac{(100 \times 5 \times 11 \times 2.5) + (100 \times 5 \times 7 \times 7.5) + (100 \times 5 \times 11 \times 12.5)}{(100 \times 5 \times 11) + (100 \times 5 \times 7) + (100 \times 5 \times 11)}$$

$$z_{NA} = 7.5 \text{ mm}$$

The composite modulus of elasticity is calculated from (12.6).

$$E_c = \frac{\sum E_i A_i}{\sum A_i} = \frac{(100 \times 5 \times 11) + (100 \times 5 \times 7) + (100 \times 5 \times 11)}{(100 \times 5 + 100 \times 5 + 100 \times 5)}$$

$$= 9.67 \text{ GPa}$$

The bending stiffness $(EI)_c$ for the whole composite is obtained from (12.8).

$$(EI)_c = \sum E_i \left(I_i + A_i z_i^2\right)$$

$$= 11\left(\frac{1}{12} \times 100 \times 5^3 + 100 \times 5 \times 5^2\right) + 7 \times \frac{1}{12} \times 100 \times 5^3$$

$$+ 11\left(\frac{1}{12} \times 100 \times 5^3 + 100 \times 5 \times 5^2\right)$$

$(EI)_c = 1405208.333 \text{ GPamm}^4$

$$I_{yy} = \frac{(EI)_c}{E_c} = 145316.2702$$

The minimum radius of gyration is obtained from

$$r_{yy} = \left(\frac{I_{yy}}{\sum A_i}\right)^{1/2}$$

$$= \left(\frac{145316.2702}{(100 \times 5 + 100 \times 5 + 100 \times 5)}\right)^{1/2} = (97.88)^{1/2} = 9.89$$

Slenderness ratio $= \dfrac{l}{r_{yy}} = \dfrac{700}{9.89} = 70.78$

Critical load for the composite laminate is obtained as below

$$P_{cr} = \frac{\pi^2 EA}{\left(\dfrac{L}{r_{yy}}\right)^2} = \frac{\pi^2 \times 9.67 \times 1000 \times 1500}{(70.78)^2}$$

$$= 28.58 \times 10^3 \text{ N}$$

12.6 DESIGN OF SINGLE SKIN PANELS

As a first approximation, the anisotropy of the material may be ignored. The average values of Young's modulus and Poisson's ratio are chosen for the given panel. For a rectangular flat plate of sides *a* and *b* (*a* >*b*) and thickness *h* subjected to a uniform pressure *p*, the maximum deflection is given by

$$w_{max} = \alpha p a^3 b \frac{(1-v^2)}{Eh^3} \tag{12.34}$$

and the maximum bending stress is given by

$$\sigma_{max} = \beta p \frac{ab}{h^2} \tag{12.35}$$

where α and β are non-dimensional parameters dependent upon the aspect ratio and the boundary conditions of the panel. Knowing the permissible values of stress and deflection, d can be calculated from (12.42) or (12.43) and from this thickness the number of plies required to form the laminate are evaluated

If $w/h > 0.5$, large deflection theory the plates is to be considered. The above formulae are modified as follows

$$p = \frac{Eh^4}{a^3 b(1-v^2)}\left[\frac{1}{\alpha}\left(\frac{w}{h}\right) + \gamma\left(\frac{w}{h}\right)^2\right] \quad (12.36)$$

$$\sigma = \frac{Eh^2}{a^2}\left[\frac{\beta}{\alpha(1-v^2)}\left(\frac{w}{h}\right) + \alpha_1\left(\frac{w}{h}\right)\right] \quad (12.37)$$

Not much data are available for the non-dimensional parameters γ and α_1, but for practical purposes a value of $\gamma = 2$ and $\alpha_1 = 1.6$ (simply supported) or 2.9 (clamped) are recommended.

The approach mentioned so far is to be treated as only approximates and first estimates, as the formulae are based on isotropic considerations. They are based on the consideration of balanced laminates (i.e., flexural properties are the same in both principal directions).

For unbalanced laminates, stiffness parameters are

$$D_1 = \frac{E_1 h^3}{12\mu} \quad \text{and} \quad D_2 = \frac{E_2 h^3}{12\mu} \quad (12.38)$$

where

$$\mu = 1 - v_{12}^2 \frac{E_2}{E_1}$$

The maximum deflection is given by

$$w_{max} = \alpha p \frac{a^3 b \mu}{E_2 h^3} \quad (12.39)$$

and maximum stresses in two directions are given by

$$\sigma_{max} = \beta_1 \frac{pab}{h^2}, \quad \sigma_2 = \beta_2 \frac{pab}{h^2} \quad (12.40)$$

Values of α, β_1 and β_2 depend on D_1 / D_2 and their calculated values are given in Ref. 12.5. The mechanical properties can also be obtained from micromechanics or mechanical testing.

12.7 DESIGN OF STIFFENED STRUCTURES

In all FRP built composite structures except for the monocoque or the single-thickness skin construction, the plate or the panel structures are supported by stiffeners. In its simplest form, a stiffener is a simple beam or other sections attached to a plate to support the applied load. The stiffener is provided to improve the bending stiffness of the plate, which then improves the structural efficiency. The stiffeners are provided in both the principal orthogonal directions. Depending on the direction of alignment, the stiffeners are generally referred to as longitudinal and transverse stiffeners. These stiffeners may be continuous or discontinuous depending upon the structural requirements.

12.7.1 Design of Composite Stiffeners

In metallic structures, panel and other structures are supported by rolled stiffeners such as T-bar, I-section, flat bar and angle bar. These stiffeners are generally regular in shape and have constant thickness. In case of laminated composite stiffeners, the individual stiffener elements can be tailored to a particular shape and other mechanical properties can be varied locally as per structural requirements. In metallic stiffeners, the bending rigidity can be improved by maximising the second moment of area of the stiffener. For example, for a rectangular beam with breadth 'b' and the thickness 'h', the second moment of area can be evaluated from the expression

$$I = \frac{1}{12} bh^3 \qquad (12.41)$$

However, for composite stiffeners, as reasons described in section 12.4.6, the bending rigidity (EI) cannot be improved in isolation of each other.

The stacking sequence of the plies for a laminated composite stiffener should be such that the lamina with high elastic modulus is to be stacked away from the panel-bending axis. The orientation of the fibre and the position of a lamina with higher strength about the neutral axis of the laminate have great influence on the bending stiffness of the stiffener. It is for this reason that unidirectional plies with fibres in the longitudinal direction of the stiffener are included in the crown or flange of the hat type stiffeners. In ship structures, woven roving and chopped strand mat plies are used in the web of the hat type stiffeners to carry the shear stresses from the crown of the hat section to the underlying panel.

12.7.2 Types of Composite Stiffeners

Composite stiffeners are classified on the basis of their geometric shapes. These shapes are built from individual plies. Some of the stiffener shapes can be produced by roll forming or pultrusion and then fastened to the panel. The regular shapes commonly used in composite structures are omega, hat type stiffeners, trapezoidal, rectangular, arc of a circle, trapezoidal with rounded crown, angle shape, T-bar, I-beam and trapezoidal supported internally by a T-section. In addition, different techniques are adopted to manufacture these stiffener sections such as layers separately moulded in shape, plies laid on foam cake in shape and pre-moulded stiffeners. The stiffeners are attached to the panel by bonding or stitching or mechanical fastening. Another method that is used for connecting a stiffener is by a concept called embedded stiffening strap. In this, a few layers of plies from the panel are opened out to embed the stiffening strap on to which an additional stiffener is positioned.

Many of the composite stiffeners mentioned have a closed section when connected to the panel. The main difference in the open versus closed section is the effect of torsional stiffness. The torsional stiffness of the closed section aids the panel in resisting the buckling deformations. The stiffener twisting resistance is very important to control the buckling deformations. In case of closed stiffeners, all parts of the stiffeners are connected to one another as against open stiffeners, which are not inter-connected at the periphery. In hat type stiffeners, shear stiffness of the web is a very important parameter, as it aids in transferring loads from the crown of the stiffener to the panel.

12.7.3 Stiffener Design Parameters

We carry out the analysis of a stiffener by taking each element that make up the complete stiffener i.e., flanges, webs, crown, and the interconnection between them separately. This is because each element has different types of lamina, fibre orientation, and their physical and material characteristics may be different. It is necessary to evaluate the contribution of each of these elements to the stiffness and strength of the panel-stiffener combinations. Also, each element has to carry out certain functions to ensure that the structural performance is completely met, for example, the web has to support the shear load and the crown has to support the bending load. The trough angle is another important parameter in the stiffener design. The trough angle is the angle between the normal and the web of the stiffener at the panel–stiffener joint. A small trough angle causes loss of flexural efficiency and emergence of unfamiliar twisting of the stiffener. Generally a trough angle of 30 degrees or more is preferred for the purpose of the design. Similarly the unsupported length of the stiffener i.e., spacing at the base of the hat-type stiffener should not be too large, as it may affect the flexural efficiency of the stiffener. The stiffener design parameters that have most relevance to the structural design are shown in Fig 12.18.

Given: Design requirements (includes loading, constraints etc.)

Aim: Design a laminate to support the load without failure during its service life

Solution: Select laminate

Decide on configuration and size

Carry out a design analysis for strength, stiffness, fatigue and Other structural responses

Result: Check for satisfactory performance; if not satisfactory, alter stacking sequence, change thickness, increase number of plies to improve the structural response measures

Fig 12.18 Stiffener design parameters

Example 12.10

A stiffener member of the cross–section shown in Fig. 12.19 has a span of 1000 m and the ends are simply supported. Find the ultimate load carrying capacity.

Plate	Dimensions	Area	z	Az	Az^2	E	$I_{base} = \frac{1}{2}bd^3 + Az^2$	$(EI)_{base}$	EA	EAZ
WR	300 × 2	600	1	600	600	11	800	8800	6600	6600
CSM	300 × 1	300	2.5	750	1875	7	1800	13300	2100	5250
CSM	300 × 1	300	3.5	1050	3675	7	3700	25900	2100	7350
WR	300 × 2	600	5	3000	15000	11	15200	167200	6600	33000

Bottom flange

CSM	140 × 1	140	6.5	910	5915	7	5926.67	41486.69	980	6370
WR	140 × 1	140	7.5	1050	7875	11	7886.67	86753.33	1540	11550

Webs

WR, WR, WR	3 × 100	300	56	16800	940800	11	2381600	26197600	3300	184800

Top crown

WR	80 × 1	80	106.5	8520	907380	11	907386.67	9981253.33	880	93720
CSM	80 × 1	80	107.5	8600	924500	7	924506.667	6471546.667	560	60200
WR	80 × 1	80	108.5	8680	941780	11	941786.67	10359653.33	880	95480

The location of the neutral axis of the stiffener is

$$z_{NA} = \frac{\sum(E_i A_i z_i)}{\sum E_i A_i}$$

$$z_{NA} = \frac{504320}{25540}$$

$$z_{NA} = 19.75 \text{ mm}$$

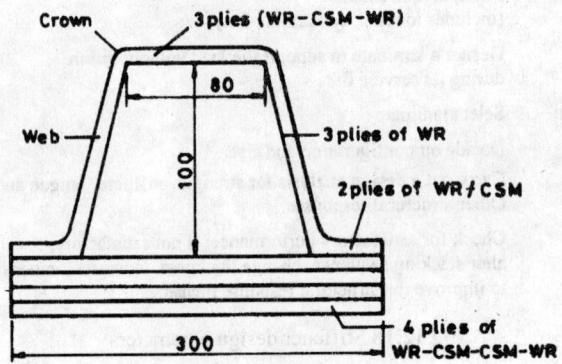

Fig. 12.19 Example 12.11

The elastic modulus of the stiffener is $E_c = \dfrac{\sum(E_i A_i)}{\sum A_i}$

$$E_c = \frac{25540}{2620} = 9.75 \text{ GPa}$$

EI about the neutral axis of the stiffener is

$$(EI)_c = \left(\sum (EI)_{\text{base}}\right) - \left(\sum EA\right) y^2$$

$$= 4.403 \times 10^7 - (25540)(19.75)^2$$

$$= 3.407 \times 10^{10} \frac{\text{N mm}^4}{\text{mm}^2}$$

The maximum bending moment at few selected plies

$$M = \frac{\sigma_{Ti}(EI)_c}{E_1 z_{NA}} = \frac{20 \times 3.407 \times 10^{10}}{11000 \times 19.75} = 686.4 \text{ Nm}$$

$$M = \frac{22 \times 3.407 \times 10^{10}}{7000 \times 19.75} = 5421.627 \text{ Nm}$$

$$M = \frac{20 \times 3.407 \times 10^{10}}{11000 \times 19.75} = 3489.9 \text{ Nm}$$

The ultimate load carrying capacity of the stiffener is

$$q_{max} = \frac{8 M_{min}}{L^2} = \frac{8 \times 686400}{(1000)^2} = 5.4912 \text{ N/mm}$$

12.7.4 Design of a Longitudinally Stiffened Panel

There are certain stiffened panels like in a ship bottom, which is designed against compressive load due to hull bending. The structure in this area is generally supported by longitudinal stiffness and relatively widely spaced transverse stiffeners. The problem is approached from the point of view of initial buckling.

A conservative estimate of the critical load of the stiffened panel is

$$Nx_{cr} = \frac{2\pi^2}{b^2} \left(H + \sqrt{D_x D_y} \right) \qquad (12.42)$$

For an orthotropic strip of width b with simply supported edges buckling will occur with a number of half waves over length a given by

$$n = \frac{a}{b} \left(\frac{D_y}{D_x} \right)^{1/4} \qquad (12.43)$$

If the edges are clamped, then

$$Nx_{cr} = \frac{\pi^2}{b} \left(2.4 H + 4.6 \sqrt{D_x D_y} \right) \qquad (12.44)$$

This will provide an upper bound estimate of the initial buckling stress,

$$n = 1.5 \left(\frac{a}{b} \right) \left(\frac{D_y}{D_x} \right)^{1/4} \qquad (12.45)$$

Longitudinal stiffeners may exhibit column like instability. Buckling stress is conservatively estimated by Euler's formula

$$\sigma_{x_{cr}} = \frac{\pi^2 D_y}{A a^2} \bigg/ \left(1 + \frac{\pi^2 D_y}{a^2 GA_s} \right) \qquad (12.46)$$

where
D_y is the flexural rigidity of the stiffener with associated effective width of the plate
A is the cross-sectional area of the stiffener including the effective width of the plate
GA_s is the shear rigidity of the stiffener web
A_s is area of the stiffener web
a is spacing of the transverse frames

REFERENCES AND SUGGESTED READINGS

12.1 P.K. Mallick, *Fibre Reinforced Composites*, 2nd Edition, Marcel Dekker Inc., USA, 1993.

12.2 J.M. Whitney, *Structural Analysis of Laminated Anisotropic Plates*, Technomic Publishing Co. Inc., USA, 1987.

12.3 R.M. Jones, *Mechanics of Composite Materials*, Second Edition, Taylor and Francis Inc., USA, 1999.

12.4 P.L.N. Murthy and C.C. Chamis, Simplified design procedures for fibre composite structural component joints, in *Composite Structures: Testing, Analysis and Design* Editors, J.N. Reddy and A.V.K. Murthy, Narosa Publishing House, India, 1992, pp. 133-153.

12.5 A.F. Johnson, Design Analysis of FRP Plates, Proceedings of Thirteenth Reinforced Plastics Congress, BPF, 1982.

CHAPTER 13

COMPOSITE JOINTS

13.1 INTRODUCTION

In this chapter, basic characteristics of composite joints are discussed. The review on the topic of composite joints is available in Ref. 13.1–13.8. There are various reasons for providing joints in composite structures. The yard where the composite structure is manufactured may not have necessary space for the production of the complete structure. In such a case, the FRP composite structure may have to be divided into components which are to be finally joined. In certain cases for inspection of the inner spaces or for the regular maintenance of the structure, access inside it is necessary. Joints may be required in the structure to meet the limitation of the transportation system. In case of damages, joints are required to be provided to the repaired structure where as many cuts are made as necessary.

The basic purpose of the joint is to transfer the load from one member to the other. It may be noted that a joint will entail an increase in the weight of the composite structure. Joints in composite structures are often the weakest link. Unlike metal structures, an FRP does not have enough ductility to redistribute the stresses that can take place. An attempt should always be made to maintain the integrity of the entire structure and to see that the structure is not weakened by the presence of joints.

13.2 CLASSES OF LAMINATE JOINTS

Two major classes of laminate joints are bonded joints (Figs. 13.1) and bolted joints (Fig. 13.2). These two types may be combined to yield what is termed as bonded–bolted joints (Fig. 13.3). Welding may be applied to thermoplastic composites which is beyond the scope of this book. The first two classes will be dealt with in this chapter.

13.3 BONDED JOINTS

When procured laminates are joined together by bonding (gluing) with a suitable adhesive, joints are so formed in the manufacturing process, in which cases the laminate and joints are to be cured together. A few common joints are shown in Fig 13.4. These joints are most appropriate for joining flat laminates and tabular members. The adhesive interlayers exist between the substrates (known as adherends). Epoxy-based composites have not presented any difficulty in bonding of polyester composite components. At present, the epoxy and acrylic based toughened adhesives are used for general application. They are easy to use, robust, durable and relatively free from toxic hazards.

13.3.1 Stress Distribution

Very often high stresses occurring at a discontinuity in a structure initiate failure. As such, in order to obtain an improved design, a reasonably good idea of joint stresses is absolutely essential

[13.9-13.11]. For efficient transfer of loads, bonded joints are most efficient when subjected to shear load and least efficient when subjected to peel load. As it is impossible to obtain loading which gives pure shear load, peel forces are to be considered.

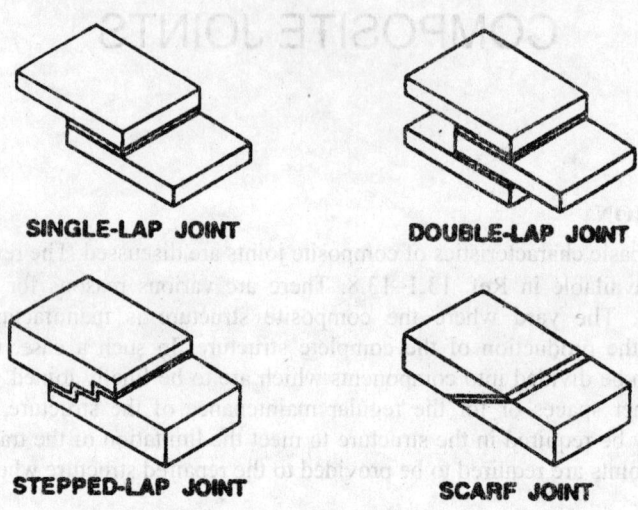

Fig. 13.1 Bonded joints

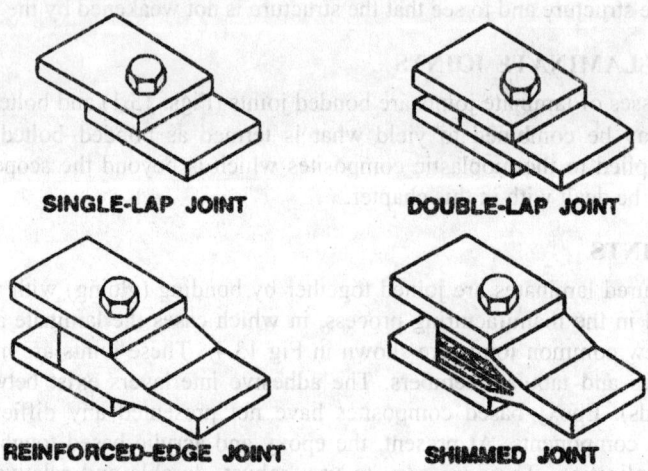

Fig. 13.2 Bolted joints

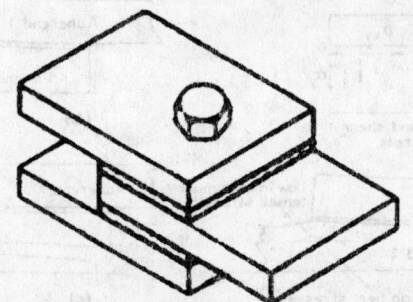

Fig. 13.3 Bonded bolted joints

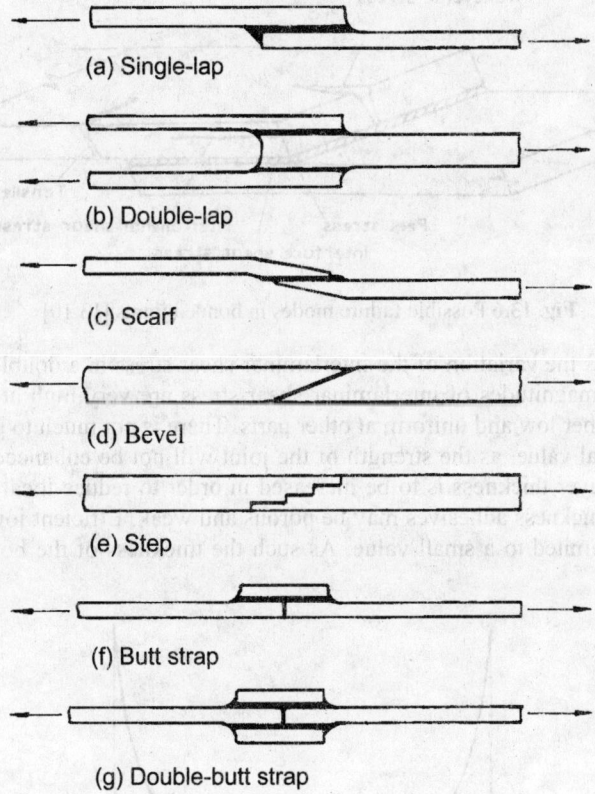

Fig. 13.4 Common bonded joints

In a single-lap joint, loads applied to the adherends are eccentric to the centre (adhesive joint), the bending resulting due to it set up a (peel) stress in the thickness direction of the adhesive (Figs 13.5 and 13.6). The combined normal and shear stress distributions exhibit high stresses at the lap ends of the adhesive layers which may adversely affect the strength of the joint. The peel stress can be reduced to a minimum by having an adequate overall length and upper limit of thickness.

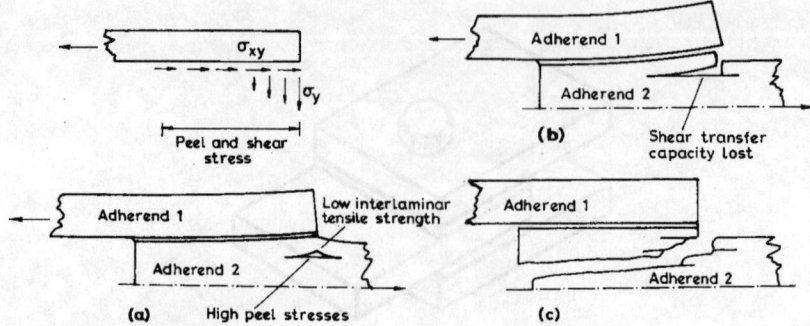

Fig. 13.5 Delamination in bonded joint

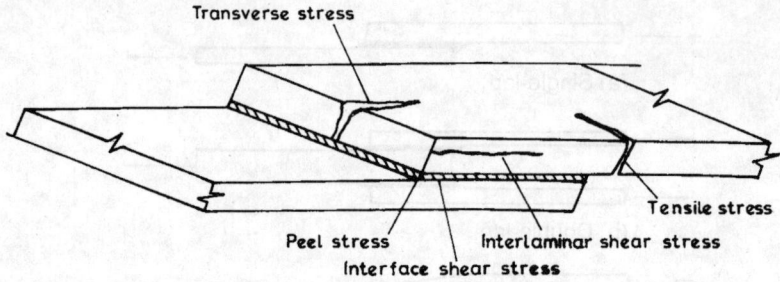

Fig. 13.6 Possible failure modes in bonded joints [13.10]

Figure 13.7 depicts the variation of the interlaminar shear stress in a double-lap joint. As can be seen in the figure, the magnitudes of interlaminar shear stress are very high at the ends of the joint whereas the stress is rather low and uniform at other parts. There is not much to gain by increasing the overlap beyond a critical value, as the strength of the joint will not be enhanced. According to Hart-Smith [13.12], adhesive layer thickness is to be increased in order to reduce local high stresses. It is to be borne in mind that thickness adhesives may be porous and weak. Efficient joints can be made only when the thickness is limited to a small value. As such the thickness of the bond may be limited to 0.1mm to 0.25mm.

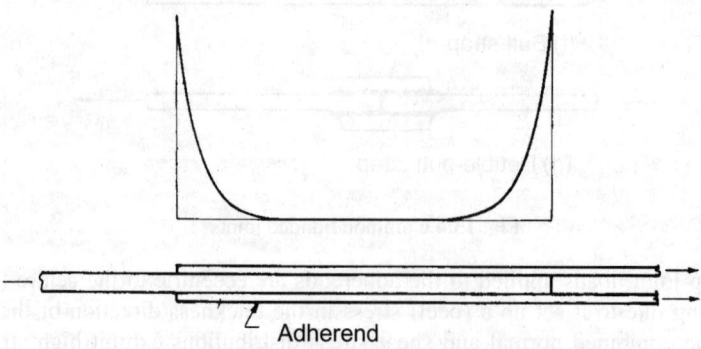

Fig. 13.7 Distribution in adhesive shear stress in bonded joints

Scarf joints between identical adherends will have a uniform distribution of adhesive shear stress and will show a higher strength than other types of joints. A stepped lap joint is also capable of reaching very high joint strengths. In practice, however, the difficulty experienced in machining the steps or stiff angles may take away some of the advantages.

13.3.2 Modes of Failure

Microcracking occurring in the adhesive layer at the interface of adhesive /adherend or in the adherends may be the starting point of the impending failure. This microcracking will ultimately lead to the failure of the joint.

There are four distinct modes of failure of adhesive bonded joints [13.10]. They are as follows

1. Failure of the adherend
2. Shear strength failure of the adhesive
3. Failure of the adhesive due to a peel load
4. Failure of fibre composite adherends by delamination

When the adhesively bonded joint is stronger than the adherend, the latter fail first. The second category of failure is due to the failure of the bond between the adhesive and the adherend. It is basically an adhesive failure.

The peel load explained earlier may lead to the failure of the adhesive. At the bonded joints, shear stresses occur with the peel stresses. Due to the peel stresses, inner laminates of the adherend splits locally which inhibits the inner and outer plies from transferring the shear to one another. The magnitude of the peel stresses can be reduced to a minimum when the overall length lies in the range of $50t$ to $100t$ (t is the thickness) with a upper limit of thickness being 1.5mm. Lastly, the laminate which forms the adherend may fail by delamination. More details can be found elsewhere [13.10].

13.3.3 Merits and Demerits of Adhesive Bonded Joints

The merits and demerits of adhesively bonded joints are given below:

1. An adhesively bonded joint distributes the load over a greater surface area. Due to this all fibres at the joint interface become effective in carrying the load so that stress concentration is reduced.

2. Due to bonded joints, very little weight is increased in the structure.

3. Adhesively bonded joints are more economical to provide.

4. The bonded joints require some environmental control during the manufacturing process, as it is affected by severe temperature, humidity and other service conditions.

5. No holes are needed in bonded joints.

6. As bonded joints are permanent in nature, cracks resulting from the initial failure will propagate easily, since there are no fibres across the joints to act as crack arresters.

7. In bonded joints, it is difficult to inspect for quality control.

13.4 MECHANICAL JOINTS

Mechanical or bolted joints are created by using three main mechanical systems – screws, rivets or bolts for joining the adherends. They are mainly used in parts where bonding cannot be implemented or uneconomic when parts are removed for inspection or maintenance. The design of FRP bolted joints are not different from that followed for metal joints. The nature of stresses occurring at these joints are basically three-dimensional, analytic prediction of which has its own

13.4.1 The Failure Modes of the Mechanical Joints

Mechanical joints are designed to resist certain select failure modes during the preliminary design stage. The principal failure modes for the mechanical joints are (1) bearing failure, (2) tension failure, (3) shear-out or cleavage failure and (4) shear failure of bolts. These failures are shown in Fig. 13.8. The combination of these failure may also occur.

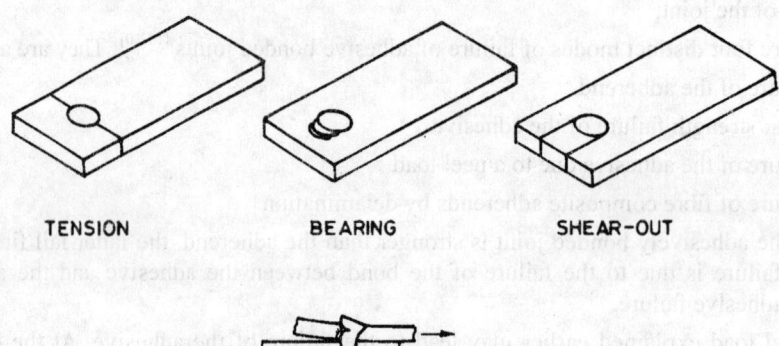

Fig. 13.8 Failure modes in bolted joints

The bearing failure of an FRP composite material depends primarily on the stacking sequence of the plies. The tensile failure modes occur due to high stress concentrations. In cleavage or shear-out failure mode, the failure is initiated by transverse tensile failure of the bolt and is followed by tensile failure. Finally, for a bad design, shear failure of the bolt cannot be ruled out.

13.4.2 Advantages and Disadvantages of Mechanical Joints

The advantages and disadvantages of mechanical joints are listed below

1. There is no need for surface preparation.
2. Quality control procedure is simple.
3. Damaged and wrongly assembled members are readily repaired or replaced without affecting the adherends.
4. It is easy to inspect the joint or other parts of the structures by removing the members.
5. The discontinuity due to the presence of holes induce high stress concentrations around the joints that may trigger the failure of the adherends. Bolted joints indicate lower efficiency of the members weakened due to the presence of holes.
6. The increase in the cost of bolted joints is proportional to its size.

13.5 PRELIMINARY DESIGN OF ADHESIVELY BONDED COMPOSITE JOINTS

The presentation of the preliminary design of adhesively bonded composite joints is based on the work of Chamis and Murthy [13.13].

A butt joint and a lap joint are shown in Fig. 13.9. Steps to be followed for the design of adhesively bonded joints are given below

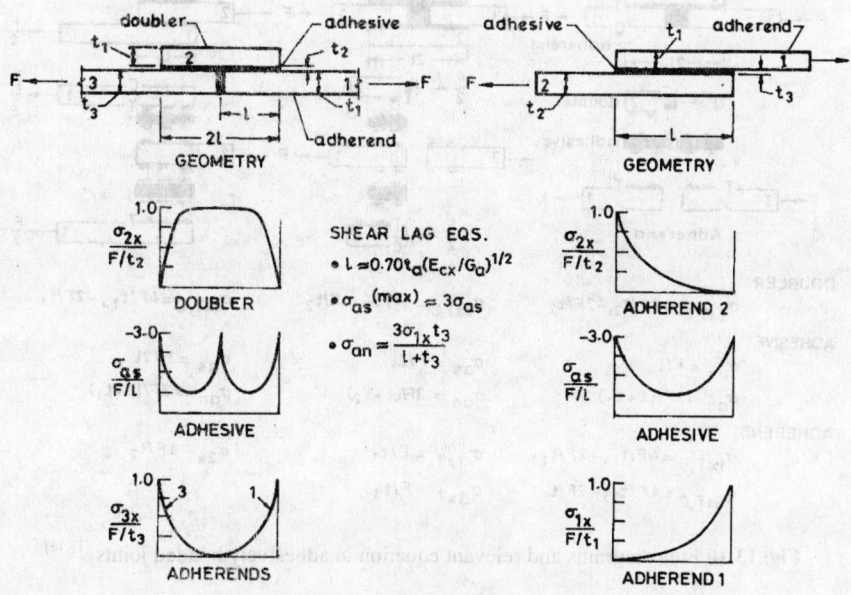

Fig. 13.9 Geometry and stress variation of adhesive joint [13.13]

1. Obtain the design requirements: laminates, adhesive, factors of safety and other particulars are required.
2. Obtain dimensions of the laminates and properties of the adherends using appropriate mechanics.
3. Obtain the properties of the adhesive. The specific typical properties required are: (a) shear strength and (b) peel-off strength.
4. Select design allowables. The choice may be limited as follows : (a) the load factor on force F to be chosen. It usually lies between 1.5 and 2.
5. The length of the joint is to be selected using the following equation

$$l = \frac{F}{\sigma_{as}} \qquad (13.1)$$

where F denotes the load (tensile/compressive/shear) in the adherends per unit width.

σ_{as} is the maximum permissible shear stress in the adhesive.

6. Check the minimum length and maximum shear and normal stresses in the adhesive [using shear lag theory equations (Fig 13.9)].
7. Determine bending stresses in the doublers and adherends using the proper equations from Figs. 13.10 and 13.11.

352 *Mechanics of Composite Materials and Structures*

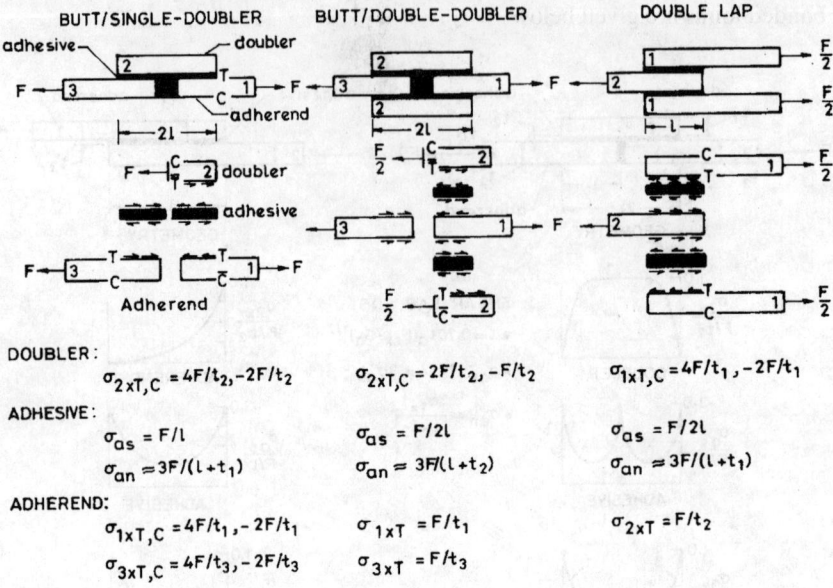

Fig. 13.10 Free diagrams and relevant equation in adhesively bonded joints [13.13]

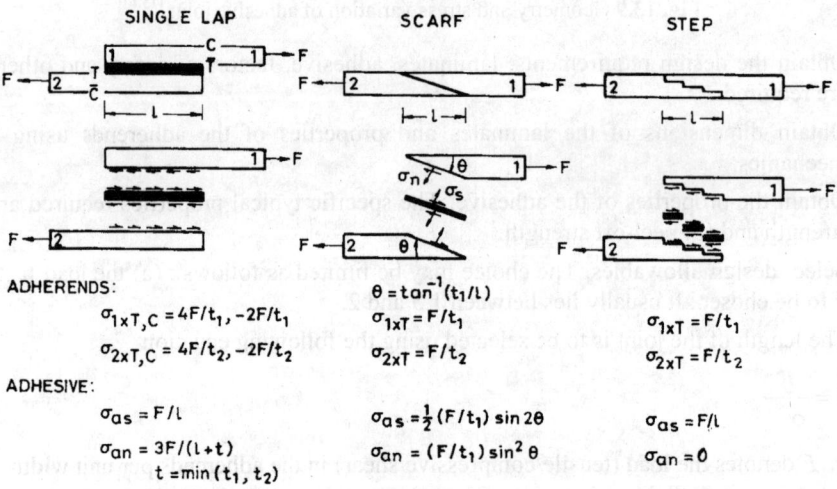

Fig. 13.11 Free digrams and relevant equation in adhesive joints [13.13]

8. The margin of safety (MOS) is to be evaluated for all computed stresses. At each step of evaluation of stresses, MOS is to be calculated and compared with permissible stresses using the following equation:

$$\text{MOS} = \frac{\text{Maximum permissible stress}}{\text{Calculated stress}} - 1 \qquad (13.2)$$

9. Calculate the joint efficiency (JE) which is defined as follows

$$\text{J. E.} = \frac{\text{Joint force transferred (F)}}{\text{Adherend fracture load } (n_{\text{cxt}_1})} \times 100 \quad (13.3)$$

Example 13.1

Butt joint with a single doubler

The step by step procedure given above is presented with an example.

1. Joint design requirements

   Loads: a tensile load of 1.5 N/mm

   Laminate: $(0|\pm 45|0|90)_s$ with thickness 1.5 mm

   Adhesive: epoxy matrix same as the laminate

   Safety factors: 1 on joint load and 0.5 on adhesive strength

   For this laminate, $\sigma_{2xt} = 5.55 \text{ N/mm}^2$, $\bar{\sigma}_{2xc} = 5.6 \text{ N/mm}^2$

2. Adhesive properties of epoxies are

   $G_a = 12.7 \text{ N/mm}^2$, $E_a = 35.2 \text{ N/mm}^2$, $\nu = 0.35$

   $\sigma_{as} = 1.00 \text{ N/mm}^2$, $\sigma_{ax} = 1.1 \text{ N/mm}^2$

3. Design permissibles

   Joint load = $1 \times 1.5 = 1.5$ N/mm

   Adhesive normal or peel-off strength = $0.5 \times 1.1 = 0.55 \text{ N/mm}^2$

   Adhesive shear strength = $0.5 \times 1 = 0.5 \text{ N/mm}^2$

4. Joint length

   $$l = \frac{F}{\sigma_{as}} = \frac{1.5}{0.5} = 3 \text{ mm}$$

   Doubler length = $2 \times 3 = 6$ mm

From Ref. 13.12 and Ref. 13.14, and assuming 0.99 as load transfer efficiency, minimum length is given by [13.14]

$$l^{\min} = 0.7 t_a \sqrt{\frac{E_{cx}}{G_a}}$$

The laminate thickness = 1.5 mm

These are 10 plies

So, thickness of each ply = $t_a = \frac{1.5}{10} = 0.15$ mm

Given, $E_{cx} = 705 \text{ N/mm}^2$

Therefore, $l^{\min} = 0.7 \times 0.15 \sqrt{\dfrac{705}{12.7}} = 0.78$ mm < 3mm

Thus, the joint length is 3 mm and doubler length is 6 mm. As it is impractical for handling maximum shear stress concentration, the length is changed to 25 mm instead of 6 mm.

$$l_{as}^{\max} = 3 \times l_{as}$$

$$\sigma_{as}^{\max} = \dfrac{3 \times 1.5}{25} = 0.18 \text{ N/mm}^2 < 0.5 \text{ N/mm}^2$$

$$\text{MOS} = \dfrac{0.5}{0.18} - 1 = 1.78$$

Average shear stress is given by

$$\sigma_{as}^{av} = \dfrac{1.5}{25} = 0.06 \text{ N/mm}^2 < 0.5 \text{ N/mm}^2$$

$$\text{MOS} = \dfrac{2.5}{0.5} - 1 = 4$$

Peel-off stress (Fig. 13.9) is given by

$$\sigma_{an} = \dfrac{3\sigma_{1x} \, t_3}{l + t_3} = \dfrac{3 \times 1.5 \times 1.5}{25 + 1.5} = 0.25 \text{ N/mm}^2 < 0.55 \text{ N/mm}^2$$

$$\text{MOS} = \dfrac{0.55}{0.25} - 1 = 1.2$$

Certain important points are to be noted.

(1) Taking into consideration the practical difficulties, the joint length has been modified to 25 mm from 6 mm, as 6 mm is rather a small value.

(2) Stresses calculated for this joint indicate that they are much smaller than the maximum permissible values, which suggests that the joint design is not efficient.

(3) Very small values of joint length is obtained using shear lag theory which suggests that the load is transferred over a small distance.

6. Bending stresses (Fig. 13.10)

Design of the doubler

$$\sigma_{2xT} = \dfrac{4F}{t_2}, \quad \sigma_{2xC} = -\dfrac{2F}{t_2}$$

$$F = 1.5 \text{ N/mm}$$

The thickness of the doubler is assumed to be the same as that of the laminate (adherends)

$$t_2 = 1.5 \text{ mm}$$

$$\sigma_{2xt} = \dfrac{4 \times 1.5}{1.5} = 4 \text{ N/mm}^2$$

$$\sigma_{2xc} = -0.5 \, \sigma_{2xt} = -2 \text{ N/mm}^2$$

Given that the permissible value of the stresses of the laminate are

$$\bar{\sigma}_{2xT} = 5.55 \text{ N/mm}^2, \quad \bar{\sigma}_{2xC} = 5.6 \text{ N/mm}^2$$

$4 \text{ N/mm}^2 < 5.55 \text{ N/mm}^2$ Therefore, OK

and $2 \text{ N/mm}^2 < 5.6 \text{ N/mm}^2$ Therefore, OK

7. The corresponding MOS are
   Tension
   $$\text{MOS} = \frac{5.55}{4} - 1 = 0.39$$
   Compression
   $$\text{MOS} = \frac{5.6}{2} - 1 = 1.8$$

It may be mentioned that the given permissible stress values of the laminate are rather approximate. Only from the analysis of the laminate can a better value of maximum allowable stress be obtained. The calculations have been made with the help of ply stress influence coefficients [13.15]. For the present laminate, the outer ply (0°–ply) stress influence coefficients which relate the principal material axis stresses to the laminate structural axis stresses, are [13.15]

Load	σ_{f1}	σ_{f2}	σ_{f12}
Inplane	1.993	−0.007	0
Bending	1.714	−0.027	−0.003

The corresponding ply stresses are calculated by using the following equation

$$\sigma_{f1} = \frac{F}{t_1}(1.993 + 3 \times 1.7114)$$

where F is the joint load = 1.5 N/mm

t_1 is the doubler thickness = 1.5 mm

$$\sigma_{f1} = \frac{1.5}{1.5}(1.993 + 3 \times 1.714) = 7.135 \text{ N/mm}^2$$

Similarly,

$$\sigma_{f2} = \frac{1.5}{1.5}[-0.007 + 3(-0.027)] = -0.088 \text{ N/mm}^2$$

$$\sigma_{f12} = \frac{1.5}{1.5}[0.0 + 3(-0.003)] = -0.009 \text{ N/mm}^2$$

From Ref. 13.15, we get the permissible values of stresses.
They are

$$\bar{\sigma}_{f1T} = 15.5 \text{ N/mm}^2$$

$$\bar{\sigma}_{f1C} = 2.46 \text{ N/mm}^2$$

$$\bar{\sigma}_{f12S} = 0.70 \text{ N/mm}^2$$

Comparing the calculated value in the outer ply of the doubler to these strengths, we have

Longitudinal tension

$$7.135 \text{ N/mm}^2 < 15.5 \text{ N/mm}^2$$

$$\text{MOS} = \frac{15.5}{7.135} - 1 = 1.17$$

Transverse compression

$$0.088 \text{ N/mm}^2 < 2.46 \text{ N/mm}^2$$

$$\text{MOS} = \frac{2.46}{0.088} - 1 = 26.95$$

Interlaminar shear

$$0.009 \text{ N/mm}^2 < 0.70 \text{ N/mm}^2$$

$$\text{MOS} = \frac{0.70}{0.009} - 1 = 76.78$$

For this joint, the approximate approach appears to be rather conservative, as the stress σ_{2xt} has an increased value.

8. Joint efficiency

    Joint efficiency (J.E) is calculated as given below

$$\text{J.E.} = \frac{F}{\overline{\sigma}_{2xT} t_1} \times 100$$

$$= \frac{1.5}{5.55 \times 1.5} \times 100 = 18 \text{ percent}$$

which is rather poor.

Due to occurrence of bending in the doubler and adherends, the joint efficiency is poor. Thus from Figs. 13.10 and 13.11, it is revealed that the efficiency of the joint can be improved by selecting a joint which will not induce any bending.

An increase of the thickness of both doubler and adherends will increase the efficiency of the joint, but at a cost of increased manufacturing complexity, time and expenses.

9. Joint design summary

Doubler	$(0\,|\pm 45\,|\,90)_s$
	(same as adherends)
Composite	T300/E
	(same as adherends)
Adherends	structural epoxy
	(same as adherends)
Length	$l = 25$ mm
	(adjusted for fabrication handling)

Stresses	Calculated σ N/mm^2	allowable σ N/mm^2	Margin of safety
Adhesive			
shear average	0.06	0.5	4.0
shear maximum	0.18	0.5	1.78
peel-off	0.25	0.55	1.2
Doubler/Adherends			
Combined tension	4.0	5.55	0.39
Combined compression	2.0	5.6	1.8
Joint efficiency = 18 percent			

13.6 PRELIMINARY DESIGN OF COMPOSITE BOLTED JOINT

The presentation of the preliminary design of mechanical bolted joints is based on the work of Chamis [13.16]. In the preliminary design phase, certain select failure modes are considered.

13.6.1 Bearing Failure

In bearing failure modes, compressive stresses caused by the bolt diameter tends to crush the composite material. Compressive stresses are developed around the loaded half of the circumference of the bolt or rivet hole.

The bearing failure stress is given by [Fig. 13.12]

$$\sigma_{cx} = \frac{F}{dt_c} \tag{13.3}$$

when

d is the bolt diameter

t_c is the laminate thickness

Maximum or allowable bearing stress of the laminate is denoted by $\overline{\sigma}_{cxc}$. An example is presented below to illustrate the bearing failure [13.16].

Example 13.2

Bearing failure of bolted joints

Calculate the average bearing stress of σ_{cx} in a $[0|\pm 45|0|90]_s$ AS/E laminate having a thickness of 1.5 mm. It has a titanium bolt 6 mm in diameter and is subjected to a load of 45 N/mm. Allowable compressive strength of the laminate is 5.6 N/mm^2.

$$\sigma_{cx} = \frac{F}{dt_c} = \frac{45}{6 \times 1.5} = 5 \text{ N/mm}^2 < 5.6 \text{ N/mm}^2$$

The margin of safety against bearing failure is

$$\text{MOS} = \frac{5.6}{5} - 1 = 0.12 \qquad \text{Therefore, O.K.}$$

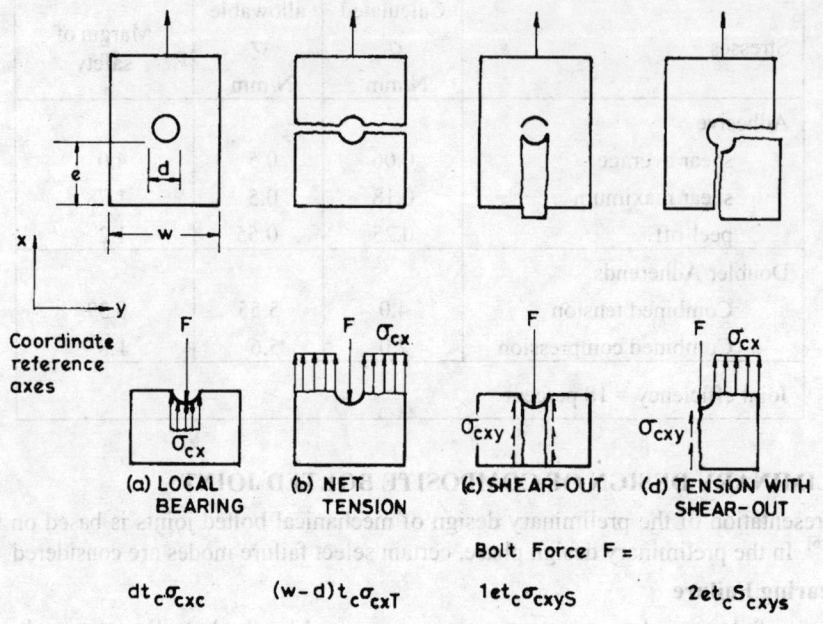

Fig. 13.12 Failure modes of composite bolted joints

13.6.2 Tension Failure

The tension failure stress can be calculated on the basis of gross area and net area. The calculation based on the net area will yield higher value of the average stress. As such design calculation for the tension failure is based on the net area.

The net tensile stress is given by (Fig. 13.12),

$$\sigma_{cx} = \frac{F}{[(w-d)t_c]}$$

where

$(w - d)$ is the net section width

t_c is the laminate thickness

F is the load.

$\bar{\sigma}_{cxT}$ is the failure stress in tension

Example 13.3

Tension failure

Calculate the net tensile stress (σ_{cx}) at the bolt hole edge for the composite bolted joint of the same laminate in example 3.1. Permissible tensile strength of the laminate is 5.6 N/mm². The width of the plate is 25 mm with a 6 mm bolt diameter.

$$\sigma_{cx} = \frac{45}{(5-6) \times 1.5} = 1.58 \text{ N/mm}^2 \quad < 5.6 \text{ N/mm}^2$$

The margin of safety is given by

$$\text{MOS} = \frac{5.6}{1.58} - 1 = 2.54 \qquad \text{Therefore, O.K.}$$

Therefore, the composite bolted joint is safe in net tension.

13.7 COMPOSITE MULTI-BOLT JOINTS

For joining two adjacent panels, multipoint composite joints are necessary. A typical joint is shown schematically in Fig. 13.13. It is assumed in the analysis that all bolts of same diameter carry equal load. This may not be the case in reality, as the first row of bolts will experience more load. But in case the first row of bolts suffer from bearing failure, then there will be redistribution of the load to the next bolt and the process will continue.

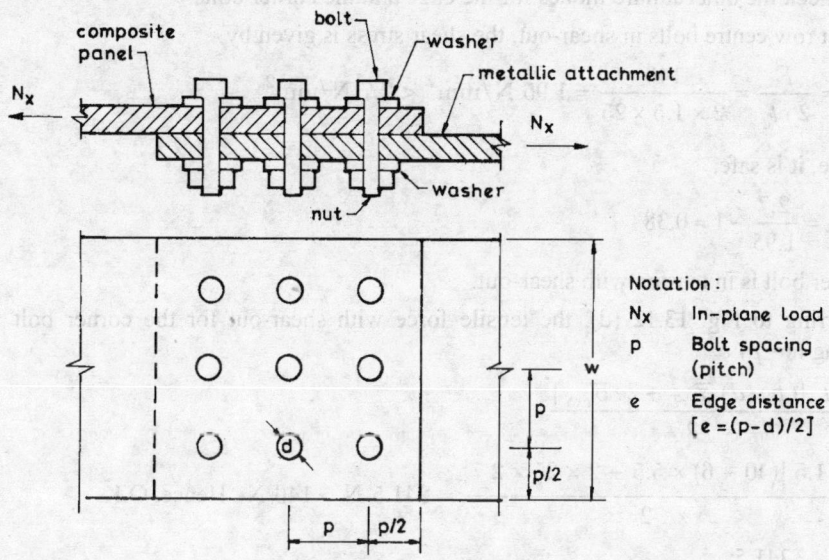

Fig. 13.13 Multi-bolt composite joint

13.7.1 Design of Composite Multi-Bolt Joints

In this section, we will discuss the design of multi-bolt joint in Fig. 13.13 in which a composite panel is to be attached to a metallic plate. Only the bolts in the composite panel are to be designed. The composite laminate particulars are $[0 \mid \pm 45 \mid 90]_s \ A_s / E$. A design tensile load of 3.65 N/mm is applied to the laminate which is 1.5 mm thick. The bolt spacing 'p' is 40 mm and the edge distance is 25mm. The permissible stresses are $\overline{\sigma}_{cxC} = 5.6 \text{ N/mm}^2$, $\overline{\sigma}_{cxT} = 5.5 \text{ N/mm}^2$ and $\overline{\sigma}_{cxyS} = 2.7$ N/mm^2.

The calculation has been presented stepwise.

Step 1: The load carried per bolt is

$$p \cdot N_{cx} = 40 \text{ mm} \times 3.65 \text{ N/mm} = 146 \text{ N}$$

Step 2: Number of bolts per bolt row is to be determined.

$$N = \frac{F}{d \cdot t_c \cdot \overline{\sigma}_{cxC}} = \frac{146}{6 \times 1.5 \times 5.6} = 2.90$$

Use 3 bolts.

Step 3: Check next for tension.

N from Fig. 13.13 is

$$N = \frac{F}{[(w-d) \, t_c \, \overline{\sigma}_{cxT}]} = \frac{146}{[(40-6) \times 1.5 \times 5.5]} = 0.52 \text{ N}$$

1 bolt is good enough to take care of net tension. For this case, more bolts are required from the consideration of bearing than net tension.

Step 4: Check the other failure modes for the edge and the corner bolts.

For the first row centre bolts in shear-out, the shear stress is given by

$$\sigma_{cx} = \frac{F}{2 \, e t_c} = \frac{146}{2 \times 1.5 \times 25} = 1.95 \text{ N/mm}^2 < 2.7 \text{ N/mm}^2$$

Hence, it is safe.

$$\text{MOS} = \frac{2.7}{1.95} - 1 = 0.38$$

Corner bolt is in tension with shear-out.

Referring to Fig. 13.12 (d), the tensile force with shear-out for the corner bolt is given by (considering $w = p$)

$$F = \frac{t_c \, [(p-d) \, \overline{\sigma}_{cxT} + 2e \, \overline{\sigma}_{cxS}]}{2}$$

$$F = \frac{1.5 \, [(40-6) \times 5.5 + 2 \times 25 \times 2.7]}{2} = 241.5 \text{ N} > 146 \text{ N} \quad \text{Hence, O.K.}$$

$$\text{MOS} = \frac{241.5}{146} - 1 = 0.654$$

13.8 OTHER APPROACHES

Till now preliminary design of both adhesive and bolted FRP composite joints have been discussed. In spite of the complexity of the problem, classical analytical reference can be made to the theoretical work by Muskhelishveli [13.17] to deduce the stress distribution around a frictionless pin-loaded hole based on the complex function method,. It was extended to the orthotropic materials by de Jong [13.18, 13.19].

Theoretical foundations for the lap-joint design were laid by de Bruyne [13.20] and Goland and Reissner [13.25] which provided initial insights into stresses and deformations in the adherends and adhesive of single and double-lap joints. It becomes obvious that classical analytical methods have their inherent limitations in analysing the complex behaviour of composite joints. Therefore, investigators took resort to numerical methods for solving problems in composite joints [13.4–13.8, 13.40–13.43]. They include the finite element method [13.4, 13.5], boundary collocation method [13.28] and the boundary element method [13.29, 13.30]. There are relative merits and demerits in all the approaches.

No doubt the finite element method is the most versatile amongst them. There are various reasons for it – its elegant treatment of the boundary is most advantageous. In order to obtain accurate stresses, care should be taken to generate fine mesh, specially around the hole. The finite element method takes into account inelastic behaviour of adherends and adhesives.

The FEM has also been applied to include other effects [13.31, 13.22]. Adams [13.33] and Vinson [13.34] have reviewed the design of single and double lap joints in metals and composite plates and tubes.

A number of investigators considered a plane stress state in a pin-loaded plate [13.35–13.38]. It is the usual practice to use two-dimensional finite element models along with applying classical lamination theory.

Perhaps the least popular of the three methods mentioned above is the boundary collocation method. It involves the use of two functions that can be expanded into a certain series. The values of the unknown coefficients are obtained by satisfying boundary conditions and imposing continuing conditions along the fastener hole for composite bolted joints.

The boundary element method has been applied for analysing mechanically fastened repairs and lap-joints [13.29]. BEM has also been applied to analyse static strength of composite joints with multiple fasteners [13.30].

For predicting the strength of composite joints, most of the methods have used two-dimensional models. It is only recently that three-dimensional models have been developed [13.39].

REFERENCES AND SUGGESTED READINGS

13.1 F.L. Mathews, P.F. Kilti and E.W. Godwin, A review of the strength of joints in fibre-reinforced plastics, Part 2, Adhesively bonded joints, *Composites*, V. 63, No. 1, 1982, pp. 29–37.

13.2 L.J. Hart-Smith, *Developments in Adhesive-2*, A.J. Kinloch (Editor), Elsevier Applied Science, London, 1981.

13.3 W.A. Less, A Review – The Design and Assembly of Bonded Composites, *Composite Structures*, I.W. Marshall (Editor), Chapter 36, Elsevier Applied Science, London, 1991.

13.4 F.K. Chang, R.A. Scott and G.S. Springer, Failure of composite laminates containing pin-loaded holes – method of solution, *Journal of Composite Materials*, V.18, 1984, pp. 255-278.

13.5 L.I. Eriksson, J. Backland and P. Moller, Design of multiple row bolted composite joints under general inplane loading, *Composite Engineering*, V.5, 1995, pp. 1051–1068.

13.6 P.P. Camhano and F.L. Mathews, Stress analysis and strength prediction of mechanically fastened joints of FRP: A review, *Composites*, Part A, 28A(6), 1997, pp. 529–547.

13.7 Dano Marie-Lasve, Guy Gendron and Andre Picard, Stress and failure analysis of mechanically fastened joints of composite laminates, *Composite Structures*, V. 50, 2000, pp. 257-296.

13.8 F.L. Mathews and R.D. Rawlings, *Composite Materials: Engineering and Science*, Woodlead Publishing Ltd., Cambridge, England, 1999.

13.9 F.L. Mathews, G.A.O. Davies, D. Hitchings and C. Soutes, *Finite Element Modelling of Composite Materials and Structures*, Woodhead Publishing Ltd. and CRC Press, U.K., 2001.

13.10 L. Hollaway, *Polymer Composite for Civil and Structural Engineering*, Blackie Academic and Professional, London, 1993.

13.11 P.K. Mallik, *Fibre Reinforced Composites*, 2nd Edition, Marcel Dekker Inc., USA, 1993.

13.12 L.J. Hart-Smith, Further developments in the design of adhesive bonded structural joints, Doughlas paper 6922, Doughlas Aircraft Corporation, Long Beach, California, April 1980.s

13.13 C.C. Chamis and P.L.N. Murthy, Simplified procedure for designing adhesively bonded composite joints, *Journal of Reinforced Plastics and Composites*, V. 10, No. 1, 1991, pp. 29–41.

13.14 R.D. Adams and W.C. Wake, *Structural Adhesive Joints in Engineering*, Elsevier Applied Science Publishers, New York, 1984.

13.15 P.L.N. Murthy and C.C. Chamis, "ICAN: Integrated composite analyser", *Journal of Composite Technological Research*, V.8, No. 1, 1986, pp. 8–17.

13.16 C.C. Chamis, Simplified procedures for designing composite bolted joints, *Journal of Reinforced Plastics and Composites*, V.9, No. 11, 1990, pp. 614–626.

13.17 N.I. Muskelishveli, *Some Basic Problems of Mathematical theory of Elasticity*, Noordhoff International Publishing, Leyden, The Netherlands, 1977.

13.18 S.G. Lekhnitski, *Theory of Elasticity of an Anistropic Body*, Holden Day, San Francisco, California, USA, 1963.

13.19 T.de Jong, Stresses around pin-loaded holes in elastically isotropic or orthotropic plates, *Journal of Composite Materials*, V.11, 1977, pp.313–331.

13.20 N.A. de Bruyne, The strength of glued joints, *Aircraft Engineering*, V.16, 1944, pp.115–118.

13.21 M. Goland and E. Reissner, The stresses in cemented joints, *Journal of Applied Mechanics*, V.7, 1944, pp. A.27–A.27.

13.22 R.N. Haddock, Joints in composite structures, *Proceedings of the Conference on Fibre Composites Vehicles Design*, AFFDL-TR-72-13, 1972.

13.23 O. Ishai and O. Gali, Direct determination of interlaminar stresses in polymeric adhesive layer, *Experimental Mechanics*, V.17, No. 7, 1977, pp. 265–270

13.24 Y.A. Bahei – Ela Din and G.J. Dvorak, New designs for adhesive joints for thick composite laminates, *Composite Science and Technology*, V.61, 2001, pp.19–40.

13.25 T.A. Collings, The strength of bolted joints in multi-directional CFRP laminates, *Composites*, V.8, No. 1, 1977, pp.43–54.

13.26 F. Nortensen and O.T. Thomsen, Analysis of adhesive bonded joint: A unified approach, *Composites Science and Technology*, V.62, 2002, pp. 100–103.

13.27 M.W. Hyer, E.C. Klong and D.E. Cooper, Effects of pin elasticity, clearance and friction on the stresses in a pin-loaded orthotropic plate, *Journal of Composite Materials*, V.21, No. 3, 1987, pp. 190–206.

13.28 E. Madenca, S. Shkarayev, B. Sergeev, D.W. Ophinger and P. Shyprykeovich, Analysis of composite laminates with multiple fasteners, *International Journal of Solids and Structures*, V.35, 1998, pp. 1793–1811.

13.29 N.K. Salgado and M.H. Alibadi, The analysis of mechanically fastened repair and lap-joints, *Fracture of Engineering Materials and Structure*, V.20, 1997, pp. 583–593.

13.30 S.T. Lie, G. Yu and Z. Zhao, Analysis of mechanically fastened composite joints by the boundary element method, *Composites*, Part B, V.31, 2000, pp. 693–705.

13.31 Y. Weitsman, Stresses in adhesive joints due to moisture and temperature, *Journal of Composite Materials*, V.11, 1977, pp. 378–394.

13.32 R.D. Adams, R.W. Atkins, J.A. Harris and A.J. Kinloch, Stress analysis and failure of carbon-fibre-reinforced – plastic/ steel double lap-joints, *Journal of Adhesion*, V.20, 1986, pp. 29–53.

13.33 R.D. Adams, Strength prediction with lap joints, especially with composite adherends, a review, *Journal of Adhesion*, V.30, 1989, pp. 219–242.

13.34 J.R. Vinson, Adhesive bonding of polymer composites, *Polymer Engineering and Science*, V.29, 1989, pp. 1329–1331.

13.35 P. Conti, Influence of geometric parameters on stress distribution around a pin-bonded hole in a composite laminate, *Composite Science and Technology*, V.25, 1986, pp. 1–19.

13.36 T. Ramamurthy, Recent studies on the behaviour of interference fit pins in composite plates, *Composite Structures*, V.13, 1989, pp. 81–99.

13.37 R.A. Naik and J.H. Crews, Stress analysis method for a clearance-fit bolt under bearing conditions, *AIAA Journal*, V.24, 1985, pp. 1348–1353.

13.38 L.I. Eriksson, Contact stresses in bolted joints of composite laminates, *Composite Structure*, V.6, 1986, pp. 57–75.

13.39 W.H. Chen, S.S. Lee and J.T. Yeh, Three-dimensional contact stress analysis of a composite laminate with bolted joint, *Composite Structures*, V.30, 1995, pp. 257–297.

13.40 N.K. Hassan, M.A. Mohamedien and S.H. Rizkallay, Finite element analysis of bolted connections of PFRP composites, *Composites* : Part B, V.27B, 1996, pp. 339–349.

13.41 W.C. Carpenter and R. Barsoum, Two finite elements for modelling the adhesive in bonded configurations, *Journal of Adhesion*, V.30, 1989, pp. 25–46.

13.42 R.H. Andruet, D.A. Dillard and S.M. Holzer, Two-and-three dimensional geometrical non-linear finite elements for the analysis of adhesive joints, *International Journal of Adhesion and Adhesive*, V. 21, 2001, pp. 17–34.

13.43 J.P.M. Goncalves, M.F.S.F. de Moura and P.M.S.T. de Castro, A 3-D finite element model for stress analysis of adhesive joints, *International Journal of Adhesion and Adhesive*, V.22, 2002, pp. 357–365.

ANSWERS TO EXERCISE PROBLEMS

Exercise 3

3.1 $E_2 = 99.283$ GPa (strength of materials)

      $E_2 = 117.06$ GPa (Halpin-Tsai)

3.2 $\sigma_1 = 1.54$ GPa, $E_1 = 208.38$ GPa

3.3 $\sigma_1 = 3.125$ GPa, $E_1 = 166.575$ GPa, $\sigma_1^u = 5$ GPa

3.4   78.51

3.5   0.47

3.6   0.4675

3.7   41.08%

Exercise 4

4.1 $\varepsilon_1 = 2.64 \times 10^{-3}$ $\varepsilon_2 = 6.99 \times 10^{-3}$ $\varepsilon_6 = -3.487 \times 10^{-3}$

4.2 $E_1 = 181$ GPa, $E_2 = 10.29$ GPa, $E_6 = 7.17$ GPa $\nu_{12} = 0.28$

4.3 $[Q] = \begin{bmatrix} 75.82 & 22.75 & 0 \\ 22.75 & 75.82 & 0 \\ 0 & 0 & 26.53 \end{bmatrix}$ GPa

      $[S] = \begin{bmatrix} 0.0145 & -0.0043 & 0 \\ -0.0043 & 0.0145 & 0 \\ 0 & 0 & 0.0377 \end{bmatrix}$ 1/GPa

4.6 $G_{12} = \dfrac{3}{16\left[\dfrac{1}{E_6} - \dfrac{9}{16 E_1} - \dfrac{1}{16 E_2} - \dfrac{3\nu_{12}}{8 E_1}\right]}$

4.7   {3.5   11.5   2.5}

4.8   {5.5   9.5   5.0}

4.11 $E_x = 14.82$ GPa $= E_y$; $E_s = 9$ GPa, $\nu_{xy} = 0.254$

       $\eta_{xs} = \eta_{yx} = 0.6872$, $\eta_{sy} = \eta_{sx} = 0.4189$

4.12 $\nu_{xy} = 0.321$

4.13 $\dfrac{1}{E_2} = \dfrac{5.99}{E_{x45}} - \dfrac{7.96}{E_{x30}}$

4.14 $\varepsilon_2' = \varepsilon_2 + \dfrac{\varepsilon_1 E_1}{E_2}$

4.15 $\varepsilon_x = \dfrac{\sigma_0}{E_x}(1 - \nu_{xy} + \eta_{xs} k)$

$\varepsilon_y = \dfrac{\sigma_0}{E_y}(1 - \nu_{yx} - \eta_{ys} k)$

Exercise 5

5.1 (a) anti-symmetric angle-ply laminate
$A_{is} = D_{is} = 0$
$B_{xx} = B_{yy} = B_{xy} = B_{sc} = 0$

(b) symmetric laminate
$[B] = [0]$; $A_{is} = D_{is} = 0$

(c) symmetric balanced laminate
$[B] = [0]$; $A_{is} = 0$

(d) symmetric cross-ply laminate
$A_{is} = 0$; $[B] = [0]$

(e) symmetric angle-ply laminate
$[B] = [0]$

(f) anti-symmetric angle-ply laminate
$A_{is} = 0, D_{is} = 0, B_{xx} = B_{yy} = B_{xy} = B_{ss} = 0$

(g) unsymmetric laminate
$A_{is} = 0$

(h) quasi-isotropic laminate
$A_{is} = 0$

5.2 $[A] = \begin{bmatrix} 9.124 & 6.726 & 0 \\ 6.726 & 9.124 & 0 \\ 0 & 0 & 7.444 \end{bmatrix}$ GPa-mm

$[B] = \begin{bmatrix} 0 & 0 & 0.344 \\ 0 & 0 & 0.344 \\ 0.344 & 0.344 & 0 \end{bmatrix}$ GPa-mm^2

$[D] = \begin{bmatrix} 0.030 & 0.022 & 0 \\ 0.022 & 0.030 & 0 \\ 0 & 0 & 0.025 \end{bmatrix}$ GPa-mm^3

5.3 $[A] = E_0 d \begin{bmatrix} 11.066 & 0.500 & 0 \\ 0.500 & 11.066 & 0 \\ 0 & 0 & 1.600 \end{bmatrix}$

$[D] = E_0 d^3 \begin{bmatrix} 3.68 & 0.167 & 0 \\ 0.167 & 3.68 & 0 \\ 0 & 0 & 0.533 \end{bmatrix}$

$[B] = E_0 d^2 \begin{bmatrix} 4.530 & 0 & 0 \\ 0 & -4.530 & 0 \\ 0 & 0 & 0 \end{bmatrix}$

5.4 $[D] = \begin{bmatrix} 0.205 & 0.042 & -0.032 \\ 0.042 & 0.059 & -0.032 \\ -0.032 & -0.032 & 0.049 \end{bmatrix}$

5.5 $[A] = \begin{bmatrix} 96.584 & 29.012 & 0 \\ 29.012 & 96.584 & 0 \\ 0 & 0 & 33.784 \end{bmatrix}$ GPa-mm

$[D] = \begin{bmatrix} 34.17 & 5.22 & 2.09 \\ 5.22 & 9.01 & 2.09 \\ 2.09 & 2.09 & 6.24 \end{bmatrix}$ GPa-mm^3

5.10 $\{\sigma_{xy}\}^{90} = \begin{Bmatrix} 0.0165 \\ -0.0043 \\ 0 \end{Bmatrix}$ GPa

5.11 $A_{xx} = A_{yy} = 373.626 \; N_0$
$A_{xy} = 87.91, \; \bar{v}_{xy} = 0.235$

Exercise 9

9.1 $\{\sigma_1 \quad \sigma_2 \quad \sigma_6\} = \{-440 \times 10^{-6} \Delta T \quad 404.82 \times 10^{-6} \Delta T \quad 0\}$

9.2 (a) 1 – direction -180×10^{-6} mm

2 – direction   0.24 mm

(b) $\sigma_3 = -37.02$ MPa

(c) $\varepsilon_{1c} = \varepsilon_{2c} = 0$, $\varepsilon_{3e} = -0.0024$

9.3 $\sigma_y = -86.28$ MPa

9.4 $\alpha_1 = 7.64 \times 10^{-6}$ /°C $\alpha_2 = 49.88 \times 10^{-6}$ /°C

9.5 $\sigma_{1f} = -0.484 \times 10^{-6}$ /°C, $\sigma_{2f} = 3.326 \times 10^{-6}$ /°C

9.6 For $-30°$ ply

$$\begin{Bmatrix} N_x^T \\ N_y^T \\ N_s^T \end{Bmatrix} = \begin{Bmatrix} 4.33 \\ 5.09 \\ 0 \end{Bmatrix} \times 10^{-6} \text{ GPa-mm}$$

9.7 $\bar{\alpha}_x = -78.6 \times 10^{-6}$ /°C, $\bar{\alpha}_y = 706.4 \times 10^{-6}$ /°C, $\bar{\alpha}_s = 0$

9.9 $\theta = 45°$, $\begin{Bmatrix} \sigma_{1e} \\ \sigma_{2e} \\ \sigma_{6e} \end{Bmatrix} = \begin{Bmatrix} -41.72 \\ 44.72 \\ 0 \end{Bmatrix}$ MPa

$\theta = -45°$, $\begin{Bmatrix} \sigma_{1e} \\ \sigma_{2e} \\ \sigma_{6e} \end{Bmatrix} = \begin{Bmatrix} -41.72 \\ 44.72 \\ 0 \end{Bmatrix}$ MPa

Exercise 10

10.1 (a) $F_{1T} = 1225$ MPa, $E_1 = 50.2$ GPa

(b) $F_{1T} = 2465.5$ MPa, $E_1 = 169.2$ GPa

10.2 35 MPa

10.3 $4m^2 n^2 \sigma_s^2 \left[\dfrac{1}{F_1^2} + \dfrac{1}{F_2^2} + \dfrac{1}{F_1 F_2} \right] + \dfrac{(m^2 - n^2)\sigma_s^2}{F_6^2} = 1$

10.4 Maximum stress theory : $\sigma_0 = 216.0$ MPa

Deviatoric strain energy: $\sigma_0 = 38.25$ MPa

10.5 $\theta = \dfrac{1}{2} \sin^{-1}\left(\dfrac{2F_6}{\sigma_0} \right)$

$$F_{yT} = \frac{F_{2T}}{\cos^2\theta} = \frac{F_0 \cos^2\theta - v_{xy}\sin^2\theta}{\cos^2\theta}$$

10.6 $\sigma_s = \left[\dfrac{1}{\dfrac{4m^2n^2}{F_1^2} + \dfrac{4m^2n^2}{F_2^2} + \dfrac{4m^2n^2}{F_1 F_2} + \dfrac{(m^2-n^2)}{F_6^2}} \right]^{1/2}$

10.7 $F_0 = \left[\dfrac{1}{\dfrac{2.25}{F_1^2} + \dfrac{0.25}{F_2^2} + \dfrac{0.25}{F_6^2} + \dfrac{0.75}{F_1 F_2}} \right]^{1/2}$

10.8 $\sigma_s = \dfrac{(f_2 - f_1) \pm \sqrt{(f_1 - f_2)^2 + 4(f_{11} - 2f_{12})}}{2(f_{11} + f_{22} - 2f_{12})}$

10.9 $F_0 = \dfrac{(f_2 - f_1) \pm \sqrt{f_1^2 + f_2^2 - 2f_1 f_2 + 4f_{11} + 4f_{22} - 8f_{12} + f_{12} + 12f_{66}}}{f_{11} + f_{22} - 2f_{12} + 3f_{66}}$

10.10 Maximum stress theory : $F_0 = 80.77$ MPa

Tsai-Hill theory : $F_0 = 79.68$ MPa

Exercise 11

11.2 Maximum stress theory: 90° ply $\bar{\sigma}_x = 44.03$ MPa

Tsai-Hill theory : 90° ply $\bar{\sigma}_x = 43.77$ MPa

Tsai-Wu theory : 90° ply $\bar{\sigma}_x = 44.77$ MPa

11.3 At ULF, $N_x = 216.19$ MPa

11.4 $N_x = 382.64$ MPa

11.5 $N_x = 480$ MPa

11.7 $\bar{\sigma}_x = 35$ MPa-mm

INDEX

aerospace	3
additives	8
aircraft	2,8,9,14,15
angle-ply	91,93,95,98,130,132,133,136,148, 151-153,313,314
anisotropic strength	260
anti-symmetric laminate	91-93, 96-98
aramid fibre	2-6
asbestos fibre	2
automobiles	12
balanced laminate	96, 97
beam	160-170, 179, 183, 184, 312, 317-319, 325, 340
beam-column	160
bending of composite plates	128
bending stiffness matrix	88
bending-stretching coupling	168
biaxial test	265, 267
bolted joints	345, 346, 349, 350, 357
bonded-bolted joints	347
bonded joints	345-349, 351-352
boron fibre	1, 2, 4, 6, 15
boundary element method	361
buckling	155,157,159,168,175
carbon fibre	1-6, 13-15
carpet plot	310, 322, 323
ceramic fibre	1, 6
ceramic matrix	1, 3
Cholsky's factorization	174
chopped strand mat (CSM)	4, 313, 318, 326
classical beam theory	161,164
classical laminate plate theory	124
closed mould system	19
coating	3, 11
coefficient of moisture expansion	232, 233, 241
coefficient of thermal expansion	231, 232, 241
column	160
compliance matrix	89, 90
composite joints	345, 350, 357, 361
composite multi-bolt joint	359
concrete	1
contact moulding	17
corrugated plate	312
coupling agents	3
coupling compliance matrix	90
coupling stiffness matrix	88
cross-ply	93, 95, 97, 99
damping	175, 178, 185
design allowable	310, 311
design criteria	308-310, 323
design spiral	309
determinant search technique	176, 177
deviatoric strain energy	261, 264
Eberline procedure	176
eigenvalue	176,177
eigenvector	176,177
E-glass	2, 4, 6, 15
engineering constants	50, 52, 55, 64
environmental criteria	308
epoxy resin	2, 6, 7, 8, 14, 15
factor of safety	320
failure theories	254-268
fibre dominated failure	254
filament winding	2, 15, 23, 24
fillers	3, 8, 9
first-ply failure	281
flakes	1, 2
finite element analysis	171-228
flexibility matrix	172,173
Fourier coefficient	129, 134
Fourier series	129, 130, 133, 136
free thermal strains	233-236, 243
frequency	168, 177, 224, 225
frontal technique	174,175
fundamental frequency	223, 224
Galerkin's method	131
Gauss elimination	174,175,05
Gauss-Lagendre rule	215
Gauss points	193,206

Gaussian quadrature	193
general anistropic material	46, 50
glass fibre	2, 4
glass transition temperature	228
hand lay-up	17, 21, 22
Hermitian matrix	197, 198
higher order shear deformation	142
high performance resin	6, 8
Holzer method	177
hybrid laminate	81
hygrothermal effects	228-253
hygrothermoelastic forces	236
inplane compliance matrix	90
inplane shear modulus	33, 34, 40
inplane stiffness matrix	88
internal damping	11
interlaminar shear	125, 138, 141, 302, 303
interlaminar stress	299, 302, 304
isoparametric element	186, 190, 191, 194, 206, 207, 210, 224
isoparametric stiffener	199
isotropic layer	49
Jacobian matrix	211
Jacobian transformation method	176
Kevlar fibre	5, 15
Lagrange	207
lamina	1, 9, 10
laminate	9, 10, 81-159
laminate strength	281-306
last-ply failure	283
layerwise theory	124
longitudinal compression	256-258
longitudinal modulus	29
longitudinal stiffness	28
longitudinal strength	28
longitudinal tension	254
macromechanics	11
manufacturing process	312
marine applications	12
mass matrix	175, 178, 194, 206
matched die moulding	18, 19, 21
maximum strain theory	260-262, 263, 269, 277
maximum stress theory	260, 262-264, 269, 277, 278
mechanical joints	350, 357, 361
metal matrix composite	2
micromechanics	11, 25-36, 231, 264
Mindlin's theory	186, 209
Myklestad method	177
Navier's solution	128
Newmark's method	178
numerical methods	25
parallel axis theorem	317
partial-ply failure	283, 284
particulate composite	1
phenolic resin	2, 6, 8
plate element	190, 191, 193, 199, 200, 206, 207
Poisson's mismatch	33, 36
Poisson's ratio	32, 33, 35, 36
polyester resin	6-8
potential energy	130, 131, 193, 215, 217
prepegs	8
principle of virtual displacement	197
pultrusion	22, 23
QR-transformation method	176
Quasi-isotropic laminate	99
reduced compliance	62
reduced stiffener	54, 62, 64
Reissner-Mindlin plate	206
residual stress	228, 229, 242, 243
restriction of elastic constants	75
Ritz method	131
rod	165
rubber (polyurethane)	1
rubber (reinforced)	1
Sander's shell theory	208
S-glass	2, 4
self-consistent model	25
shape function	191, 192, 208
shear deformation	138, 142, 186, 195, 224
shear coefficient	220

shear coupling	197
shear locking	193, 194, 205
shell	173, 206-210, 224
simultaneous iteration method	177
specially orthotropic material	48, 50, 62, 75, 93, 99
sporting goods	12, 15
spray-up	17
strain energy	147
stacking sequence	81, 92
stability	152, 172, 176, 178
static analysis	175
stiffener	194, 199, 200, 202, 203, 204, 205, 206, 312, 339, 340, 341, 342
stiffness matrix	171-178, 181-185, 193, 194, 198, 199, 205
Stodola's method	176
Taylor's series	215, 216
tension members	321
thermoplastic	1, 2, 6, 7
thermoset	1, 6, 7
torsional rigidity	204
transfer matrix method	177
transformation method	176
transformation of elastic constants	61
transformation matrix	181, 202
transformation of strain	59-61
transversely isotropic material	49
transverse modulus	30-32, 35
transverse normal stress	124
transverse shear strain	124, 192, 217
transverse shear stress	124, 220
transverse tension	259
truss element	181
Tsai-Hill theory	261, 264-266, 277
Tsai-Wu theory	261, 266-268, 277
T-bar	340
ultimate laminate failure	281, 283
variational approach	25
vector iteration method	176
vessel	2, 14
vibration	
free	148, 151, 175, 176, 223-225
transient	175
vinyl ester resin	6-8
virtual work	197, 199, 205
volume fraction	26, 27, 29-31, 34
warpage	228, 243, 244
weight fraction	26, 27
woven roving	4